ROUTLEDGE HANDBOOK ON GENDER IN TOURISM

This comprehensive handbook delves into the multifaceted dimensions of the role of gender in tourism, spanning education, research, and practice.

With 40 international contributions from leading thinkers in the field, this book brings together diverse themes such as entrepreneurship, mobility, sustainability, and sexuality. In doing so it shatters traditional boundaries and dissects how gender influences perceptions, experiences, and opportunities, advocating for equality and challenging entrenched power dynamics. Informed by the United Nation's Gender Equality goals, this handbook champions the potential of gender-aware tourism to reshape the world by fostering inclusivity, empowerment, and understanding. It adopts diverse insights, encompassing feminist and queer perspectives, challenging norms, and exploring marginalised voices. By dissecting gender in educational, entrepreneurial, and research contexts, it unveils hidden dynamics. This book empowers readers to grasp the breadth of gender's role and equips them with tools to foster equality and reshape the tourism landscape, while making suggestions for future research agendas.

This book is intended for scholars, educators, researchers, government officials and practitioners in the fields of gender studies, tourism, education, entrepreneurship, employment, mobility, research, sustainability, and sexuality.

Magdalena Petronella (Nellie) Swart is an Associate Professor in Tourism at the University of South Africa (Unisa) and a Certified Meeting Professional. She holds a DCom in Leadership Performance and Change. Nellie authored and co-authored journal articles, books, book chapters, and several conference proceedings and has organised local and international conferences. Recently she co-edited two tourism handbooks and is a regular speaker at tourism industry events. She is the Chair of the Tourism Educators South Africa (TESA) and the programme leader for the Executive Development Programme for Women in Tourism at Unisa. In 2022 the International Hospitality Institute recognised Nellie as one of the 100 Most Powerful People in Africa Hospitality. She is also recognised as the G100 City Chair for Johannesburg.

Wenjie Cai is an Associate Professor in Tourism at the University of Greenwich, United Kingdom. Wenjie gained his PhD in Tourism from the University of Surrey, United Kingdom. Wenjie's area of expertise and research interests include digital well-being, social inclusion, knowledge creation, technology use, and consumer behaviour. Wenjie aims to advocate digital well-being and support marginalised and disadvantaged groups through his research. Wenjie has more than 30 research outputs, including publications in world-leading journals in the field. His research has been reported globally by more than 500 media.

Elaine Chiao Ling Yang is a Senior Lecturer in Tourism at Griffith University. Elaine's work focuses on the empowerment of marginalised groups in tourism, including women, children and migrants, as travellers, entrepreneurs, and workers. Most of her work entails an intersectionality lens that foregrounds the intertwined gender, race, and cultural identities. She also specialises in solo travel, Asian tourism, and visual research methods. Elaine is an associate editor of *Frontiers in Sustainable Tourism* (Social Impact of Tourism) and editorial board member of *Tourism Management Perspectives*. Elaine has received multiple research awards, including the CAUTHE Fellows Award in 2023.

Albert Nsom Kimbu is a Reader and Head of the Tourism and Transport Department in the School of Hospitality and Tourism Management, University of Surrey, Guildford, United Kingdom. He is also the co-founder of the Gender, Entrepreneurship and Social Policy Institute (GESPi). He is also a Senior Research Fellow at the School of Tourism and Hospitality, University of Johannesburg, South Africa. Dr Kimbu researches gendered entrepreneurial pathways, development-led, and inclusive tourism. He has received funding from the British Academy, Newton Fund, UKRI, UNWTO, UNIDO among others, to consult and undertake research on these topics, resulting in publications in leading tourism management journals, including *Annals of Tourism Research, Tourism Management, Journal of Travel Research, Journal of Sustainable Tourism*, industry reports, and edited books.

ROUTLEDGE HANDBOOK ON GENDER IN TOURISM

Views on Teaching, Research and Praxis

*Edited by Magdalena Petronella (Nellie) Swart,
Wenjie Cai, Elaine Chiao Ling Yang
and Albert Nsom Kimbu*

Routledge
Taylor & Francis Group

LONDON AND NEW YORK

Designed cover image: gettyimages.ca/komunitestock

First published 2024
by Routledge
4 Park Square, Milton Park, Abingdon, Oxon OX14 4RN

and by Routledge
605 Third Avenue, New York, NY 10158

*Routledge is an imprint of the Taylor & Francis Group,
an informa business*

British Library Cataloguing-in-Publication Data
A catalogue record for this book is available from the British Library

Library of Congress Cataloging-in-Publication Data
Names: Swart, M. P. (Magdalena Petronella), editor. | Cai, Wenjie, editor. | Chiao Ling Yang, Elaine, editor. | Kimbu, Albert N., editor.
Title: Routledge handbook on gender in tourism : views on teaching, research and praxis / edited by Magdalena Petronella (Nellie) Swart, Wenjie Cai, Elaine Chiao Ling Yang, and Albert Nsom Kimbu.
Other titles: Handbook on gender in tourism
Description: First edition. | New York : Routledge, 2024. | Includes bibliographical references and index.
Identifiers: LCCN 2023042879 (print) | LCCN 2023042880 (ebook) | ISBN 9781032261348 (hbk) | ISBN 9781032261409 (pbk) | ISBN 9781003286721 (ebk)
Subjects: LCSH: Tourism--Sociological aspects. | Women in tourism.
Classification: LCC G156.5.S63 R69 2024 (print) | LCC G156.5.S63 (ebook) | DDC 306.4/819082--dc23/eng/20231025
LC record available at https://lccn.loc.gov/2023042879
LC ebook record available at https://lccn.loc.gov/2023042880

ISBN: 978-1-032-26134-8 (hbk)
ISBN: 978-1-032-26140-9 (pbk)
ISBN: 978-1-003-28672-1 (ebk)

DOI: 10.4324/9781003286721

Typeset in Sabon LT Pro
by KnowledgeWorks Global Ltd.

CONTENTS

FIGURES

TABLES

CONTRIBUTORS

Marina Abad Galzacorta is an Associate Professor in Tourism at the University of Deusto in the Faculty of Social and Human Sciences (Basque Country). She is accredited as teaching staff by Unibasq (Agency for Quality of the Basque University System) and member of "Leisure, Culture and Tourism for Social Transformation" research group which was recognised by the Basque Government (IT 1457-22). She has collaborated on research projects with several institutions and authored and co-authored journal articles, book chapters, and several conference proceedings. Her research areas are tourism, technology and innovation, sustainable and responsible tourism and tourist behaviour analysis.

Núria Abellan-Calvet is a Professor in Tourism at CETT Barcelona School of Tourism, Hospitality and Gastronomy, affiliated with the University of Barcelona. She holds a Gender Studies MA at the University of Sussex (United Kingdom) and is currently developing her PhD in University of Lleida, focusing on lesbian tourism. She is a researcher at the Research Group in Tourism, Culture, and Territory (TURCiT), and her work focuses on tourism and gender, LGBTIQ tourism, and cultural tourism.

Husna Zainal Abidin is a lecturer at the Center for Tourism Research, Wakayama University, Japan. She received her PhD in Tourism Management from the University of Surrey, United Kingdom. Husna is an early career researcher who actively publishes in academic journals such as Tourism Management and non-academic publications such as industry reports and blogs. Her research revolves around destination management, focusing on technology in tourism and Muslim tourism. With a unique background spanning both Asia and the West, as well as having a Muslim perspective, Husna offers a diverse and multifaceted approach to her research.

Issahaku Adam is a Professor at the Department of Hospitality and Tourism Management of the University of Cape Coast, Ghana. Prof Adam's research interests Inclusion in tourism and leisure, sustainable tourism, gendered entrepreneurial pathways in tourism, and tourist behaviour and experience. He has been involved in numerous projects and consultancy works on the different aspects of tourism development in Africa. Prof. Adam has published widely on his research interests in leading tourism and hospitality journals, including Annals of Tourism Research, Journal of Travel Research, Tourism Management, International Journal of Hospitality Management, and Journal of Travel and Tourism Marketing amongst others. He is currently the Regional Editor for Africa for Leisure Studies and a member of the editorial board of the Journal of Travel Research, and Tourism Planning and Development.

Ogechi Adeola is a Professor of Marketing at Lagos Business School, Pan-Atlantic University, Nigeria. In addition, she serves as a member of the Governing Board of University of Kigali, Rwanda. Her multi-dimensional research focuses on advancing knowledge across the intersection of marketing, entrepreneurship, tourism, and gender. She is the editor of the book "Indigenous African Enterprise: The Igbo Traditional Business School (I-TBS)," published in 2021, as well as the two-volume book titled "Casebook of Indigenous Business Practices in Africa" – Volume 1 (Apprenticeship, Craft, and Healthcare) and Volume 2 (Trade, Production, and Financial Services) – both published in 2023 by Emerald Publishing Limited.

Ewoenam Afua Afenyo-Agbe is a Senior Lecturer at the Department of Hospitality and Tourism Management of the University of Cape Coast, Ghana. She holds a BSc in Tourism, an MPhil, and a Ph.D. in Tourism Management all from the University of Cape Coast, Ghana. Her principal research interests are in community tourism, ecotourism, tourism management, tourism entrepreneurship, gender in tourism, and tourism-related livelihood issues. She has published in international tourism journals, including Annals of Tourism Research, Tourism Management, and Tourism Planning and Development. She has recently co-edited a book on the prospects and challenges of community-based tourism.

Hugh Bartis is a Principal Lecturer in the Tourism Department at Nelson Mandela University in South Africa. He holds a PhD in Tourism Management and has published several articles in accredited journals, book chapters, and conference proceedings. Hugh has also been a Non-Executive Director at Nelson Mandela Bay Tourism, the Eastern Cape Tourism Board, the Southern African Tourism Services Association and is an executive member of the Tourism Educators of South Africa. He has extensive experience of curriculating Tourism and Hospitality programmes.

Vanessa S. Bernauer is a doctoral researcher and lecturer at the Chair of Human Resource Management at Helmut Schmidt University Hamburg in Germany. She conducts research on equality, diversity, inclusion, and identity in organisations with a focus on service work in the luxury segment, alternative forms of organising in digitalised workplaces, and sexual orientation. Her work has been published in leading international journals such as Human Resource Management Journal or Equality, Diversity and Inclusion. As a guest editor she co-edits Special Issues with (inter)national colleagues, e.g., in EDI and she conceptualises and conducts courses and workshops, e.g., for the Academy of Management or specialists' and executives' further education.

Brenda Boonabaana is an Assistant Professor in the Department of Geography and the Environment, the University of Texas at Austin, United States. She has expertise in Gender and Sustainable Tourism, qualitative research, and has co-authored several publications in this field. She is on the Editorial Board for Gender, Place and Culture Journal, and has provided technical expertise to numerous institutions, including Cornell University, UNWTO and IFPRI. She has worked for the UNWTO as a Regional Field Expert for Africa, towards producing the Second Global Report on Women in Tourism and Tourism Lecturer at Makerere University, Uganda.

Laiara Amorim Borges, founder of the Quilombo Aéreo Organization, Brazilian black woman researcher and flight attendant. Research the favelas and outskirts in relation to mobilities through aviation. She is a postgraduate in social project management. She is a Member of research groups "Gender, race and intersectionalities in tourism" (UFSCar) and "Gender and Performance" (Quilombo Aéreo).

Alexa Bufkin is an Adjunct Professor in the Department of Tourism, Event, and Sport Management at Indiana University She holds a Bachelor of Science and Master of Science in Tourism, Events and Sport management. Alexa has co-authored journal articles and book chapters and has spoken at different seminars locally and domestically. Her research interests are based on high-risk niche tourism ranging from human trafficking prevention and hazards within travelling to adventure activities.

Amanda K. Cecil, Ph.D., CMP, is a Professor in the Department of Tourism, Event, and Sport Management and directs the Events & Tourism Institute at Indiana University. works on a variety of projects in event tourism with other faculty members and students. Her research interest involves linking travel, tourism and event trends and impacts. Additionally, she has scholarly interests in experiential and engaging learning models. She has consulting experience in instructional design for educational programs in customer service, business travel management, strategic meeting management, and sports travel management. She has published in many tourism journals, including The Journal of Convention and Event Tourism, The Journal of Teaching in Travel & Tourism, and The Journal of Sustainable Development. She serves as the she serves as the Editor-In-Chief for the Journal of Convention & Event Tourism.

Maria Cendoya Garmendia graduated in Tourism at the University of Deusto in the Faculty of Social and Human Sciences (Donostia, Basque Country). She focused her Degree Thesis Project on "Solo travellers Women's emotions in the XXI century" and she did her internships in Centre Touristique et Culturel Casamance (Senegal). In her internship she collaborated in the Center management but also she cooperated with Kassumay Foundation in a project with local entrepreneurship women in Diakene Ouoloff (Casamance, Senegal).

Frederick Dayour (PhD) is an Associate Professor in Tourism and Hospitality and the Head of Department for the Department of Hospitality and Tourism Management at the SD Dombo University of Business and Integrated Development Studies, Ghana. Frederick obtained his BSc. and MPhil degrees in Tourism Management at the University of Cape Coast and PhD in Tourism and Hospitality at the University of Surrey, United Kingdom. He is also currently a Senior Research Fellow at the School of Tourism and Hospitality, University of Johannesburg, South Africa. He has published in Annals of Tourism Research, Journal of Travel Research, and Tourism Management.

Natália Araújo de Oliveira is a Brazilian woman researcher, PhD in Sociology (Federal University of Rio Grande do Sul). She holds a Master's in Social Science (University of Vale do Rio dos Sinos), Bachelor in Tourism (Mato Grosso State University). She researches tourism and racial relations. She is a Member of research groups "Gender, race and intersectionalities in tourism" (UFSCar) and "Labor Movens: Working Conditions in Tourism" (UnB).

Gabriela Nicolau dos Santos is a Brazilian woman researcher. Postdoctoral fellow at the Center for Languages, Literatures and Cultures of the University of Aveiro (Portugal) since 2020. Completed the PhD in Advanced Studies in Social Anthropology in 2016 at the University of Barcelona. She researches tourism, gender, and intangible heritage. Member of research groups "Gender, race and intersectionalities in tourism" (UFSCar) and "Gender and Performance" (University of Aveiro).

Ikechukwu O. Ezeuduji is full Professor of Tourism Management at the University of Zululand, South Africa. He obtained his PhD degree from the BOKU-University of Natural Resources and Life Sciences, Vienna, Austria. He has more than 12 years of teaching experience in Higher Education Institutions. He has widely published in broader areas of tourism development (rural and events tourism), tourism management (strategic tourism management and tourism entrepreneurship) and tourism marketing (brand essence, brand competitiveness, brand image and brand loyalty).

Xiangli (Sally) Fan is an Associate Professor in College of tourism, HuaQiao University in China. She is interested in gender and tourism research, such as female human resource management in tourism and hospitality industry, and women's safety perceptions on tour amongst others.

Cassiana Panissa Gabrielli is a Professor at the Federal University of São Carlos, where she leads the research group "Gender, Race and Intersectionalities in Tourism," in addition to teaching and developing research on gender inequalities in the tourism sector. She holds a PhD in Interdisciplinary Studies on Women, Gender and Feminism (Federal University of Bahia) and a Master's degree in Culture and Tourism (Santa Cruz State University). For more than ten years, she has been seeking to strengthen discussions on gender in Brazilian tourism, authoring some articles and book chapters and teaching short courses and specific disciplines on the subject.

Vizak Gagrat is a PhD student at the University of Queensland, Brisbane, Australia. Vizak researches LGBT+ events, focusing on marginalised groups and social inclusion. This focus complements his contributions to various LGBT+ groups and organisations within the Brisbane area, strengthening sense of community for these groups. Vizak also lectures for James Cook University, Brisbane campus in the areas of Tourism, Hospitality, and Business.

Maria Gebbels is an Academic Portfolio Lead for Hospitality and Tourism at the School of Management and Marketing, University of Greenwich. Her main research focuses on gender issues and career development in hospitality, professionalism, in-prison fine dining, hospitality in adventure tourism, and critical hospitality as a lens to understand social relations. Maria works closely with the hospitality industry advising on broader ED&I concerns. She also co-authored "Adventure Tourist: Being, Knowing, Becoming" and has been collaborating on research projects with colleagues from Malaysia, New Zealand, the Netherlands, and Australia.

Eylla Laire M. Gutierrez is the Research Manager at the Asian Institute of Management's Dr. Andrew L. Tan Center for Tourism and a doctoral student at the Ritsumeikan Asia Pacific University. She also served as the Gender and Sustainable Tourism Consultant of the Philippine Center for Environmental Protection and Sustainable Development, Inc., Senior Consultant at Warwick & Rogers, and the Sustainable Tourism Specialist of the Masungi Georeserve, Foundation Inc. She has been engaged in several commissioned research projects in partnership with the University of Auckland, Konrad Adenauer Stiftung, and the Philippine APEC Study Center Network, among others. Her research interests include sustainability, women's studies, community development, and tourism management.

Ireena Nasiha Ibnu received her PhD degree in Migration Studies from the University of Sussex, United Kingdom. She is currently a senior lecturer and coordinator (Liberal Communication) at the Faculty of Communication and Media Studies at the Universiti Teknologi Mara (UiTM), Malaysia. Ireena authored and co-authored journal articles, books, book chapters, and several conference proceedings and has organised local and international conferences. Her research interests include intercultural communication, transnational migration, international students' experience, and cultural anthropology.

Md. Tariqul Islam holds a Master of Science (by research) in Tourism from Universiti Putra Malaysia, Malaysia. He graduated with distinction in Airlines, Tourism, and Hospitality from Lovely Professional University, India. Tariqul has published several research articles in ABDC-listed and Scopus-indexed journals and presented the findings of his research at various national and international conferences. His area of research includes consumer behaviour and technology adoption.

Bingjie "Becky" Liu-Lastres, Ph.D., is an Assistant Professor in the Department of Tourism, Event, and Sport Management at Indiana University. Her main research interests include risk and crisis communication/management in tourism and hospitality, tourist safety and security, social media in tourism and hospitality, and tourism management. The goal of Liu-Lastres's research agenda is to promote safe travel and to ensure the health and well-being of tourists, organisations, and other key stakeholders within the tourism and hospitality industries.

Nompumelelo Nzama is a Lecturer in the Department of Tourism and Events Management, Cape Peninsula University of Technology, South Africa. She holds a Master's of Tourism degree, obtained from the University of Zululand, South Africa. Her research interests are in the areas of gender studies in tourism and tourism entrepreneurship. Nompumelelo has authored and co-authored journal articles, book chapters, and conference papers. She is currently on an academic journey to obtain a PhD in Tourism and Recreation.

Uma Pandey is an Assistant Professor in the Department of Tourism and Airlines at Lovely Professional University, India. She holds a Ph.D. in Tourism Management. She has a deep belief in the transformative power of tourism. She believes that tourism can be a force for good in the world, and she is committed to using her research and teaching to promote sustainable tourism development. Her research interests include forms of tourism, ecotourism, consumer behaviour, gender equality, and tourism education.

Haili Qin received her education in tourism management from the Huaqiao University. She is currently serving as a teacher of tourism management at Wuzhou University. She is broadly interested in the area of gender issues in tourism research. Her areas of research include cultural tourism, feminist leisure and minority female development.

Marta Salvador-Almela is a Professor at CETT Barcelona School of Tourism, Hospitality and Gastronomy, affiliated with the University of Barcelona. She holds a Master's in Anthropology at the University of Sussex (United Kingdom) and is currently doing her PhD at the University of Lleida. Her research areas are tourism, anthropology, culture, ethics, and gender. She is a member of the Research Group in Tourism, Culture, and Territory at CETT-UB (TURCiT) and the co-editor of the Tourism & Heritage Journal.

K Thirumaran specialises in tourism and hospitality management at James Cook University Singapore. His research focuses on service excellence in hospitality and cultural and heritage tourism. He is interested in questions related to modern developments and their impact on emerging destinations' cultural values and the use of virtual and augmented reality to improve training programs in the industry. Thiru coined the term "affinity tourism." See, Thirumaran, K. (2009). "Renewing bonds in an age of Asian travel: Indian tourists in Bali," in Tim Winter, Peggy Teo and T.C. Chang, eds. Asia on Tour, pp. 127–137. New York: Routledge. Thirumaran has co-edited several books related to shared services business model, societies in tropical constraint environments, and service excellence.

Shireen van Zyl is a lecturer in tourism, at a tertiary institution. She is involved in the supervision of Master's and Doctoral level research. She is also involved in engagement activities with the tourism private and public sectors as well as community stakeholders. She is currently the Head of Department: Tourism, at the Nelson Mandela University, in Gqeberha. Her highest qualification obtained is a doctoral qualification in Development Studies, with the focus of her thesis and articles published being in field of the responsible and sustainable development of tourism.

Jiamei Zhang is a postgraduate student at College of Tourism, Huaqiao University, China. Her main research interests include gender and tourism, rural tourism with a particular focus on the rural revitalisation and rural development.

Fan Zhong is a postgraduate student of Tourism at the School of Management, HuaQiao University, China. She is interested in solo female travellers' behaviour and the situation of women travel practitioners.

FOREWORD

Tourism, Gender, and Empowerment: A Path to Equity

In the ever-expanding multi-disciplinary field that is tourism studies, there remains a crucial imperative to critically explore the multifaceted dynamics of gender. After all, gender, as a fundamental aspect of human identity, plays a pivotal role in shaping the experiences, practices, opportunities, and challenges encountered within the tourism sector. Moreover, it has become increasingly obvious to me that understanding and promoting gender equity is crucial to the flourishing of humanity on this planet. However, by comparison with other dimensions and topics in tourism, gender remains underserved and undervalued. As the editors of this collection note, taking some kind of gendered lens, such as a feminist or queer perspective, to tourism enquiry, contests' "malestream" ways of thinking, and asymmetric power relations shifts the focus to women and beyond, to "marginalised communities, suppressed research areas, and silenced voices." This important and timely collection seeks to address these and other challenges by examining the interplay between gender and tourism, shedding light on the often overlooked and underexplored dimensions that influence the lives of individuals engaged in tourism, both as tourists and as workers.

Why is research on gender issues in tourism studies of such paramount importance? The answer partly lies in the recognition that gender is not a binary construct but a spectrum that encompasses a diverse range of identities, experiences, and inequalities. It is important to emphasise that gender research has more than a female-centric focus and includes research on/with men and masculinities and on/with lesbian, gay, bisexual, transgender, queer, intersex, asexual, and others (LGBTQIA+) communities. By examining the gendered dimensions of tourism, we can uncover the ways in which power structures, social norms, and cultural expectations shape the opportunities and constraints faced by individuals across the globe. In practice, gender research in tourism encompasses, amongst other topics: gender identities; sexualities; the (re)production, performance, and construction of gender; embodiment; stereotyping, discrimination, and gender-based harassment in travel and the workplace; the gender pay gap; unpaid, emotional, and aesthetic labour, leadership and entrepreneurship. All of these and others are addressed in this handbook.

By conducting rigorous research on these and other gender issues in tourism studies, we can identify the mechanisms through which gender-based discrimination, violence, and marginalisation persist within the industry. This knowledge empowers us to develop strategies and policies that promote gender equality, social justice, and inclusivity within tourism. Tourism, as a transformative force, has the potential to challenge and disrupt existing gender norms, but it can also perpetuate and reinforce inequalities, something recognised by the United Nations Sustainable Development Goal #5: Gender Equality and Empower all Women and Girls, and the World Tourism Organisation, which has recently been emphasising its significance in the tourism context, particularly as women represent the majority of workers in the sector. Understanding the gendered dimensions of tourism is thus crucial not only for academic scholarship but also for practitioners, policymakers, and industry stakeholders. Moreover, the adoption of feminist ethics of care, as discussed by Angela Kalisch and Stroma Cole, is central to addressing the key challenges now facing humanity, notably the climate crisis and the transition to a low-carbon future. By recognising the diverse needs, preferences, and aspirations of tourists, we can create more inclusive and sustainable tourism experiences. Similarly, by addressing the gendered challenges faced by workers and entrepreneurs in the tourism sector, we can foster fair and equitable employment practices, ensuring the well-being and empowerment of those who contribute to this significant industry.

This handbook brings together a collection of insightful research that explores the role and importance of a range of these gender issues in tourism studies. It showcases the work of scholars from diverse disciplines, regions, and perspectives, offering a comprehensive and nuanced understanding of the subject matter. The handbook is truly global and its chapters are authored by a diverse range of English-speaking and non-native English-speaking researchers from 16 countries across four continents, including several contributions from the Global South, typically under-represented in such collections. The contexts and topics are wide-ranging and its four sections cover the following: feminist epistemologies and methodologies (including practical guidelines for research design, and data collection and analysis); teaching and learning gender in tourism, and student experiences; tourism development, responsible tourism policies, women's empowerment, and community development; the gendered tourism workforce; racism and sexism in the airline industry; gender mobilities; women tourism entrepreneurs.

Taking a feminist approach (such as intersectionality, self-reflexivity and positionality, power dynamics, and feminist standpoint theory) challenges hegemonic and traditional knowledge production, with a specific aim of social change and empowering marginalised communities. Despite the work in this handbook, there remains much more to do in this area, and many of these areas of future enquiry are mapped in the interviews with leading tourism gender researchers in the handbook's conclusion. Although we have seen a positive development in gender research in tourism in recent years, tourism gender studies remains poorly connected to wider debates in feminist and gender studies and, compared with other research areas in tourism, is under-represented in the classroom and in terms of discussion on teaching and learning approaches. We need more critical and alternative voices, further empirical insights from the world's overlooked regions and countries, and more scholarship on the intersectionality of race, gender, and sexual orientations, which foregrounds the importance of multiple worldviews and paradigms and shifts tourism knowledge away from Anglo-centric traditions and philosophies.

This handbook addresses these omissions by offering a detailed and critical understanding of the discussions related to teaching and learning, and gendered tourism experiences and practices. In addition to being a novel contribution to contemporary conversations on paradigmatic standpoints, knowledge production, tourism education, the tourism workplace, entrepreneurship, etc., its format helps to develop the scope and directions for future research. By proposing practical guides, its contributors offer transferable "takeaways" for future researchers, while its conclusion presents an agenda for future enquiry. As you read and engage with tourism and gender through the pages of this handbook, embrace the transformative power of research. By delving into the gendered dimensions of tourism, we can and must challenge existing paradigms, advocate for change, and pave the way to a more inclusive and equitable future. Gender studies in tourism has already had an impact beyond academia; findings are effectively influencing policies, raising awareness amongst a range of stakeholders and the public, and transforming future generations and workforces. May this book serve as a catalyst for further dialogue, reflection, and action, inspiring researchers, practitioners, and policymakers to embark on a collective journey towards a gender-just tourism industry and a more equitable world.

Professor Nigel Morgan
University of Surrey
6 September 2023

PREFACE

Gender studies in the field of tourism has witnessed a remarkable growth and transformation in recent years, yet it necessitates a continuous refinement of our investigative approaches in teaching, research, and practices to unpack the array of nuances across the sector. *The Routledge Handbook on Gender in Tourism: Views on Teaching, Research and Praxis* fills a notable gap in the current body of knowledge, by offering a well-rounded understanding of the diverse discussions related to gender in tourism and how this is understood, researched, and practised. This handbook is the first scholarly work to combine a diverse range of topics related to teaching and learning, research, and gendered tourism practices in one publication. It covers a variety of topics related to education, entrepreneurship, employment, mobility, research, sustainability, tourism development, and sexuality to name a few. The unique harmonisation of the tourism sector with these topics aims to broaden the appeal of the handbook.

Several mentorship and developmental initiatives were started by non-native English-speaking editors since the conceptualisation of the handbook in 2021. In adhering to our ethos, the editors and reviewers mentored emerging scholars in the developing regions of the Global South, on how to refine their academic writing and publication skills to craft quality chapters according to international publication standards. Specific guidance revolved around the focus and structure of the chapters, the development of conceptual/theoretical frameworks, the use of well-developed methodologies, and "key takeaways" to ensure significant contributions were made to the current conversations on gender in tourism. Before acceptance, each chapter was double peer-reviewed by scholars with strong academic credentials who are experts in the field, which in some cases involved four rounds. All contributions had to be original, and authors were not allowed to replicate work previously published, which was assessed against similarity reports. Proceeds from this publication will be re-invested in tourism development programmes for women and youth in Africa.

Following the call for chapters, 39 abstracts were received of which four abstracts were rejected, as they did not sufficiently emphasise the gender in tourism element of the call. During the first round of chapter submissions, 23 chapters were received, and another 4 chapters were rejected as the authors did not address the identified shortcomings following the review of abstracts. Two more rounds of chapter reviews and submissions followed, where four chapters were not resubmitted and a further three chapters were withdrawn. This rigorous peer-review process concluded with 16 chapters, written by 37 mostly non-native English-speaking authors from 16 countries. Together with the introduction and conclusion chapters, this collection consists of 18 chapters, organised into six sections.

This handbook begins with an introductory chapter that explores and unpacks the six themes related to teaching and learning, research, and gender practices. Part I explores current discussions related to teaching and learning gender in tourism and the associated student experience, while Part II focuses on researching gender in tourism. Practising gender in tourism consists of four sections, starting with Part III on tourism development, Part IV provides insights on the gendered tourism workforce, Part V highlights gendered mobility challenges through the experiences of female travellers, while nuances on gender and entrepreneurship are unpacked and explored in Part VI. The handbook concludes with a reflection on the editorial journey, takeaways from the authors, conversations with esteemed tourism gender scholars on how they foresee the future evolution of the topic, and final forward-looking views of the editors.

Magdalena Petronella (Nellie) Swart
Department of Applied Management,
College of Economic and Management Sciences,
University of South Africa, South Africa

Wenjie Cai
Tourism and Marketing Research Centre,
Greenwich Business School,
University of Greenwich,
London, United Kingdom

Elaine Chiao Ling Yang
Griffith Business School,
Department of Tourism,
Sport & Hotel Management,
Griffith University,
Nathan, Australia

Albert Nsom Kimbu
Faculty of Arts & Social Sciences,
School of Hospitality & Tourism Management,
University of Surrey, Guildford, United Kingdom;
School of Tourism and Hospitality,
University of Johannesburg, South Africa

Introduction

1

UNVEILING THE GENDER LENS IN TOURISM

Wenjie Cai, Albert Nsom Kimbu, Magdalena Petronella (Nellie) Swart, and Elaine Chiao Ling Yang

Abstract

Gender profoundly influences all aspects of tourism teaching and knowledge production and practices, yet it lacks the critical and reflexive understanding it deserves. In response the *Routledge Handbook on Gender in Tourism: Views on Teaching, Research and Praxis* aims to fill the gap in gender studies in tourism, offering critical insights, diverse perspectives, and practical guidance for future research and policymaking. This chapter provides an oversight on how this handbook delves into the profound influence of gender on aspects of tourism teaching and learning, research and practices. It explores how adopting a gendered lens challenges conventional knowledge and benefits marginalised communities, suppressed research areas, and silenced voices. Across six parts, the book covers a wide range of gender-related topics in the tourism context, such as gender identities, sexualities, discrimination, workplace challenges, and entrepreneurship.

Keywords

Teaching gender in tourism, Researching gender in tourism, Practising gender in tourism, Tourism development, Tourism workforce, Gendered mobilities, Entrepreneurship.

Introduction

Gender, as an essential dimension of social lives, has been deeply rooted and reflected in every aspect of tourism knowledge production and practices. However, without critical and reflexive understandings, such an important dimension is yet to receive the attention it deserves. From an epistemological aspect, a gendered lens brings alternative perspectives and challenges conventional knowledge productions and ways of thinking. Such a lens (e.g., feminist and queer approaches) does not only benefit research on women but also shifts focus to marginalised communities, suppressed research areas, and silenced voices. It also opens doors for critically understanding power dynamics and challenging unequal power structures. In practice, gender serves as an umbrella term for topics such as gender

DOI: 10.4324/9781003286721-2

identities, sexualities, and the (re)production, performance, and construction of gender. Issues related to these topics play an important part in the tourism context, such as bodied and gendered tourist experiences (Small, 2016), stereotypes, discrimination and gender-based violence in travel (particularly towards solo female and LGBTQ+ travellers) (Eger, 2021; Usai et al., 2022) and tourism workplace (Dudley et al., 2022), gender pay gap and the glass ceiling in the sector (Carvalho et al., 2019), unpaid labour in family tourism businesses (United Nations World Tourism Organization [UNWTO], 2019), workplace harassment and mistreatment (Cheung et al., 2018; Zhou et al., 2021), entrepreneurial leadership (Kimbu et al., 2021), and structured constraints and challenges for women entrepreneurs (Khoo et al., 2023; Ribeiro et al., 2021).

Following UN Sustainable Development Goal #5: Gender Equality and Empower all Women and Girls, the United Nations World Tourism Organisation (UNWTO) has been emphasising its significance in the tourism context, particularly as women represent the majority of workers in the sector. UNWTO (2023) believes that through job provision and entrepreneurship, tourism can empower women to be the leaders of society. Recognising several constraints, such as unpaid work in family businesses and low pay and low status compared with their male counterparts (UNWTO, 2019), UNWTO started several initiatives working with strategic partners globally to empower women in the tourism sector. Although the COVID-19 pandemic has been proven to foster gender inequality in tourism, particularly in employment (Claudio-Quiroga et al., 2022), UNWTO believes the pandemic provides a golden opportunity to tackle gender-related issues. For instance, Kalisch and Cole (2022) suggest a Feminist Alternative Tourism Economics approach emphasising the Feminist Ethic of Care, Social and Solidarity Economy, and Human Rights-Based Economy in the post-COVID recovery and tourism transformation. Discussing gender in practice and knowledge production is thus invaluable in the post-COVID era when setting future research directions.

It is important to stress that gender issues cover much wider issues than women's issues. Gebbels et al. (2020) emphasised the importance of engaging men in the process of gender reform and equality in the context of tourism and hospitality. Such arguments shift away from a female-centric focus to the relational aspect of gender. In addition, as an umbrella term, gender studies also cover research on lesbian, gay, bisexual, transgender, queer, intersex, asexual, and other (LGBTQIA+) communities, especially how they negotiate their sexualities and gender identities outside their familiar environment, particularly for transgender individuals (Monterrubio et al., 2021). Shedding light on the overlooked sphere of men and masculinities in tourism, Thurnell-Read and Casey's (2014) edited work is a great initiative to bring in a wide range of empirical insights on the interdependencies of masculinities and travel. In terms of methodology, two recently edited books, *Masculinities in the Field* (Porter et al., 2021) and *Femininities in the Field* (Porter & Schänzel, 2018), provided excellent platforms for researchers to engage in reflexive discussions on gender, and its influences on the tourism fieldwork. It is encouraging to see these initiatives broadening the scope of gender studies.

Although we have seen a positive development in gender research in tourism in recent years, Figueroa-Domecq and Segovia-Perez (2020) pointed out that tourism gender studies rarely consulted their parent disciplines of feminist and gender studies. In the same vein, Aitchison (2005) emphasised the importance of tourism gender studies to engage and critically appraise the origin, particularly feminist empiricism, standpoint feminism, and poststructural feminism. In addition to the epistemological aspect, gender studies in tourism also have a significant real-life impact; the findings have effectively influenced policies, raised awareness for the general public, and are transforming future generations and work-forces. Therefore, we, in this handbook, seek to advance this knowledge further theoretically and empirically.

In addition, compared with other research areas in tourism, the current body of knowledge is still lacking in terms of discussion on teaching and learning, research approaches, and the practices of gender in tourism, particularly in the Global South. We need more critical and alternative voices, empirical insights from the overlooked regions, first-hand reflections on the pedagogical practices and research methods, and more investigations on the intersectionality of race, gender, and sexual orientations. This handbook thus aims to fill this notable gap by offering a detailed and critical understanding of the discussions related to teaching and learning, research and gender practices. The book is a novel contribution to the current conversations related to how gender is perceived in tourism. Furthermore, the handbook format helps to develop the scope and directions for future research. By proposing practical guides, contributors in this handbook offer transferable "takeaway points" that future researchers can apply. In this handbook, we also aim to provide a holistic picture, including the cycle of paradigmatic standpoints, knowledge production, tourism education, workplace, and entrepreneurship. We also aim to cover both macro (wider issues on pre-, during-, and post-COVID-19 impacts and tourism development and policy) and micro (individual experiences) perspectives.

An Idea Is Born

In 2016, the then Deputy Minister of Tourism in South Africa, Ms Thokozile Xasa, stated that "….there was a low percentage of women at board and executive management levels of large enterprises in the tourism sector" (South African Government, 2015). To transform the sector and to grow the tourism economy it was deemed necessary to build more capacity amongst women leaders and executives, especially amongst previously disadvantaged communities. The Executive Education of black women was prioritised and institutions of higher learning could tender to offer a programme on behalf of the National Department of Tourism. The University of South Africa won the bid to offer the one-year programme under the leadership of Nellie Swart for the next five intakes. Since 2016, 124 students graduated from the programme. The programme's success is evident in the regular promotions of graduates, not only in tourism but also in related sectors. In 2019, the UNWTO recognised this programme as a good practice in tourism education and training (UNWTO, 2019. However, results from the annual student surveys indicated the majority had challenges in maintaining a work-life balance, with concerns of little support from their families and employers. These alarming revelations motivated a wider investigation into the challenges experienced by individuals in the tourism industry based on their gender or gender identity.

The impetus for starting the book began with Nellie Swart's suggestion to Routledge on 30 November 2020, to work on a "Routledge Handbook on Women in Tourism" based on her involvement in the Executive Development Programme for Women in Tourism, in South Africa. Concerns raised by the students in the programme motivated the conceptualisation of the book. Faye Leerink, Commissioning Editor from Routledge, responded positively but suggested expanding the scope to include gender in tourism. Nellie shared the idea with Wenjie Cai on 2 December 2020, and Wenjie proposed inviting Elaine Yang and Albert Kimbu to join the editorial team. The first editorial meeting took place on 16 February 2021. A handbook proposal was submitted to Routledge on 15 March 2021 and was resubmitted to address reviewers' comments on 3 June 2021. Once the editorial contracts were signed with Routledge, the call for chapters was distributed on 9 November 2021.

All four editors represent the broad demographics of scholars in tourism and are non-native English speakers but work in Anglo-centric academia. Such experience provides editors with a particular lens in editing the book. Although trained in Anglo-centric institutions, all four editors' own research work has been emphasising the importance of alternative worldviews and paradigms in developing tourism knowledges away from Anglo-centric traditions and philosophies. We also believe in the importance of representation and giving platforms to the marginalised voices in knowledge production to challenge dominant discourses. Being non-native English speakers ourselves, we are also taking a more empathetic and developmental approach in the process of review and editing.

The Editorial Process

In adhering to the academic and research integrity of the collection, a rigorous peer-review process was followed. During the review process authors were commended for their interesting and well-presented contributions. Through the mentorship process, reviewers provided positive feedback on the chapters, acknowledging the authors' efforts and highlighting the potential of their research. The author's honesty and vulnerability were appreciated when sharing their autoethnographic experiences. The reviewers recognised the importance of the topics addressed in the chapters and believed they could make valuable contributions to their respective fields. Several recommendations were made by the reviewers to the authors to improve the chapters. They suggested clarifying the focus and structure, aligning the content with the aims of the edited collection, strengthening the literature review with more recent references, and ensuring theoretical foundations are well-developed. The reviewers emphasised the need for clearer connections between theory and findings, incorporating intersectionality as a lens and providing practical takeaways for readers. Furthermore, the reviewers recommended enhancing the methodology sections by justifying research approaches, providing more details on data collection and analysis, and addressing issues of translation and interpretation. They also advised revising the findings to better align them with gender perspectives, including more participant quotes, and discussing the implications of the research. In terms of writing style, the reviewers recommended improving the flow, grammar, and proofreading for consistency. Overall, the feedback acknowledged the chapters' potential contributions while providing constructive suggestions to enhance their quality, focus, and relevance to the edited collection. By implementing these recommendations, the authors have strengthened their chapters and increased their impact in their respective fields.

In the editorial process, it is with regret that we must reject certain papers submitted for inclusion in the handbook. The double peer-review process ensured the academic credibility and rigour of the chapters and also ensured all of the submissions had substantial contributions to gender studies and practices. Unfortunately, some submissions did not meet the necessary criteria. The reasons for rejection varied, but common issues included a lack of emphasis on the gender element, insufficient engagement with existing literature, unclear practical contributions, and a misalignment with the book's scope. Additionally, the rejected chapters lacked well-defined purposes, theoretical foundations, and the necessary gendered insights. Problems with methodology, weak literature reviews, insufficient use of data, and writing style and structure also contributed to the rejection. We did provide constructive and actionable feedback for authors to improve their works for future submissions. It is essential for authors to address these shortcomings by providing stronger foundations, improving the structure and clarity of their work, and adopting a more focused and gender-centric approach in future submissions.

In this handbook, we have contributions and contexts from all around the world. A wide geographical range of gendered topics captured in the 16 chapters is authored by a diverse range of 37 English-speaking and non-native English-speaking researchers. Authors are represented by 16 countries (Australia, Brazil, China, Germany, Ghana, India, Japan, Malaysia, Nigeria, Portugal, Singapore, South Africa, Spain, Uganda, the United Kingdom (UK), and the United States of America (USA)) across four continents, to enhance the international appeal and circulation of the handbook. Although we never aim to represent every corner of the world, the contextual findings and insights from different continents do offer a comprehensive understanding of gender issues across the world. In particular, we have several contributions from the Global South, which provide an alternative, yet important contribution to knowledge. We also feel it is important to provide a platform for scholars from the Global South to have their voices heard. In addition to the vast geographical representations, chapters in this book also make up a full knowledge production and application cycle, which consists of discussions on philosophical discussions, fieldwork and pedagogical reflections, and student experiences. In addition, our authors discuss both how gender is a tool of empowerment and the challenges and negotiating strategies. Some chapters also bring in novel perspectives and new understandings, either through intersectional insights or by challenging conventional understandings.

Structure of the Handbook

Six themes emanated from the 16 chapters, which are summarised in the following sections.

Teaching and Learning Gender in Tourism

Maria Gebbels (Chapter 2) and Husna Zainal Abidin and Ireena Nasiha binti Ibnu's (Chapter 3) chapters provide perspectives from both educators and students in Western Higher Education Institutions. Gebbels' chapter emphasises the need to design a gender-conscious curriculum as the key to addressing gender issues in the wider society and future workforce. Reflecting on her own teaching and learning practice, and using the principles of heutagogy, the chapter recommends educators reflexively evaluate their teaching philosophies by turning to the critical feminist theories or critical pedagogy literature, co-creating a collective and inclusive learning space where gender issues can be openly discussed, challenged, and reflected, and using innovative techniques to facilitate dialogue and conversations. From the student's perspective, Abidin and Ibnu interviewed female Muslim students from Malaysia and Indonesia studying tourism programmes in the UK and USA universities. Their study found that female Muslim students have positive experiences in a conducive learning environment, in which they feel respected and accepted by their lecturers and peers. However, the study also indicates some challenges, such as a lack of awareness regarding prayer times in the teaching terms and limited representations in the course contents. Recommendations are given from both insightful studies. Teaching and learning serve a primary role in knowledge dissemination. Discussions on gender and intersectionality should be embedded in the curriculum design and further implemented in an open, supportive classroom and beyond.

Researching Gender in Tourism

Two methodological chapters provide rich insights into researching LGBTQ+ communities. From a more theoretical level, Núria Abellan Calvet and Marta Salvador-Almela's work (Chapter 4) explores the potential of feminist research when applied to the study of tourism, specifically, the inclusion of gender non-conforming identities in the tourism sector. Investigating a case study of non-binary, genderfluid, and genderqueer identities in Barcelona, the chapter aims to fill a knowledge gap encompassing tourism and gender non-conforming tourists. The feminist approach (such as intersectionality, self-reflexivity and positionality, power dynamics, and feminist standpoint theory) challenges hegemonic and traditional knowledge production, with a specific agenda on social change and giving voices to the marginalised. By proposing some excellent questions, the chapter offers a checklist for those who would like to carry out feminist studies in tourism. From a more practical perspective, Vizak Gagrat's study (Chapter 5) discusses the challenges and negotiating strategies when conducting research with bisexual and transgender participants. Adopting an autoethnographic approach, the author discussed challenges such as difficulties in recruiting participants, sensitivity issues when researching vulnerable communities, and participants' fear of biphobia and transphobia. By reflecting on the fieldwork, Gagrat emphasised the benefits of attending events and snowballing to recruit participants and building a trusting relationship with respect and positive affirmations. The author also emphasises the importance of maintaining ethical boundaries and being mindful of sensitive issues for vulnerable participants. It is worth pointing out that both chapters contribute to the overlooked literature on transgender, bisexual, and gender non-confirming studies in tourism. By offering practical guidelines for research design, data collection, and data analysis, these two chapters are invaluable contributions to this handbook for future knowledge generations in this area.

Practising Gender in Tourism I – Tourism Development

Tourism developed in a responsible and sustainable manner is often considered a pathway to women's empowerment, achieving gender equity and overall community development, especially in marginalised societies. This is a topic explored in three chapters of this collection. A starting point to achieving gender equity and community development is through the formulation and implementation of the right policies and strategies. Consequently, in Chapter 6, Shireen van Zyl and Hugh Bartis analyse gender inclusion in South Africa's responsible tourism policies, identifying the positive changes induced by extant legislations. They equally articulate the need for more improvements in terms of alignment of terminology within policy documents, and importantly, the interpretation and implementation of these policies for women to be fully empowered. In Chapter 7, Eylla Laire M. Gutierrez further adopts an empowerment lens to examine how the experiences of Filipino women's participation in tourism activities within their communities contribute to their psychological, economic, and political empowerment while at the same time enabling them to become active agents and facilitators of community development through tourism. They equally highlight the critical importance of right supporting frameworks. Taking a focus on the role of Ubuntu in leadership and tourism development in Africa, Ogechi Adeola and Albert Nsom Kimbu in their conceptual work (Chapter 8) contend that Ubuntu can provide a foundation for socially and environmentally sustainable tourism development that fosters gender equity, promotes the well-being of local communities, and preserves natural and cultural resources. They conclude that if well understood and practised in tourism, Ubuntu could lead to the revitalisation and promotion of African traditions and values that prioritise collective well-being and community development. Collectively, these chapters provide a snapshot of the importance of policies and societal values in enabling tourism to be a driver of women's empowerment and community development.

Practising Gender in Tourism II – Gendered Tourism Workforce

The section on the gendered tourism workforce encapsulates two chapters (Chapters 9 and 10) that address different aspects of the tourism and hospitality industry in different countries, India and Brazil, with a focus on the perceptions and empowerment of female workers. The travel and tourism sector in India is described as highly profitable, generating substantial foreign revenue and creating numerous employment opportunities. Within this context, Chapter 9 by Md. Tariqul Islam and Uma Pandey provides valuable insights into the perceptions of Indian female undergraduate students who aspire to work in the rapidly expanding tourism and hospitality industry. These students' perceptions are influenced by factors such as social status, career prospects, and the work environment and benefits offered by the hospitality industry. It calls on policymakers and stakeholders in the hospitality industry, to modify their policies and work environments to cater to the needs of female workers in the industry. In Chapter 10, Cassiana Panissa Gabrielli, Natália Araújo de Oliveira, Gabriela Nicolau Santos, and Laiara Amorim Borges report the lack of attention given to racism and sexism within the Brazilian airline industry and proposes an analysis of the empowerment of black women workers in this context. Quilombo Aéreo (2022), a collective of black aeronauts formed to combat racism in aviation, is highlighted as a relevant initiative. Four dimensions of female empowerment (cognitive, psychological, political, and economic) are the focus of the analyses which support the emphasis on the intersectionality of race and gender in the airline industry. Chapter 10 highlights the need for black Brazilian women crewmembers to develop strategies based on their experiences and collective efforts, which contribute practically to their empowerment and that of other black women in the industry. Overall, both studies shed light on the experiences and perspectives of female workers in the tourism and hospitality industry in their respective countries, but they differ in their specific contexts and foci.

Practising Gender in Tourism III – Gendered Mobilities

As gender mobilities examine and explore how gender roles intersect with class, race, and sexuality to shape patterns of travel in the broader cultural, economic, and social processes, we share two chapters on this perspective. A study on "Solo female travellers' emotions. An analysis of specialist bloggers' narratives" (Chapter 11) by Marina Abad Galzacorta and Maria Cendoya Garmendia highlights the booming trend of solo female tourism worldwide and the emotions experienced by women during their solo travels, including feelings of freedom, spontaneity, empowerment, anger, and fear. The chapter analyses the profiles, emotions, and motivations of solo female travellers, focusing on Spanish-language travel blogs that emphasise emotional and safety aspects. It is evident that women's travel narratives have evolved, with expressions of emotions being a dominant theme in the weblogs of solo female travellers. Bingjie Liu-Lastres, Alexa Bufkin, and Amanda Cecil (Chapter 12) explore the role of female business travellers in the tourism and travel industry, with a focus on risk perceptions and safety concerns. A risk profile of female business travellers is developed, focusing on their perceived risks, safety concerns, and travel willingness before and after the COVID-19 pandemic. A shift in concerns towards health and safety due to the pandemic is highlighted in this chapter. Perceived safety is found to mediate the relationship between risk perception attitude variables and travel willingness during the pandemic. Additionally, self-efficacy and perceived severity significantly predict travel intentions. The study emphasises the importance of a gendered approach in understanding female business travellers' risk perception and provides implications for crafting effective marketing messages to encourage their return to business travel post pandemic. Both chapters contribute to the understanding of women's experiences in travel and tourism, examining their emotions, concerns, and evolving dynamics in the context of solo travel and business travel, respectively.

Practising Gender in Tourism IV – Gender and Entrepreneurship

The chapters in this section unpack the role of gender in shaping women tourism entrepreneurs' experiences, highlighting the opportunities, challenges, and constraints faced by women tourism entrepreneurs. Evidence increasingly suggests that women entrepreneurs are creating and managing successful businesses, but failure rates remain consistently high among women entrepreneurs (Figueroa-Domecq et al., 2022). However, Nompumelelo Nzama and Ikechukwu O. Ezeuduji (Chapter 13) in exploring gender nuances in tourism-related business performance in Durban, South Africa, uncovered that even though gender contributed to enabling access to start-up capital, it played no significant role on business performance and success. Rather management, networking, and marketing capabilities were pivotal in engendering entrepreneurial success which if deployed correctly could increase entrepreneur's performance. This viewpoint is supported by Brenda Boonabana in Chapter 14 examining women's entrepreneurship in rural Uganda where she suggests adopting a gender transformative approach to address gender constraints and underlying gender inequalities faced by tourism women entrepreneurs in Uganda.

In a similar vein, Ewoenam Afua Afenyo-Agbe, Issahaku Adam, Albert Nsom Kimbu, and Frederick Dayour (Chapter 15) adopt a migrant entrepreneurship lens to explore the role of gender and ethnicity in shaping migrant tourism entrepreneurs' experiences in Ghana. Their qualitative study not only unpacks the role of gender and ethnicity in shaping migrant tourism entrepreneurs' business experiences but equally provides key insights into their mobility decision-making process while underscoring the role of the family, as well as their experiences of dealing with and living in host communities. Similar findings are shared by Xiangli (Sally) Fan, Haili Qing, Jiamei Zhang, and Fan Zhong in Chapter 16. In this study, they unpack the critical role played by women in revitalising rural communities in China through tourism entrepreneurship. Their quantitative study not only evidences the dual disadvantages (rural and gender discrimination) faced by rural Chinese women, but, more importantly, the chapter equally highlights the critical influence of family support (psychological, capability, and financial) in helping rural Chinese women to overcome these challenges and establish tourism enterprises, thus gaining respect and recognition from their families, communities, and institutions. With this in mind, in reviewing literature on the motivations, barriers, non-governmental strategies, and government policies on women tourism entrepreneurship, Magdalena Petronella (Nellie) Swart, Vanessa S. Bernauer, and K. Thirumaran's systematic literature review in Chapter 17 contributes to debates by different stakeholders (researchers, public, and industry stakeholders) to better understand critical success factors for women tourism entrepreneurship and reflect on how these could be maximised and challenges minimised, enabling women entrepreneurs to thrive and flourish in hospitality and tourism.

Conclusion

This handbook on gender studies in tourism sheds light on the critical and reflexive understanding of gender as an essential dimension of social life that impacts every aspect of teaching and learning, tourism knowledge production and practices. Gender studies in tourism have the potential to influence policies, raise awareness, and transform future generations and workforces. The handbook aims to advance this knowledge both theoretically and empirically, encouraging more critical and alternative voices and empirical insights from overlooked regions. It also emphasises the intersectionality of race, gender, and sexual orientations in tourism studies, encouraging future research to address these aspects. Through rigorous double-blind peer reviews, the book maintains academic integrity, ensuring contributions are substantively valuable to gender studies and practices. The book is structured around six themes: teaching and learning gender in tourism, researching gender in tourism, practising gender in tourism regarding tourism development, the gendered tourism workforce, gendered mobilities, and gender and entrepreneurship. Each chapter contributes valuable insights from different parts of the world, offering a comprehensive understanding of gender issues in tourism. It covers diverse contexts and foci, providing a holistic picture of gender's impact on the tourism field. Overall, the book fills significant gaps in the current body of knowledge as it offers practical guidelines and transferable takeaways and contributes to setting future research directions. The diverse geographical author representation ensures a comprehensive understanding of gender issues worldwide, with contributions from scholars in 16 countries across four continents. Moreover, the book emphasises representation and giving platforms to marginalised voices in knowledge production, while challenging dominant gender discourses.

References

Aitchison, C. C. (2005). Feminist and gender perspectives in tourism studies: The social-cultural nexus of critical and cultural theories. *Tourist Studies*, 5(3), 207–224.

Carvalho, I., Costa, C., Lykke, N., & Torres, A. (2019). Beyond the glass ceiling: Gendering tourism management. *Annals of Tourism Research*, 75, 79–91.

Cheung, C., Baum, T., & Hsueh, A. (2018). Workplace sexual harassment: Exploring the experience of tour leaders in an Asian context. *Current Issues in Tourism*, 21(13), 1468–1485.

Claudio-Quiroga, G., Gil-Alana, L. A., Gil-López, Á, & Babinger, F. (2022). A gender approach to the impact of COVID-19 on tourism employment. *Journal of Sustainable Tourism*, 31(8), 1818–1830.

Dudley, K. D., Duffy, L. N., Terry, W. C., & Norman, W. C. (2022). The historical structuring of the US tourism workforce: A critical review. *Journal of Sustainable Tourism*, 30(12), 2823–2838.

Eger, C. (2021). Gender matters: Rethinking violence in tourism. *Annals of Tourism Research*, 88, 103143.

Figueroa-Domecq, C., de Jong, A., Kimbu, A. N., & Williams, A. M. (2022). Financing tourism entrepreneurship: a gender perspective on the reproduction of inequalities. *Journal of Sustainable Tourism*. https://doi.org/10.1080/09669582.2022.2130338

Figueroa-Domecq, C., & Segovia-Perez, M. (2020). Application of a gender perspective in tourism research: A theoretical and practical approach. *Journal of Tourism Analysis: Revista de Análisis Turístico*, 27(2), 251–270.

Gebbels, M., Gao, X., & Cai, W. (2020). Let's not just "talk" about it: Reflections on women's career development in hospitality. *International Journal of Contemporary Hospitality Management*, 32(11), 3623–3643.

Kalisch, A. B., & Cole, S. (2022). Gender justice in global tourism: Exploring tourism transformation through the lens of feminist alternative economics. *Journal of Sustainable Tourism*, 1–18.

Khoo, C., Yang, E. C. L., Tan, R. Y. Y., Alonso-Vazquez, M., Ricaurte-Quijano, C., Pécot, M., & Barahona-Canales, D. (2023). Opportunities and challenges of digital competencies for women tourism entrepreneurs in Latin America: A gendered perspective. *Journal of Sustainable Tourism*, 1–21.

Kimbu, A. N., de Jong, A., Adam, I., Ribeiro, A. M., Adeola, O., Afenyo-Agbe, E., & Figueroa-Domecq, C. (2021). Recontextualising gender in entrepreneurial leadership. *Annals of Tourism Research*, 88.

Monterrubio, C., Mendoza-Ontiveros, M. M., Rodríguez Madera, S. L., & Pérez, J. (2021). Tourism constraints on transgender individuals in Mexico. *Tourism and Hospitality Research*, 21(4), 433–446.

Porter, B. A., & Schänzel, H. A. (Eds.). (2018). *Femininities in the field: Tourism and transdisciplinary research*. Channel View Publications.

Porter, B. A., Schänzel, H. A., & Cheer, J. M. (Eds.). (2021). *Masculinities in the field: Tourism and transdisciplinary research*. Channel View Publications.

Quilombo Aéreo. (2022). *Quilombo Aéreo*. https://quilomboaereo.com.br/#

Ribeiro, A. M., Adam, I., Kimbu, A. N., Afenyo-Agbe, E., Adeola, O., Figueroa-Domecq, C., & de Jong, A. (2021). Women entrepreneurship orientation, networks and firm performance in the tourism industry in resource-scarce contexts. *Tourism Management*, 86.

Small, J. (2016). Holiday bodies: Young women and their appearance. *Annals of Tourism Research*, 58, 18–32.

South African Government. (2015). *Deputy Minister Tokozile Xasa: Launch of Executive Development Programme for black women managers in tourism sector*. Available from: https://www.gov.za/speeches/deputy-minister%E2%80%99s-speech-during-occasion-launch-executive-development-programme-black-women

Thurnell-Read, T., & Casey, M. (Eds.). (2014). *Men, masculinities, travel and tourism*. Springer.

United Nations World Tourism Organization (UNWTO). (2019). *Global report on women in tourism* (2nd ed.). UNWTO. Available from: https://www.e-unwto.org/doi/epdf/10.18111/9789284420384

United Nations World Tourism Organization (UNWTO). (2023). *Tourism for SDGs*. https://tourism4sdgs.org/sdg-5-gender-equality/

Usai, R., Cai, W., & Wassler, P. (2022). A queer perspective on heteronormativity for LGBT travelers. *Journal of Travel Research*, 61(1), 3–15.

Zhou, Y., Mistry, T. G., Kim, W. G., & Cobanoglu, C. (2021). Workplace mistreatment in the hospitality and tourism industry: A systematic literature review and future research suggestions. *Journal of Hospitality and Tourism Management*, 49, 309–320.

PART I

Teaching and Learning Gender in Tourism

2
TOWARDS A GENDER-CONSCIOUS TOURISM CURRICULUM

Lessons from the Classroom

Maria Gebbels

Abstract

Promoting gender equality and discussing gender issues in tourism openly and inclusively can be facilitated by moving towards a gender-conscious tourism curriculum. University students need to be equipped with knowledge, skills, and experiences that will enable them to navigate the ever-changing world in a purposeful and meaningful way. This chapter shares reflections on teaching and learning, drawing on the author's classroom experience, to illustrate the challenges and opportunities of incorporating gender in tourism programmes. By critically drawing on the principles of heutagogy, this chapter offers insights on the value of "revisiting teaching philosophy," "co-creating a collective and inclusive learning space," and promoting "learning and teaching through conversation" as essential pillars to teaching and learning gender. In doing so, this chapter contributes to the ongoing discussions on the need to design a gender-conscious curriculum bringing to attention gender issues as being mainstream in tourism education and beyond.

Keywords

Tourism education, Self-reflection, Heutagogy, Learning and teaching, Self-determined learning

Introduction

Despite increased research on gender and tourism in recent years, little is known about teaching and learning gender in tourism. Yet, according to Segovia-Perez et al. (2019), tourism education can play a significant role in removing stereotypes and gendered assumptions, as well as promoting good practices in relation to equity. Furthermore, tourism employment is considered a source of empowerment for women and minority groups and achieving gender equality and empowering women and girls features prominently as the fifth United Nations Sustainable Development Goal (UN SDG). Yet, as pointed out by

DOI: 10.4324/9781003286721-4

Alarcón and Cole (2019, p. 903), "without gender equality, there can be no sustainability," further concluding that gender equality is the foundation for all 17 UN SDGs. Academia and education, because they produce and disseminate knowledge, have an important role to play in shaping ideas and discourses to help achieve the SDGs (Dashper et al., 2022) in the classroom environment and through relevant curriculum changes.

Our responsibility as educators has got to be to engage in teaching practice which enables students to not only learn their core subjects and better themselves but also equip them with knowledge, skills, and experiences that will allow them to navigate the ever-changing and uncertain (post)modern world. Tourism education, especially in the United Kingdom (UK) where vocational elements of programmes continue to be highly valued, is an important component of the reproduction of the new workforce. Future managers and leaders are, arguably, students in tourism, and it is during their time at university that introducing them to current debates and misconceptions on the ever-present gender stereo-types and inequalities needs to take place (Gebbels et al., 2020). It is this new generation of future leaders who are indispensable to eradicating gender inequality in tourism and beyond. As Jeffrey (2017) urges us all teaching gender equality and female empowerment in tourism helps to promote these issues on a larger scale and raise awareness for our future leaders. Gebbels et al. (2020) have also called for the need to address gender issues within educational institutions in order for students to enter into the work environment with minimal gender bias.

This chapter contributes to the first part of this handbook: *Teaching and learning gender in tourism*, with the aim of sharing reflections on teaching and learning, in particular drawing on the author's classroom experiences, to illustrate the challenges and opportunities of incorporating gender in tourism programmes in the UK. By critically drawing on the principles of heutagogy, which advocates learner agency and is defined as a self-determined learning where the learner takes centre stage in the learning process, this chapter offers insights into the value of *revisiting our own teaching philosophy*, *creating a collective and inclusive learning space*, and promoting *learning and teaching through conversation* as essential pillars of teaching and learning gender. In doing so, this chapter contributes to the ongoing discussions on the need to design a gender-conscious curriculum bringing to attention gender issues as being mainstream in tourism education and beyond.

Gender and Curriculum: Contested Terms

Our societies are governed by unwritten rules, norms, and expectations, many of which are culturally conditioned. For instance, the societal norms of gendered roles, labour division, and patriarchy are easily translated into the classroom environment. In that space, they should be questioned rather than unintentionally reinforced. Within the context of tourism and hospitality education, underpinned by strong practical and vocational elements, Tribe (2002) has called on academia to educate and raise graduates as *philosophic practitioners*. In short, tourism programmes should incorporate both liberal and vocational education equipping students with the ability to reflect and act (Inui et al., 2006). To better reflect the changes in higher education and industry employment, such graduates are now being characterised as those who can understand and critique higher order academic knowledge and skills, apply and critique higher order practical knowledge and skills, develop a critical self with a critical understanding of the world, formulate reasoned visions of a better tourism world, and participate in activities for a better tourism world (Tribe & Paddison, 2021, pp. 11–12). Developing such future leaders will not be possible without a strong commitment from teachers, in this case study called the facilitators of learning to encapsulate the principles of heutagogy, who create a classroom environment which encourages critical thinking, transformation, and identity formation (Lalendle & Msila, 2020). Such progressive teachers are also known to be very aware of the importance of discussing issues of gender, race, or class alongside traditional content of tourism curriculum in order to raise awareness of current debates among future employees in the industry (Segovia-Perez et al., 2019).

Before going any further, it is essential to clarify that the definitions of the two key terms used here, gender and curriculum, are contested. The definition of gender is different across varying theoretical, research, and practice domains, including also very diverse interpretations of "gender," what it means, how we study it, and how it mediates our understanding of the world and our experiences of it. Bradley (2013) considers gender as a lived experience operating and influencing three areas of social life, which are "production," "reproduction," and "consumption." For other thinkers, like Judith Butler, a postmodern feminist, gender is fluid and therefore doing gender is underpinned by performativity, whereby sex and gender are inextricably linked and are played out by male and female identity in everyday lives. This symbolises a move away from gender as dichotomous.

Similar to gender, the curriculum is also a contested term, positioned differently depending on the theoretical and ideological leanings of the scholars working within the field. A classic definition of curriculum comes from Lawton (1975, p. 6) who argued that rather than it being "that which is taught in classrooms" curriculum is "essentially a selection from the culture of society ...certain aspects of our way of life, certain kinds of knowledge, certain attitudes and values are regarded as so important that their transmission to the next generation is not left to chance." As such, the curriculum is influenced by social and cultural values, knowledge and skills that are deemed necessary for young people to know to prepare them for life and their future work. Therefore, I want to argue that like gender, the curriculum is not a fixed but a dynamic entity. It is influenced by the ideological perspectives of lawmakers and government leaders, the changes in economies and societies as well as the beliefs, traditions, and values of those who teach and of those who learn. Indeed, cultural and social contexts are both at the heart of Vygotsky's theorisation of learning, whereby the possibility of learning cannot be separated from its social context, giving rise to social constructivism (Shabani & Ewing, 2016). This results in the need for academics to ensure that their students are aware of the most recent debates and emerging paradigms which are related to tourism both directly and indirectly. One way to do that is by including the highly debated subject of gender in the tourism curriculum using the principles of heutagogy.

Principles of Heutagogy

In order to incorporate the subject of gender successfully into the tourism curriculum, first, we need to consider the teaching strategy that will help us to achieve that. Pedagogy, which is developed on theories of cognitivism, constructionism, and behaviourism, primarily focuses on teaching children, is subject-centred, delivers knowledge objectively, and therefore follows a very uniform step-by-step progression (Halupa, 2015).

Rethinking education within the context of adult learning and higher education implies a shift from self-directed, or student-directed learning which is at the core of andragogy to self-determined learning, key in heutagogy. As simply explained by Halupa (2015), pedagogy is faculty-centred education, andragogy is student-centred education, and heutagogy is self-determined and transformative. Yet, these three modes of learning are now being seen as applicable during a person's entire intellectual learning and development journey which is life-long, and the choice of any of the modes is based on a larger context and what is being learned (Paine, 2021). Hase (2015) acknowledges the difference between self-directed learning and self-determined learning, which are often mistaken for each other, by stating that self-directed learning is "a subset of self-determined learning (…) it is a quality and a process of self-determined learners." Therefore, heutagogy is often referred to as a teaching strategy or a methodology for adult learners built on dialogue, self-reflection, and exchange (Kenyon & Hase, 2001; Lalendle & Msila, 2020). First defined by Hase and Kenyon (2013) as self-determined learning, heutagogy has become a strategy that helps individuals to know how to learn as the key skill for the 21st century (Advance HE, 2020). Heutagogy builds on the self-directed principles of andragogy in which students develop their own learning skills. As such, it is underpinned by the philosophies of humanism and constructivism (Hase & Kenyon, 2013) whereby the learner is central to the educational process of learning (see Rogers, 1969), and the learner constructs their own reality using past and present experiences and thus becomes actively engaged in learning (see Dewey, 1938; Piaget, 1973; Vygotsky, 1978).

As it calls for self-determined learning, the process rather than the outcome of learning is emphasised (Kenyon & Hase, 2001). Therefore, when teaching a complex subject such as gender, the key principles of heutagogy can be applied. These are self-reflection and double-loop learning, based on the theory of action. They emphasise the importance of questioning personal values and assumptions during learning (Akyıldız, 2019).

Heutagogy is prospective in approach, whereby knowing how to learn is recognised as a fundamental skill (Snowden & Halsall, 2014). It does this by providing a learner-centred environment by helping students define their learning path. The role of the educator is less prominent than that of a student, and learning becomes a negotiated experience rather than a formal exchange of ideas. Critical to heutagogy in the context of vocational education and training is being able to recognise that the learner is key to all areas of the learning process (Kenyon & Hase, 2001). McPherson (2016) contents that collaboration is a powerful tool in heutagogy because students learn much from one another as they do from the facilitator of learning. Snowden and Halsall (2014, p. 4) cite research that concluded the heutagogical approach as being key in helping students to control their learning through reflective practice, resulting in enhancing their professional development. Heutagogy changes traditional teaching as we know it by addressing the past inactivity of the learner who is now expected to work with the teachers/facilitators of learning as co-creators of knowledge.

The essence of heutagogy has all the elements that seek to free the learner who is a trusted player in the education process, as it emphasises the idea of knowledge as a shared experience (Lalendle & Msila, 2020). As Halsall et al. (2016) explain, heutagogical approaches to education emphasise the importance of holism, self, capability, community, and societal needs. Within that, Snowden and Halsall (2014) propose two key collaborative strategies which support heutagogy: solution-focused approach to teaching and learning, and mentor-assisted learning. The former is considered a real-world approach to pedagogy. It develops critical consciousness, collective identity, and solution-orientated strategies for change because it is based on the discovery of challenging beliefs, values, and solutions, introducing the learner to concepts such as social injustice, oppression, inequality, and domination. It is therefore based on creating transformative learning and teaching experience, activating learners to become committed, engaged citizens, and recognising that development requires change at individual, societal, and cultural levels (Snowden & Halsall, 2014). Mentor-assisted learning, established by providing guidance and support, is based on developing strong mentor and mentee relations which will culminate in a learning landscape for the student that recognises and helps to battle anxieties, promotes ways to navigate university systems and processes, and offers encouragement and motivation (Halsall et al., 2016). Therefore, heutagogy places less importance on the traditional teacher and more on students who are guided to take responsibility for designing their learning pathways. In doing so, they are being equipped with the skills and capabilities to become autonomous and lifelong learners (Advance HE, 2020).

Educating the Future Workforce

As already established earlier in the chapter, education plays a crucial role in (re)producing the new workforce and many of our future leaders in tourism and hospitality are or have been students at further education or higher education institutions. For hooks (1994), the classroom should be the most radical space of possibility in the academy; exciting and never boring. It is there that status quo needs to be questioned. This bringing about change can be facilitated by taking on a liberal approach which aims to bring to attention to a broad range of issues relevant to tourism and hospitality curriculum, including the subject of gender (Tribe, 2002). It should be clear to us as educators that the new generation of women and men are indispensable to eradicating gender inequality in the industry (Gebbels et al., 2020).

Therefore, a shift to a gender-conscious curriculum design should be encouraged in order to create a commonplace where gender (in)equality can be talked about. It is our responsibility to encourage students to learn about and reflect upon gender issues and challenges present in the hospitality and tourism industry. There are two ways in which this can be achieved. Jeffrey (2017) proposed two methods of incorporating the subject of gender in tourism curriculum (Gebbels et al., 2020; Jeffrey, 2017). Jeffrey (2017) recommends that the subject of gender can be discussed throughout any module/course, including gender-related topics called "gender mainstreaming." There can also be a dedicated module/course which covers contents focusing on specific issues of gender (in)equality known as "gender specialising." The following self-reflective accounts will illustrate the challenges and suggestions about including gender in the tourism curriculum drawing on the author's personal experiences of teaching this subject.

Examples from the Classroom: Becoming a Self-Reflexive Practitioner

Using principles of auto-ethnography, I will reflect on my practice as an educator (Ryan, 2012), drawing on my own experiences in the classroom by recounting my efforts of incorporating the subject of gender in learning and teaching activities delivered to postgraduate students in a post-1992 university in London.

Methodology

As a useful feminist method, auto-ethnography allows for critical reflection by giving us a space to engage in internal conversation (Ryan, 2012). As a valuable research method, it is both a process and a product as the researcher draws on autobiography as well as ethnography to write the auto-ethnography (Edwards, 2021; Ellis et al., 2011). Questioning the possibility of ethical auto-ethnography, Edwards (2021) explains that this method can be undertaken with temporal congruence or as it was here, retrospectively. This means that I engaged in doing auto-ethnography much later after the teaching took place. I recalled and recounted the events, experiences, and feelings of delivering the session on gender issues from memory. But because auto-ethnography epiphanies happen as part of our everyday lives and are not located in the actual space of research, I never actually took any notes right after the teaching experience. I reflected on the teaching experience and the feedback that I received and used that as an opportunity to make future changes to the way that session could be delivered (Edwards, 2021; Ellis et al., 2011). Although aware of the emerging practice of collaborative auto-ethnography, I decided to reflect on the teaching experience as a teacher alone rather than also involving my students in the process (see Reyes et al., 2020). Therefore, engaging in retrospective auto-ethnography enabled me to focus on and selectively write about and analyse a particular teaching experience, which, like epiphanies for autobiographers, further allowed me to question, dissect, and understand the culture and practices of a higher education institution (Edwards, 2021; Ellis et al., 2011).

It was previously explained that principles of heutagogy are grounded in (social) constructivism. In particular, feminist social constructionism sees what we know and how we know it as socially constructed through interactions with others (Small et al., 2011). Also, the two key terms used throughout this chapter, gender and curriculum, are contested and there is a lack of universal definitions. Therefore, it was necessary for me to acknowledge that how students define gender and curriculum for themselves during and after my teaching would be largely driven by their own ontological and epistemological perspectives, influenced by their own experiences, upbringing, and worldviews. The same could be said for me. Thus, my role as the facilitator of learning, practising heutagogy, was to help them challenge their existing preconceptions and biases. For that, it was important to me to draw on our already well-established working relationship based on respect, trust, kindness, and mutual understanding. I try to achieve such a relationship from the beginning of all my classes, by facilitating an in-depth discussion about the expectations I have of students on this module, as well as asking them to outline their expectations of teaching and learning they will be undertaking.

In the following case study, I will reflect on how my own (re)thinking of gender has changed by showcasing a gender-specialising learning activity, informed by the principles of heutagogy, and how the taken for granted aspects of student feedback can act as the basis for critical self-reflection and discussion.

Case Study: Self-Reflective Account of Teaching Gender

I have been a full-time academic for six years and have taught on undergraduate and postgraduate modules. Part of my role over the last three years has been also as a postgraduate programme leader looking after tourism and hospitality management students.

I have found that there are many very effective ways to achieve a gender-conscious curriculum which I demonstrate in this case study. Guiding students through collaborative learning and helping them to become independent, self-determined learners has been key. Therefore, when teaching gender in tourism, practising the principles of heutagogy including also helping students to challenge their existing preconceptions and biases enables the development of shared vision on a topic whilst paying attention to individual voices and opinions.

Critical pedagogy, coined by Paulo Freire, advocates inclusive teaching by examining issues of power in the classroom (Saunders & Wong, 2020) and can be embraced in the classroom by beginning to question our values, beliefs, and assumptions, and by translating knowledge into practice through challenging the biases based on gender, as well as race, ethnicity, or class, and questioning the existing, oppressive power structures that inhibit student learning (Brookfield, 1995; Fullagar & Wilson, 2012). It is about helping students make sense of and respond to key issues and for us as educators to move out of our comfort zones (Brookfield, 1995).

It has always been my practice to begin certain lectures by laying out the definitions of the key terms. I tend to do that more so in the first one or two lectures; when I introduce a new module; in foundation modules delivered to first-year students especially when they are being taught a wide range of subjects; and in modules where the focus is on critical tourism and hospitality. I find that beginning with two or three definitions has many benefits, including a short discussion on which sources are appropriate to cite when giving definitions and why deciding on the most relevant definition shows critical analysis. Another benefit is to question the gendered assumptions of definitions, which I think far too often can be taken for granted as gender neutral. For instance, when discussing the subject of entrepreneurship, it is necessary to spend some time challenging the gendered entrepreneurial discourse. To move further with a gender-conscious curriculum, presenting the definition of entrepreneurship without a critical perspective is not enough. One may choose to plan an activity where students deliberate whether entrepreneurship is a gender-neutral term, leading to a realisation that the dominant discourse of entrepreneurship is masculinist, and capitalistic, assuming gender neutrality, and positioning men as the proxy of normative practice (Marlow & Martinez Dy, 2018).

We, the educators, need to continue to question the gendered assumptions, statements, and definitions in all of our teaching and allow our students to reflect on their assumptions and preconceived ideas in the safety of the classroom environment. Therefore, "gender mainstreaming" can be a very effective method in continuing to eradicate gender inequalities.

I decided to try out "gender specialising" for the first time, consciously, in one of my postgraduate modules. Student feedback indicated that they wanted to discuss current themes in tourism and hospitality. Co-designing teaching and learning activities is one example of incorporating heutagogy into the curriculum. Also, I wanted to make the subject of gender the entire focus of the lecture and tutorial learning and teaching activities. The purpose of that chosen module is to develop students' knowledge of the characteristics and the role of critical tourism and hospitality studies within wider society emphasising the point that the two sectors are both a manifestation of society and culture and a means by which these are experienced. Since we have been strongly encouraged to practise research-informed teaching, this module fitted very well as the space where I could share my latest research on gender issues in tourism and hospitality, co-authored with two other colleagues, both male. This was also the first year of running this particular module, and I wanted to use this time to introduce new topics which I knew where not being covered on other modules on this programme. The cohort was made up of 22 students, largely female, with the large majority of international students (students who came from outside of the UK and the European Union).

I was quite confident about presenting to my students on this subject. By then, the students and I had gotten to know each other quite well and we managed to establish a good working relationship based on respect, kindness, and a sense of mutual understanding. Students also got to know me as their programme leader. My self-confidence was also based on the fact that I am the first author of the journal article, I am a female, with dual nationality, I presented this paper to other audiences before, including hospitality industry professionals in the UK. I basically felt I knew my stuff.

I decided to structure my two-hour session beginning with a discussion on the topic: "gender inequality in the tourism industry belongs to the past," for which students were asked to prepare in advance by researching gender issues in hospitality and tourism from their countries of origin using official government websites, research publications, and other relevant resources.

To embrace the principles of heutagogy based on learner's agency and equip students with tools to become self-determined learners, I wanted the students to lead this session from the start, and for my research to act as a much-needed conduit between students' findings and the recommendations that I was planning on sharing with them. I also added further learning and teaching resources on the virtual learning environment for students to engage with in their own time after the session.

From the beginning of the session, it became clear that only about half of the class came prepared for the discussion, so I split the cohort in such a way to ensure that both groups had equal chances to do well in the discussion, by having students who prepared in advance split between the two groups. Both groups did their best in the discussion by presenting a range of arguments, citing research in their native language, and demonstrating critical analyses. After 30 minutes, they reached a conclusion that more needed to be done to eradicate gender in-equalities in their own home countries and beyond. This is when I was able to share my latest research, from the UK context. I began by sharing an overview of the key themes on gender in tourism and hospitality, including precarious working environment, being highly gendered, and male-dominated in its values and boardrooms. I outlined the research methodology and stressed that findings are derived from thematic analysis of the discussion which had taken place during a public seminar. I then moved on to the recommendations proposed in the paper on reducing gender inequalities presented under three headings: listening environment, education, and bringing men into the equation. I finally concluded my presentation by quot-ing prof. Nigel Morgan whom I once heard stating that "gender is everyone's problem," with a hope to encourage students to continue discussions on gender and questioning their own positionality, preconceptions, unconscious bias, and stereotypes.

After the two-hour session came to an end, I felt the activities I had planned went well based on the overall good student engagement, and positive classroom atmosphere. Two students, both male, decided to stay behind to speak to me after everyone else left. They wanted to offer some feedback and I listened. Although they enjoyed the session, they felt that as the minority gender, they could not fully engage in the discussion. They felt uncomfortable and somewhat marginalised, and much of the discussion was focused on how highly male-dominated the industry is, and on female underrepresentation in senior leadership roles. They felt that men, and therefore them personally, were being blamed for all aspects of gender inequality, which we know it is not correct. Clearly, I placed too lit-tle emphasis on the importance of male allies during my lecture, not enough on structural barriers, and possibly came across as judgmental. My own positionality and prejudice took the better of me, and I failed to establish a safe and non-judgmental space for all students to feel comfortable in. Since this feedback took me by surprise, I had to think on my feet. I asked the two students what I could have done differently. They suggested that having one of the male authors of the research presenting with me would have made them feel less in-timidated. They also suggested a different structure to a discussion to ensure that the more knowledgeable and confident students were not taking over the activities, and everyone would have a chance to contribute.

Recommendations and Advice for Future Practice

The following section on recommendations and advice for future practice results from my further reflections on the above case study. Also, the three pillars to teaching and learning gender, based on the principles of heutagogy, have enabled me to better structure these practical recommendations and enhance my own teaching practice of working towards a gender-conscious curriculum in tourism. By no means is the list an exhaustive one. My intention is to share what I have considered a helpful set of practical advice and suggestions largely derived from my own experiences and reading and research on the topic. In doing so, I hope to contribute to the discussions on how to contribute towards designing a gender-conscious curriculum bringing into attention gender issues as being mainstream in tourism education and beyond.

1 Revisit Your Teaching Philosophy

Heutagogy, as previously discussed, leads to a transformative learning and teaching experience. An effective teacher who takes on a new identity as the facilitator of learning will incorporate the two collaborative heutagogical strategies, solution-focused approach to teaching and learning, and mentor-assisted learning (Snowden & Halsall 2014). Before this can be achieved, however, my experience of teaching gender has helped me to realise how important it is to revisit our own teaching philosophy and, by doing so, continue to develop our critical self. For instance, turning to critical feminist theories or critical pedagogy literature can assist us in (re)examining our own values, goals, and beliefs about teaching and learning. This is also a good opportunity to consider our own (implicit) bias, how we exert or contain power in the classroom, and our ability to try out different teaching styles successfully. Practical strategies include but are not limited to starting with writing down our teaching philosophy and revisiting it on a regular basis, keeping a teaching journal, recording voice notes after teaching sessions using a reflective framework to aid the reflection, sharing teaching experiences and practices with other educators, and attending teaching and learning conferences.

Doing so may not only help to protect and promote interests of our female students and minority ethnic groups within our classrooms but also to highlight research-informed teaching heavily promoted and encouraged in higher education.

#2 Co-Create a Collective and Inclusive Learning Space

Another practical suggestion to move towards a gender-conscious curriculum design and practice heutagogy is to co-create a space where students feel safe, included, and have a sense of belonging. When designing my learning and teaching activities for the session on gender issues in tourism and hospitality, I had not considered the importance of the learning space and those who will inhabit it with me. Only upon reflection I began to realise that this was missing. One method to create such a space is to remove the possibility of unintentionally creating parallels between our own students and examples of gender-related issues. Vignettes have been called a research method tool, a methodological paradigm, a reflective writing tool, and can be written or visual in form (Langer, 2016). Despite their many research purposes, "a common purpose for using vignettes is to elicit information through inviting responses, encouraging discussions, and probing for understandings to gain insights to participants' beliefs, emotions, judgments, attitudes and values about the particular phenomenon that lies at the heart of the research" (Skilling & Stylianides, 2020, p. 542).

Using vignettes allows students and teachers to self-reflect and question personal values and assumptions during learning, which are also key principles of heutagogy. By creating a fictional character or an employment organisation, students can begin to provide solutions, recommendations, and action points to resolve an issue at hand around the subject of gender, whilst drawing similarities and differences between their own experiences and those of the created individual. An example of a short vignette about Alex is below.

Example

"Alex": head receptionist, satisfied with her career prospects; has worked in luxury hotels for ten years; a single, young professional, passionate about dancing. Her family does not understand her professional life choices and her passion for hospitality; they think that as a woman she should be looking for a husband and have children. In her culture, women still tend to hold positions that are extensions of their traditional domestic roles; and women are not being represented where true power exists.

Her passion for dancing has led her to consider becoming a qualified dance teacher. However, she realises that her career in hotel operations has often come first before her hobbies because so far any informal requests to change her working hours were not met. Yet she decides to pursue her dream of becoming a dance teacher.

After she qualifies as a dance teacher, she wants to do more teaching, but she also does not want to give up her work which she loves. She decides to put in a formal request for flexible working, be it job sharing, working from home on some of the days of the week, or compressed hours. She has been made aware that her request may be declined again if it is not in line with the business needs.

#3 Learn and Teach through Conversation

Dialogue becomes a critical component when teaching using heutagogy and can facilitate social transformation of the student and teacher experience (Msila, 2021). However, effective dialogue can also be a challenging element to include in teaching for reasons to do with students' knowledge of the subject; feeling reserved about discussing the subject of gender based on their cultural or other predispositions; and feeling challenged or uncomfortable to express themselves freely and confidently about such a subject in front of the class. As discussed in the case study, some students may feel that the discussion is dominated by one or two other students, and therefore it becomes our collective responsibility to mitigate such an issue.

The example of a vignette shared earlier can act as the starting point for conversation and larger discussions, which can be further supported by the techniques discussed below. There are many innovative, creative, and cost-effective techniques that can be used in teaching complex subjects such as gender that further facilitate an inclusive learning space (Gebbels et al., 2020). The key is creating dialogue as a collective approach where each student is given time to reflect, express their feelings using these creative methods, and finally arrive at a shared vision (McCusker, 2019).

- LEGO® Serious Play®, based on building models using LEGO® bricks, is a playful method to discuss contested issues, research ideas, and co-create solutions and recommendations for a challenging subject such as gender in tourism (Wengel, 2020). It facilitates collaborative group discussion environments where students, as stakeholders from diverse backgrounds, can share their opinions, and unlike other more traditional methods such as focus groups, LEGO® Serious Play® does not emphasise homogeneity (McCusker, 2020). To facilitate this method about gender in tourism, students can be presented with a topic "the impact of gender inequalities on tourism" and asked to build a model that represents their ideas. Leaving the topic rather vague is intentional to enable the students to interpret the key words: impact, gender inequalities, tourism, based on their own knowledge and conceptualisation (McCusker, 2020). Each student is then asked to explain their models in more detail, sharing about the significance of using different colours or figures. Students are then asked to work together to combine their models into a single model of a shared vision on the topic.

- Hands-on tools for creative group work include Mandala, originally known as a form of art, and refers to drawing a set of circles whilst becoming more self-aware (Potash et al., 2016), and Ketso which is a growing in popularity technique that promotes inclusivity and gender equality and helps to overcome barriers (Ketso.com, 2019). Ketso method is particularly valid when discussing gender because it simultaneously facilitates individual voice and group analysis (McIntosh and Cockburn-Wootten, 2016). Dashper et al. (2022) used Ketso to stimulate open and honest discussion about gender inequalities in tourism academia and to encourage active involvement and break down barriers in teams. Each of the four Ketso workshops is based on a tree analogy, and the leaves represent ideas offered by each participant. This allows for a constructive dialogue, ability to contribute to a discussion, and freedom to voice opinions in a non-hierarchical and inclusive environment.
- Designed to be used alongside existing curriculum and teaching methods, drama-based techniques based on principles of drama-based pedagogy help to overcome racial or gender stereotypes (Dawson & Lee, 2016). Designed to be used alongside the existing curriculum, drama-based pedagogy includes tools such as activating dialogue, theatre games as a metaphor, image work, and role work (DBP, 2022).

Conclusion

Integral to all social institutions and interactions is the subject of gender, and gender equality is the foundation for all 17 UN SDGs. It should, therefore, be an integral part to learning and teaching, research, and critical reflection in all educational settings. Yet, the subject of gender has been neglected within tourism curriculum leaving students unprepared for dealing with the realities of future workplaces and ill-equipped to contributing to global debates on eradicating gender inequalities.

The aim of this chapter was to share reflections as well as challenges and opportunities of incorporating gender in tourism programmes in the UK. Having established that both, gender and curriculum, are contested terms, which meaning is influenced by the ontological and epistemological perspectives of those defining them, the key principles of heutagogy were discussed as foundations to teaching gender. Placing less importance on the teacher and more on students who are guided to take responsibility for designing their learning pathways, heutagogy is the teaching strategy fit for a 21st-century student and future manager, leader or business owner in tourism. It aims to equip students with skills and capabilities to become autonomous and lifelong learners (Advance HE, 2020).

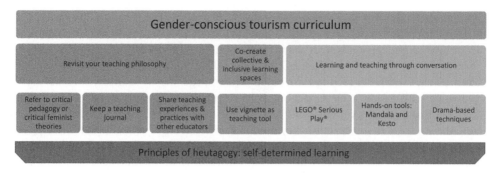

Figure 2.1   Key-takeaway recommendations.

In order to share her own experiences of incorporating gender in the tourism curriculum, the author reflected on her teaching practice by discussing examples of gender specialising and gender mainstreaming, which led to practical recommendations for future practice, founded on the principles of heutagogy. Firstly, educators are encouraged to evaluate their teaching philosophies by turning to critical feminist theories or critical pedagogy literature, which can assist in (re)examining one's values, goals, and beliefs about teaching and learning. Secondly, co-creating a collective and inclusive learning space can act as an enabler for challenging discussions about gender, and bringing in vignettes allows students and teachers to self-reflect and question personal values and assumptions. Thirdly and finally, learning and teaching through conversation can facilitate social transformation of student and teacher experience. LEGO® Serious Play®, Mandala, Ketso, and drama-based techniques have been proposed as effective tools to create dialogue in a collective way by giving each participant time to reflect, express their feelings in creative ways and to arrive at a shared vision (Figure 2.1).

As highlighted in this chapter, learning from and acting on student feedback becomes invaluable for the educator to reflect on the teaching experience and their practice. One limitation of this study is related to relying on memory work due to a lack of written or voice notes. Therefore, in the future taking notes and scheduling in feedback from students about their experiences of heutagogical learning could aid in further improvements to one's teaching practice.

Collaborative auto-ethnography is becoming an emerging practice. Researchers are encouraged to engage in this form of auto-ethnography, which would allow for the unveiling of shared experiences of teaching gender that will further inform the teaching practice and offer additional insights on the subject itself.

References

Advance HE. (2020). *HEA to Z: Heutagogy*. Retrieved from https://www.advance-he.ac.uk/knowledge-hub/heutagogy

Akyıldız, S. T. (2019). Do 21st century teachers know about heutagogy or do they still adhere to traditional pedagogy and andragogy? *International Journal of Progressive Education*, *15*(6), 151–169. https://doi.org/10.29329/ijpe.2019.215.10

Alarcón, D. M., & Cole, S. (2019). No sustainability for tourism without gender equality. *Journal of Sustainable Tourism*, *27*(7), 903–919. https://doi.org/10.1080/09669582.2019.1588283

Bradley, H. (2013). *Gender*. Cambridge.

Brookfield, S. (1995). *Becoming a critically reflective teacher*. Jossey-Bass.

Dashper, K., Turner, J., & Wengel, Y. (2022). Gendering knowledge in tourism: Gender (in)equality initiatives in the tourism academy. *Journal of Sustainable Tourism*, *30*(7), 1621–1638. https://doi.org/10.1080/09669582.2020.1834566

Dawson, K., & Lee, B. K. (2016). *Drama-based pedagogy activating learning across the curriculum.* The University of Chicago Press.

DBP. (2022). *Drama-based pedagogy.* Retrieved from https://dbp.theatredance.utexas.edu/about

Dewey, J. (1938). *Experience and education.* Macmillan Company.

Edwards, J. (2021). Ethical autoethnography: Is it possible? *International Journal of Qualitative Methods, 20,* 1–6. https://doi.org/10.1177/1609406921995306

Ellis, C., Adams, T. E., & Bochner, A. P. (2011). Autoethnography: An overview. *Historical Social Research/Historische Sozialforschung, 36,* 273–290.

Fullagar, S., & Wilson, E. (2012). Critical pedagogies: A reflexive approach to knowledge creation in tourism and hospitality studies. *Journal of Hospitality and Tourism Management, 19*(1), 1–6. https://doi.org/10.1017/jht.2012.3

Gebbels, M., Gao, X., & Cai, W. (2020). Let's not just 'talk' about it: Introducing an action plan for transforming gender relations in tourism and hospitality. *International Journal of Contemporary Hospitality Management, 32*(11), 3623–3643. https://doi.org/10.1108/IJCHM-05-2020-0385

Halsall, J. P., Powell, J. L., & Snowden, M. (2016). Determined learning approach: Implications of heutagogy society based learning. *Cogent Social Sciences, 2*(1), 1–11. https://doi.org/10.1080/23311886.2016.1223904

Halupa, C. (2015). Pedagogy, andragogy and heutagogy. In C. Halupa (Ed.), *Transformative curriculum design in health sciences education* (pp. 143–158). IGI Global.

Hase, S. (2015). *Self-directed learning and self-determined learning: An exploration.* Heutagogy Community of Practice. Retrieved from https://heutagogycop.wordpress.com/2015/12/05/self-directed-learning-and-self-determined-learning-an-exploration/

Hase, S., & Kenyon, C. (2013). *Self-determined learning: Heutagogy in action.* Bloomsbury Publishing.

hooks, b. (1994). *Teaching to transgress education as the practice of freedom.* Routledge.

Inui, Y., Wheeler, D., & Lankford, S. (2006). Rethinking tourism education: What should schools teach? *Journal of Hospitality, Leisure, Sport and Tourism Education, 5*(2), 1–12.

Jeffrey, H. L. (2017). Gendering the tourism curriculum whilst becoming an academic. *Anatolia, 28*(4), 530–539.

Kenyon, C., & Hase, S. (2001). Moving from andragogy to heutagogy in vocational education. In *Proceedings of the Australian Vocational Education and Training Research Association (AVETRA) conference,* 4th, Adelaide, Australia, March 28–30.

Ketso.com. (2019). *Ketso is empowering women around the world.* Ketso Insights. Retrieved from https://ketso.com/ketso-is-empowering-women-around-the-world/

Lalendle, L. L., & Msila, V. (2020). Heutagogy, decolonisation and rethinking knowledge: Voices of university teachers. In V. Msila (Ed.), *Developing teaching and learning in Africa: Decolonising perspectives* (pp. 165–186). African Sun Media.

Langer, P. C. (2016). The research vignette: Reflexive writing as interpretative representation of qualitative inquiry—A methodological proposition. *Qualitative Inquiry, 22*(9), 735–744. https://doi.org/10.1177/1077800416658066

Lawton, D. (1975). *Class, culture and the curriculum.* Routledge.

Marlow, S., & Martinez Dy, A. (2018). Annual review article: Is it time to rethink the gender agenda in entrepreneurship research? *International Small Business Journal, 36*(1), 3–22. https://doi.org/10.1177/0266242617738321

McCusker, S. (2019). Everybody's monkey is important: LEGO® serious play® as a methodology for enabling equality of voice within diverse groups. *International Journal of Research Method in Education, 1,* 1–17. https://doi.org/10.1080/1743727X.2019.1621831

McCusker, S. (2020). Everybody's monkey is important: As a methodology for enabling equality of voice within diverse groups. *International Journal of Research & Method in Education, 43*(2), 146–162. https://doi.org/10.1080/1743727X.2019.1621831

McIntosh, A., & Cockburn-Wootten, C. (2016). Using Ketso for engaged tourism scholarship. *Annals of Tourism Research, 56,* 148–151. https://doi.org/10.1016/j.annals.2015.11.003

McPherson M. L. (2016). Who gives a hot about heutagogy? Self-determined learning in an online Master of Science program in palliative care [Final paper]. University of Maryland.

Msila, V. (2021). Heutagogy and teaching: Toward liberatory methods. *The International Journal of Humanities Education, 18*(1), 1–18. https://doi.org/10.18848/2327-0063/CGP/v18i01/1-18

Paine, N. (2021). *Four stages to embedding a self-determined learning culture.* Training Zone. Retrieved from https://www.trainingzone.co.uk/lead/culture/four-stages-to-embedding-a-self-determined-learning-culture

Piaget, J. (1973). *To understand is to invent.* Grossman.

Potash, J. S., Chen, J. Y., & Tsang, J. P. Y. (2016). Medical student mandala making for holistic well-being. *Medical Humanities, 42*(1), 17–25. http://dx.doi.org/10.1136/medhum-2015-010717

Reyes, N. A. S., Carales, V. D., & Sansone, V. A. (2020). Homegrown scholars: A collaborative autoethnography on entering the professoriate, giving back, and coming home. *Journal of Diversity in Higher Education, 14*(4), 480–492. https://doi.org/10.1037/dhe0000165

Rogers, C. (1969). *Freedom to learn: A view of what education might become.* Charles Merill.

Ryan, I. (2012). A career journey: An auto-ethnographic insight. *Gender in Management: An International Journal, 27,* 541–558. https://doi.org/10.1108/17542411211279724

Saunders, L., & Wong, M. A. (2020). *Instructions in libraries and information centers.* Windsor & Downs Press. https://doi.org/10.21900/wd.12

Segovia-Perez, M., Figueroa-Domecq, C., Fuentes-Moraleda, L., & Munoz-Mazon, A. (2019). Incorporating a gender approach in the hospitality industry: Female executives' perceptions. *International Journal of Hospitality Management, 76,* 184–193. https://doi.org/10.1016/j.ijhm.2018.05.008

Shabani, K., & Ewing, B. F. (2016). Applications of Vygotsky's sociocultural approach for teachers' professional development. *Cogent Education, 3*(1), 1–10. https://doi.org/10.1080/2331186X.2016.1252177

Skilling, K., & Stylianides, G. J. (2020). Using vignettes in educational research: A framework for vignette construction. *International Journal of Research & Method in Education, 43*(5), 541–556. https://doi.org/10.1080/1743727X.2019.1704243

Small, J., Harris, C., Wilson, E., & Ateljevic, I. (2011). Voices of women: A memory-work reflection on work-life dis/harmony in tourism academia. *Journal of Hospitality, Leisure, Sports and Tourism Education, 10*(1), 23–36.

Snowden, M., & Halsall, J. (2014). Community development: A shift in thinking towards heutagogy. *International Journal of Multi Disciplinary Comparative Studies, 1*(3), 81–91. Retrieve from http://www.ijmcs-journal.org/

Tribe, J. (2002). The philosophic practitioner. *Annals of Tourism Research, 29*(2), 338–357. https://doi.org/10.1016/S0160-7383(01)00038-X

Tribe, J., & Paddison, B. (2021). Degrees of change: Activating philosophic practitioners. *Annals of Tourism Research, 91,* 1–13. https://doi.org/10.1016/j.annals.2021.103290

Vygotsky, L. S. (1978). *Mind in society: The development of higher psychological processes.* Harvard University Press.

Wengel, Y. (2020). LEGO® serious play® in multi-method tourism research. *International Journal of Contemporary Hospitality Management, 32*(4), 1605–1623. https://doi.org/10.1108/IJCHM-04-2019-0358

3

GENDER AND RELIGION IN TOURISM EDUCATION

Experiences of Female Muslim University Students Studying Tourism in the West

Husna Zainal Abidin and Ireena Nasiha binti Ibnu

Abstract

This book chapter explores the experiences of female Muslim students studying tourism in tertiary education in the West. While extensive research explores the discrimination and marginalisation of female Muslims in their everyday lives, there is limited study investigating how they experience and perceive their university experience in tourism education. This study contributes to the less researched area of gender, religion, and tourism education. A qualitative approach was adopted, and six female Muslims studying in Western countries were interviewed. The research findings indicate that Muslim women in tourism education, to a certain extent, have had an overall positive experience. Such experiences may be attributed to Islamic teachings and national culture that guides the behaviours and decision-making among female Muslim tourism students. Despite the positive experience, female Muslim students also experience some challenges, mainly due to the limited awareness of teaching staff towards their needs and the lack of representation in the teaching curriculum. This study has practical implications for tourism tertiary education institutions.

Keywords

Muslim, Islam, woman, tourism education, hijab, marginalisation

Introduction

A report by the National Union of Students (NUS) in the United Kingdom (UK) identified that one in every three Muslim students constantly feels worried about Islamophobic experiences on campus (National Union of Students, 2018). Specifically for Muslim women, the head veil quickly identifies them as Muslims, exposing them to a greater risk of discrimination (Nagra, 2018). Seggie and Sanford (2010), who examined the perceptions of female veiled Muslim students on university campuses in the United States of America (USA), identified that, particularly for female Muslim students who are veiled, instances of

DOI: 10.4324/9781003286721-5

marginalisation, prejudice, and discomfort within class settings are prevalent. The research by Seggie and Sanford (2010) did not differentiate between female Muslim students of different degree programs suggesting that the extent of marginalisation between varying university programmes requires further examination. As such, with Muslim female inclusivity emerging as an essential topic for the tourism education sector in the West, this book chapter responds to the research call by Seggie and Sanford (2010) to explore the programme-specific experiences of female Muslim students studying in Western universities by focusing on those who pursue tourism programmes. This chapter examined whether such discriminatory experiences occur and how universities can better improve the learning experiences of female Muslim students.

Past studies have indicated that Muslim women, in general, often face undesirable experiences in the West. For example, Nagra (2018) reported that they often face public abuse, such as yelling, pushing, and hitting. At the workplace, Sekerka and Yacobian (2018) found evidence that Western employers displayed intolerance towards their employed Muslim women's beliefs and practices. A recent study reported that Muslims are considered one of the most discriminated groups in the West, with individuals from the middle and upper classes reported to be more inclined to have discriminatory attitudes towards Islamic beliefs than those from the working classes (Jones and Unsworth, 2022). Therefore, it is not a surprise that the National Union of Students discovered such a high number of concerned female Muslim students.

Meanwhile, from a tourism tertiary education perspective, there is an increasing trend for predominantly Muslim countries to encourage their youth to further their studies in tourism. Muslim countries such as Saudi Arabia, Indonesia, and Malaysia are focusing on enhancing their tourism sector (Mir & Kulibi, 2023) by providing scholarships to pursue overseas studies in tourism that will fulfil the country's long-term tourism goals. As most of the top tourism and hospitality universities are in the West, such as in the UK and the USA, Western countries will see an increase in international Muslim students coming to complete their tertiary education. As such, Western universities need to be prepared to embrace, welcome, and better understand the needs of these new students. Although research literature is abundant on female Muslims in the West, these are mainly on the discrimination and marginalisation they experience in everyday lives, with limited research exploring their experiences within tertiary education.

This chapter will first delve into the basic tenets of Islam and how education is perceived from an Islamic lens. The historical and theoretical literature relating to Islam, female Muslims, and their experiences in higher education will be reviewed. Subsequently, the qualitative method approach will be critically discussed, followed by the presentation of results and discussion of findings. Lastly, the book chapter will end with the main conclusions of the research, limitations and areas for future research.

Education, Tourism, and Decision-Makings from an Islamic Lens

Before delving into the experiences of Muslim women in Western societies, it would first be appropriate to understand how education is perceived within the Islamic religion briefly.

Recite in the name of your Lord Who created – Created humans from a clinging clot. Recite, and your Lord is the most Generous – Who taught by the pen – Taught humanity what they knew not.

[Quran, 96: 1–5]

From the above verse, instruction to recite is understood in Islam to be synonymous with the requirement to seek knowledge. Thus, education has always been an integral aspect of Islam, as demonstrated by the above verse from the Holy Quran. This verse was the first verse (ayat) revealed to the Prophet Muhammad (peace be upon him, PBUH). This verse emphasises the importance of reading and seeking knowledge, i.e., being educated. Therefore, for practising Muslims, seeking knowledge irrespective of discipline is an integral aspect of life. Islam further encourages its followers to travel the earth as part of seeking knowledge and to be educated, as demonstrated in the following verse; *"And God made the earth a spread for you. That you may travel its diverse roads."* [Quran 71: 19–20]. Thus, to travel the world is a virtue in Islam which links to the need for tourism education.

For Muslims, Islam is not merely a religion but a complete code of life that provides guidance in all aspects of life; social, economic, spiritual, political, cultural, and even for education. The Islamic principles offer an ethical and moral compass for Muslims in their decision-making, including those related to their education. Therefore, the curriculum content that a Muslim student is willing to engage with and the types of activities and interactions will also heavily depend on Islamic teachings.

To understand how Muslims (males and females) perceive and understand the world, it would be necessary first to understand the basic tenets of Islam. The basic principles of Islam are sourced from the Holy Quran and Hadith. Hadiths are sayings and actions of the Prophet Muhammad PBUH and act as a complementary source of guidance alongside the Quran. In this section, we will discuss two tenets concerning life purpose and Islamic rulings (the basis for decision-makings) as these are the most relevant for the current discussion. For Muslims, this life is a means for achieving goodness in the next life, i.e., the hereafter. The actions and behaviours in this current life determine the outcomes of the afterlife. Thus, conduct and behaviours must conform to Islamic tenets and rulings. Firstly, in dealing with the complexities of life, Muslims are encouraged to practice *husnudzon* (حُسْنُ الظَّنّ) as far as possible. This is an important concept, as explained by Yucel (2014). According to Yucel (2014), the word *husnudzon* encompasses having *"good intentions, thought and action towards God, the universe and human beings. Moreover, it also carries the meaning of optimism and of holding good opinions and thinking well about others."* Therefore, Muslims strive to behave and think positively in accordance with Islamic values and principles in order to achieve a good outcome in the afterlife. Practising *husnudzon* will be more likely to lend Muslim students to feelings of less resentment towards others when faced with non-ideal life situations.

Secondly, in decision-making, actions and behaviours are based on core rulings known as *Shariah*. *Shariah* applies equally to both men and women but may have specific nuances for the different gender. These rulings can be divided into two main categories; *halal* (permissible) and *haram* (forbidden). Within the *halal* category, it can be further divided into four sub-categories; *wajib* (obligatory), *sunnah* (recommended), *mubah* (allowed), and *makruh* (to be avoided). For example, studying and being educated is considered *wajib*, i.e., obligatory upon every Muslim, in accordance with the Shariah as indicated by the Quranic verse [Quran, 96: 1–5] and further supported by the hadith in Sahih Muslim, 2699a, *"... he who treads the path in search of knowledge, Allah would make that path easy, leading to Paradise for him..."*. On the other hand, working at the bar to make and serve alcoholic drinks is considered *haram* due to the handling of alcohol. Therefore, Muslim students, in general, will have positive experiences during their tertiary education where curricular activities are aligned with Islamic principles and will have negative experiences (e.g., psychological dilemmas) where activities are misaligned.

As previously mentioned, *Shariah* equally applies to both women and men and may only have minimal nuances between the genders. Therefore, when considering the experiences of female Muslims studying tourism tertiary education, it will not be significantly different to that of Muslim men from the perspective of Islamic practices. Nonetheless, the outward appearance of practising Muslim women may expose them to more challenging experiences than Muslim men.

In Islam, the Quran and hadith provide an ethical and moral compass that remains universal and timeless, on which Muslims base their decisions. What is considered moral and ethical is generally a Westernised view of ethics, which can be regarded as socially constructed and ever-changing through time. Therefore, to better understand the experiences of female Muslim students studying tourism in the West, one cannot shy away from analysing how they see the world through an Islamic lens and evaluating their personal decisions based on their own code of ethics.

Women, Islam, and Education

Stereotypes and Challenges of a Muslim Woman

As the world's second-biggest religion, Islam makes up 24.9% of the global population in 2015. This constitutes almost 1.8 billion people and is projected to grow further in the decades ahead (Pew Research Center, 2015). Countries such as Indonesia, Pakistan, and India have the largest number of Muslims. Despite being the second-biggest religion, Western society is still prone to a distorted view of Islam as oppressive, particularly concerning Muslim women, i.e., a general lack of knowledge and understanding about Muslims and Islam (Nagra, 2018). The body of scholarly literature to date has concerned itself with the experiences of Muslim women related to the issues of gender equality (Chapman, 2018; Galloway, 2014), identity (Kabir, 2016; Steele et al., 2021; Zempi, 2016), the veil (Abdul Fatah, 2019; Evolvi, 2019; MacMaster, 2020; Zempi, 2019;), and piety movements and Islam (Ibnu, 2022). Interestingly, studies indicate how, due to changes in modern societies and globalisation, the roles of Muslim women are also transforming depending on political, economic, and societal needs (see Haddad & Esposito, 1998). Such changes are suggested to also affect their everyday experiences, particularly in non-Muslim countries.

Jeffery and Qureshi (2022) argued that the negative constructions of Muslim women and their tenacious stereotypes make it important for academic scholars to investigate the various challenges Muslim women encounter in their daily lives. In addition, the stereotypical portrayals of Muslim women in Western media as "Muslim-as-terrorist" or "agents" of terrorism cause them to be more vulnerable and expose them as targets of violence (Freedman, 2007; Perry, 2014). This phenomenon is known as Islamophobia, a term described as *"a fear or hatred of Islam and its adherents that translates into individual, ideological and systemic forms of oppression and discrimination"* (Zine, 2003). Adopted from Runnymede Trust (1997), the original definition of Islamophobia, Awan and Zempi (2018) suggested that Islamophobia can be expressed as:

A fear, prejudice and hatred of Muslims or non-Muslim individuals that lead to provocation, hostility and intolerance by means of threatening, harassment, abuse, incitement and intimidation of Muslims and non-Muslims, both in the online and offline world. Motivated by institutional, ideological, political and religious hostility that transcends into structural and cultural racism which targets the symbols and markers of being a Muslim.

Unfortunately, in Western countries such as Canada, the USA, France, and Australia, Islamophobia is not a new occurrence. In fact, the fear and hate towards Islam have intensified since the 9/11 incident. This incident instigated hate crimes against Muslims, with veiled Muslim women often being the first targets, as the hijab was presumed as a sign of danger and aggression by the West (Nagra, 2018). Therefore, the experiences of Muslim women have often been discussed within the topic of gendered Islamophobia. Nonetheless, Dagkas and Benn (2006) highlighted that not all Muslim women behave in the same way, and they have different approaches to resolving cultural and religious obligations. In breaking the stereotypes towards Muslim women, Margaretha (2019) argued that Muslim women in Dutch societies have challenged the "oppressed Muslim woman" stereotype through their own behaviour. They do so by adopting an ambassadorial role in their everyday encounters with non-Muslims in public places such as at the office, on campus, or in shopping malls. Meanwhile, in Canada, numerous approaches by Muslim women, including activism, media awareness campaigns, and increased community engagement, have been adopted to respond to Islamophobia (Khokhar, 2022). Both Margaretha (2019) and Faiza's (2022) findings demonstrate that female Muslims negotiate their multiple and complex layers of identity differentially.

In a world which faces a rising issue of media-fuelled Islamophobia by the West, it is imperative to examine the experiences of similar groups in other diaspora communities, such as female Muslim students in higher education (Brown, 2008, 2009; Ibnu & Azman, 2022, 2013).

Intersectionality – Female Muslim Experiences in Higher Education

The term intersectionality can be understood as a form of discrimination and subordination that takes place when identity variables such as race, class, gender, ethnicity, and age are present simultaneously (Collins & Bilge, 2016; Crenshaw, 1991). Other scholars have applied this concept to understand the interactions of multiple identity variables in different contexts, such as the shifting power and oppression relations in colonialism, geopolitics, and cultural configurations relations (Rice et al., 2019). In this research, we focus on the intersectionality of gender and religious identities by exploring the participants' female Muslim experiences, specifically those studying tourism education in Western tertiary education. As Karaman and Christian (2020) note, in the West, Muslim women are frequently portrayed as repressed or dangerous individuals due to the process that essentialises religious, gendered identity. Edward Said has critically discussed the negative portrayal towards Islam and Muslim women in the West in Orientalism 1978, which is the concept of the "other" or the "orient," the knowledge which constructs the East as the "other" (Ashcroft et al., 1995). According to Said, some approaches have been used in literature to make the East or "other" appear underdeveloped and uncivilised to Westerners (Said, 1978). Such an approach represents a truly Eurocentric perspective in knowledge development. Hence, it is essential to understand how the intersection of gender and religious identities influences Muslim women's everyday experiences in particular types of contexts (Aziz, 2012; Selod, 2018).

In Islamic-majority societies, Muslim women's access to education has proven crucial to their survival. Roudi-Fahimi and Moghadam (2003) report that education is vital to empowering women to become contributing members of society. With the growth of globalisation and internationalisation, there is an increase in global student mobility in studying in the West (Anderson, 2020; Ding, 2016). However, the rapid growth of international students arriving at higher education institutions overseas may cause disappointment among the students if their needs are not fulfilled. Unfortunately, previous studies pertaining to female Muslim experiences on campus demonstrate how they often experience a sense of exclusion (Dimandja, 2017; Schatz, 2008; Seggie & Sanford, 2010) and gender discrimination (Al-deen, 2019; Chen et al. 2019; Dimandja, 2017; McGuire et al., 2016). Chen et al. (2019) argued that Muslim students frequently struggle with a range of issues, such as the feeling of social exclusion and educational prejudice. In Canada, female Muslim students who donned the hijab reported a higher likelihood of difficulties in interacting with other non-Muslim students and faculty members (Khosrojerdi, 2015).

While these reports are alarming, they often do not differentiate between the types of degree programmes that female Muslim students are enrolled in. As Seggie and Sanford (2010) suggested, far less is known about the experiences of Muslim women in different degree programmes, such as tourism education. Different degree programmes may attract different students depending on the present different environmental and social contexts. For example, an ethnographic finding on female Malaysian Muslim students studying Finance and Accounting in the UK felt disappointed as they were segregated from the local British and international European students in their courses. Thus, they limited their intercultural communication experiences while studying abroad (Ibnu, 2022).

Meanwhile, tourism education is characterised by a high degree of people-centeredness. Thus, it could be considered that a female Muslim student may be exposed to far greater risks of potential discrimination and challenges, as highlighted by Sekerka and Yacobian's (2018) study on female Muslims working in the tourism industry. Furthermore, a study by Bontenbal and Aziz (2013), who surveyed 80 tourism and hospitality students to understand students' perceptions and attitudes towards tourism jobs in Oman's Tourism Industry, identified negative perceptions as a major issue. The study found that tourism employment in a Muslim context is often seen as something negative, particularly for women exposed to gambling, eating pork, and drinking alcohol while at work and interacting with "strangers" in hotels. Nonetheless, experiences and expectations of female Muslims in tourism may differ between Muslim and non-Muslim countries. Fewer studies have examined the perception and experiences of female Muslim students in non-Muslim countries such as the West. Thus, this research is expected to contribute to a greater understanding of female Asian Muslims studying tourism in Western higher education, especially since Asia-focused research continues to lag behind non-Asian research in the field of tourism (Yang & Ong, 2020).

Furthermore, beyond the limitations in degree-specific experiences, there are far fewer studies analysing such a topic from a Muslim perspective by Muslim researchers. As previously discussed, current knowledge production in tourism education is commonly derived and produced from a Eurocentric view (Wijesinghe et al., 2019). Therefore, this research responds to the call by Seggie and Sanford (2010), who highlight the need for research to explore the experiences of Muslim women in different university disciplines, as analysed by two Muslim women. As such, this research will explore the experiences of female Muslim students studying tourism in tertiary education in the West as examined by two female Muslim researchers.

Methods

For this study, the authors conducted qualitative exploratory research to explore and understand the experiences of past and current female Muslim students studying tourism in universities in the UK and the USA. The research focuses on female Muslims from Malaysia and Indonesia. Geographically, both countries comprise the largest proportion of Muslims in the world. Culturally, Malaysia and Indonesia have similar cultural traditions and religious subscriptions, similar to that of the researchers, making data interpretations more authentic. Data collection and analysis were underpinned by a constructivist philosophical view with an interpretivist epistemology (Gray, 2013; Saunders et al., 2012). Both authors believe that reality is created by the interpretation of the researcher. This research positions itself through the understanding that the experiences of female Muslim tourism students are derived from the interactions between other students and staff and the meanings they create from these interactions (Stenbacka, 2001).

Reflecting on the positionality of the two researchers, both researchers are female Muslims from Malaysia who undertook their tertiary education in the West. As highlighted by Ndimande (2012), the researcher's understanding of the participant's background and personal experiences contribute greatly towards having better insights into the community/phenomenon being explored, allowing a similar analytical lens on the phenomena. The topic of understanding female Muslim experiences in the West was mainly derived from the literature gap (Seggie & Sanford, 2010) and further motivated by their religious identity and personal experiences. Both researchers developed qualitative research skills through their interdisciplinary and multinational institutions. Their positions on reflexivity are firmly rooted in subjectivist or social constructionist paradigms. Being Muslim women with educational backgrounds in tourism and migration studies, the authors hoped to reveal how the intersection between gender and religious identities influences the lives of female Muslim students studying tourism.

In this study, participants were purposely sampled based on the following criteria: (1) female Muslim, (2) from Malaysia or Indonesia, (3) is studying or has studied at a university in the West in the past decade, (4) is studying or has studied tourism at the undergraduate, Masters, or PhD level, and (5) can speak and write in English. Including PhD students in the interview allowed the researchers to gain a broader understanding of the varying or similar experiences of female Muslim students in tertiary education. A list of participants is illustrated in Table 3.1. Participants were first purposely sampled based on the criteria, and later, a snowball sampling method was adopted. A structured interview with open-ended questions was chosen to give participants the freedom to relay their thoughts while

Table 3.1 Research participants

Participant	Study level	Country of study	Country of origin	Hijab
F_01	Masters	UK	Indonesia	Yes
F_02	Masters and PhD	UK	Malaysia	Yes
F_03	Masters	UK	Indonesia	No
F_04	Masters	UK	Indonesia	No
F_05	Undergraduate	UK	Malaysia	Yes
F_06	Undergraduate	USA	Malaysia	Yes

ensuring all participants received similar objective questions. The structured interview was divided into three main sections; demographics of participants, university experience, and recommendations. The interviews aimed to understand the overall experiences of Muslim students studying in the West, the challenges and opportunities they faced and suggestions on how their university experience could be improved.

Participants were given two options, i.e., to participate in an online call structured interview or an email structured interview. These two options were offered due to the extreme time difference and availability of some participants. Additionally, due to time constraints, email-structured interviews allowed the researchers to access multiple participants simultaneously. Therefore, these options allowed a greater opportunity for the researchers to access the participants. Online call-structured interviews lasted between 30 and 60 minutes, and email-structured interviews were completed within two weeks per participant. Structured interviews were analysed based on thematic analysis. This research was not aimed to achieve generalisability but rather as an initial exploratory study to initiate discussions on this less-discussed topic of religion and tourism education and identify areas for future research. An inductive approach in the coding of empirical data was used. The analysis focused on the key impactful events in the participants' tourism degree programme and future recommendations for universities to learn from.

Findings and Discussion

Theme 1: Respect and Acceptance

Respect was one of the main themes that emerged from the structured interviews. In contrast to the common expectation that female Muslims are discriminated against, marginalised and othered (Nurein & Iqbal, 2021; Seggie & Sanford, 2010), our study demonstrates that specifically for tourism education, female Muslims in tertiary education in the West have had to a certain extent, a positive overall experience. Students highlighted how they enjoyed their tourism degree as they felt a sense of respect and acceptance even though they had different faiths and perspectives compared to their peers. Hence, the positive experience and openness towards Western culture by these female Muslims were shown to be strong enough to counter the Western orientalist misrepresentation, undermining the orientalist view of the oppressed, passive Muslim woman. This view was demonstrated by a participant, "*personally, I enjoyed myself throughout the course. A sense of acceptance despite different religious beliefs is what makes me comfortable going to university daily over there.*" (F6_USA).

Furthermore, participants had no issues related to their religious attire, such as wearing the veil in class. All students had a positive experience as they reported feeling safe and comfortable wearing the hijab, even when they were the only students in their class to wear it. As one of the participants explained, "*I was the only Muslim and hijabi in my course year and was treated with respect by both the staff and peers. Nobody questions my wearing the hijab, and I've never felt unsafe or uncomfortable*" (F05_UK). Another participant concurred with the theme of respect by saying, "*Alhamdulillah [praise be to God], to be honest, I never had the real odd situation as a female Muslim student. People surrounding me respect me as a Muslim*" (F03_UK). Such respect and acceptance of female Muslims within tourism education can be seen when two participants were selected as their course representatives within their tourism cohort (F01_UK and F02_UK). Therefore, suggesting that female Muslim tourism students are accepted amongst their peers and experience limited to no discrimination. A contrasting finding to Khosrojerdi's (2015) study. This may be *"due to how the tourism school community establishes their belief. Celebrating differences and respecting one another"* (F06_UK).

Different education disciplines most likely attract different characteristics of students and teaching staff. Self-selection could be an influential factor determining a female Muslim's university experience. Tourism is a highly people-centric discipline and thus is open and accepting in nature. It is expected that tourism students and staff would tend to be those who are keen on experiencing and learning more about different cultures, traditions, and religions of others. Where differences may be a hurdle in other disciplines, differences may be celebrated and well-accepted in the tourism education sector. Therefore, not surprisingly, our study indicates that female Muslim students in tourism education do not experience educational prejudice, which contrasts with Chen et al. (2019). Discrimination and prejudice were rather, an experience that happened outside of the tourism programme. One participant highlighted a traumatic experience of being verbally harassed by a group of men while waiting at the university bus stop (F02_UK). Little was known about the men, but such experiences resonate with past studies that female Muslims are more at risk of harassment and discrimination in their everyday lives (Nurein & Iqbal, 2021).

Theme 2: Representation within Teaching Content

While tourism education within universities has generally been positive for female Muslim students, all students have highlighted the lack of representation as one of the major limitations in their tourism programme. Despite the growing Muslim population and the Halal tourism market, the findings indicate a lack of representation of Muslims in the teaching content. As a consequence, participants felt hesitant to initiate a discussion on a new topic in class.

> *Our course content focused mainly on European tourism and European travellers. As far as I can remember, there weren't many occasions where the lecturers touched on the subject of Muslims/Islam... I guess I had hoped for more diverse topics in the course instead of just European and Western tourism case studies...Still, it would have been nice to have more diversity, learning about Halal tourism and Muslim travellers or Muslim destinations. As for my part, I didn't really feel comfortable initiating those subjects myself. Partly because I'm a minority in the course and partly because I didn't feel confident enough about my knowledge. Looking back, I think it would've been fine to do a couple of presentations about Halal tourism or Muslim-majority destinations. I think many would be interested in a bit of diversity.*
>
> *(F05_UK)*

Similarly, another participant had similar thoughts on the need for more representation. The participant felt that having Muslim representation in teaching content would help empower her to voice her opinions (F01_UK). *"The Muslim tourists are growing a lot. I wish they would include Halal tourism or Muslim Tourism so I could say something about it"* (F01_UK). As reported by Papastathopoulos et al. (2020), current tourism trends indicate a significant growth of Muslim travellers. In fact, a growing number of young female Muslim tourists are educated, have higher purchasing power than their previous generations, and are culturally curious (Mastercard-CrescentRating, 2019). In response to the growing

Muslim traveller community, Muslim countries such as the Middle East, Malaysia, and Indonesia are currently focusing on developing more tourism products and services catering for the halal tourism market. Non-Muslim countries, such as Korea and Japan, are also delving into Muslim-friendly tourism, indicating the growing area of Muslim tourism and the need to incorporate such trends into the teaching content. The concern of the students on the lack of inclusiveness on halal tourism within tourism education is valid and an important one to be addressed by tertiary institutions.

Meanwhile, another participant mentioned that the representation of Muslims in teaching content would provide an opportunity for her to explain the misconceptions about Islam and tourism, which she currently did not have the chance to do so. *"Halal tourism isn't just about the food. But it's also about hygiene, about family. I would like to explain the common misunderstandings of Halal tourism that people may not know"* (F04_UK). Therefore, suggesting that for some female Muslim students, they may take on an "ambassadorial" role in raising awareness of their faith and overcoming misconceptions associated with it (Seggie & Sanford, 2010). Such findings are similar to that of Margaretha (2019), where female Muslims challenge the stereotypes of Islam in everyday encounters with non-Muslims.

Despite feeling the lack of tourism education content and curriculum, none of the participants had the courage to voice their thoughts and concerns on this matter. Since all participants were from countries which have previously been colonised by the West, the phenomenon as described by Alatas could be on display here, i.e., how the colonised remain in their *"captive minds"* through education, even beyond the years of their countries becoming independent (Wijesinghe et al., 2019; Winter, 2009). Subconsciously, *"designed products"* (such as curriculum content) from past colonisers are perceived to be superior and of higher intellectual status and thus creating a lack of confidence to make suggestions to curriculum content. This suggests the need for tourism educators to consider how they could instil in their female Muslim students a sense of confidence in their identity via the inclusion of topics relevant to them. To be fair, tourism education has generally been produced from a Eurocentric view (Wijesinghe et al., 2019). Therefore, Muslim-related tourism may not be a natural concern for the curriculum developers, who are predominantly from a Western and non-Muslim background.

Theme 3: Awareness towards Muslim Practices

While participants generally highlighted a good level of acceptance within tourism education, they highlighted challenges regarding the awareness of Muslim practices and beliefs. For example, students experienced the class schedule sometimes conflicts with their prayer times. Students felt it was difficult and challenging to raise the issue of prayer during class. *"It was difficult for me to find time to pray when classes were back-to-back. But still, in class, I don't know how to tell the lecturer, and I feel uncomfortable saying that I have to pray and everything"* (F04_UK) and *"the schedule of classes sometimes clashed with the prayer times. So, I should really be wise to organise my schedule to catch up with both"* (F03_UK). Muslim women may face greater challenges here than men due to the time required to prepare for prayer (Chen et al., 2019). Thus, female students have undertaken diverse approaches to resolve cultural and religious obligations (Dagkas & Benn, 2006). While some students may easily *"take a quick break to pray in an empty room"* (F05_UK), others may skip their prayers, which often results in psychological dilemmas for some.

This is because prayers are *wajib* (obligatory) for every Muslim. Again, most participants explained their hesitancy in voicing their needs, similar to the case of the previously high-lighted educational content.

The intersectionality among culture, gender, and religion further exacerbates such ex-periences. The cultural element of female Muslims in Southeast Asia is that they tend to take extra precautions to avoid causing inconveniences to others. The culture of Southeast Asia women is such that they are generally shy. Malay Muslim women, in particular, are expected to be reserved and to avoid attention when in public, especially in the company of men (Collins & Bahar, 2000). According to Collins and Bahar (2000), being reserved is a positive characteristic in the Malay culture, whereas similar concepts are not as promi-nent in Western societies. From a Muslim perspective, such positive shyness is also seen in the concept of "*hayā*'", which carries the meanings of conscientiousness and modesty (Elshinawy, 2023). Female Muslims, in particular, are enjoined to always have *hayā*' in their demeanour, and it is a sunnah (recommended) to behave so. Part of such behav-iours includes the etiquette of how one should speak and deal with others. Thus, it fur-ther promotes reservations about voicing contradictory ideas during class. Furthermore, in Islam, teachers are highly respected and are positioned as a place of authority and, therefore, should already be aware of the considerations of their students. For students, they do not want to inconvenience others on their behalf, suggesting a cultural element of female Muslims in Southeast Asia. Therefore, it would be appropriate for future staff training to include more cultural and religious awareness relating to female Muslim students.

Overall, while the findings of this study generally presented positive experiences, it does not necessarily indicate that there are no discriminatory and challenging experiences by female Muslim tourism students. When interviewed, all participants primarily focused on the positive aspects. Expressions such as "*I never had the real odd situation as a female Muslim student*" (F03_UK) can be interpreted that the student has previously experienced some challenges but not to a great extent. As further explained by another female student, "*Even if I had encountered them [i.e., a challenging situation], it might be less than a per-cent... Honestly, no matter where one is going, there will always be a small fraction of us facing this issue-- downfall on the need for help/feeling like being discriminated against. I believe it comes down to the student outlook.*" As previously highlighted, female Muslims are taught to practise *husnudzon*, i.e., having good intentions of the universe and thinking well of oneself and others. Such practices are considered a *sunnah* (recommended) as per the *Shariah*. Therefore, suggesting that female Muslim students generally view the world from a more positive perspective, thus, which allows them to be more resilient and accepting when facing difficult situations.

In addition, the mere characteristics of these female Muslim students who have chosen to move away from their families to live in a foreign country by themselves already sug-gest that these female students are mentally determined. It could be suggested that they are more likely to see experiences more positively. For example, one participant indicated a traumatic experience of being shouted at by a group of local boys for no apparent reason (F02_UK). Despite this experience, it did not deter her from continuing her studies again in another Western country. Female Muslims who travel understand their need to adapt and do not expect a non-Muslim country such as the UK and USA to have similar perspectives to that of Muslim countries. The findings of Muslim experiences in Muslim countries will likely differ from those in non-Muslim countries. Nonetheless, it is important to note that the tourism discipline attracts those drawn towards diversity, which helps develop a conducive learning environment for female Muslim students.

Conclusion

This research sought to explore the experiences of female Southeast Asian Muslim students studying tourism in tertiary education in the West. This research contributes to the research gap on gender, religion, and education and responds to the research call by Seggie and Sanford (2010), who proposed the need for research to explore the experiences of Muslim women in different university disciplines. Therefore, this study explores the experiences of female Muslims from Indonesia and Malaysia who studied tourism education, specifically in the UK and USA.

Contrary to the common understanding that Muslim women are marginalised, neglected and "*othered*" in Western countries, Muslim women in tourism education, to a certain extent, have had an overall positive experience with some minor challenges regarding cultural and religious awareness and representation. The female Muslim tourism students perceive having a high level of acceptance in their tourism education which could be attributed not only to their Islamic practice of *husnudzon* but also to the nature of those interested in tourism, i.e., the tourism programme attracts people who accept diversity.

Despite the positive experience, female Muslim students also experience challenges and negative experiences, particularly when curricular activities are misaligned with the Islamic principles of the *Shariah*, causing psychological dilemmas. The intersectionality of gender, cultural background, and religion is identified to sometimes hinder female Muslim students, particularly from Southeast Asian cultures, from fully living up to their religious expectations.

The findings from this research provide implications for current practice in tourism education. We suggest practical recommendations, particularly in the area of representation and identity strengthening. Representations should be implemented in three different areas; teaching content, timetable scheduling, and academic staff. First, tourism education should consider more representation of Muslim tourism in their teaching content, such as topics on Halal tourism and the experiences of female Muslim tourists.

Second, we suggest that the management and administration of universities consider the design of inclusive curriculum and teaching plans which could include the consideration of prayer times in the timetable. Universities should additionally consider raising awareness of the cultural and religious challenges faced by female Muslim students. While both male and female Muslim students are obliged with similar expectations as laid out in the *Shariah*, the degree of challenges experienced by female Muslim students may vary slightly due to the cultural influence on how a female is expected to behave, i.e., being more reserved and shy in addition to the preparation time required to perform certain religious obligations.

Third, more female Muslim scholars should be hired as part of the academic staff teaching tourism education. Having female Muslim teaching staff could contribute to a more inclusive and diversified knowledge production of tourism education. As highlighted by Winter (2009), including the philosophical perspectives of regional scholars should be the heart of tourism scholarship. In terms of identity strengthening, academic staff and the tourism curriculum should consider ways in which they can encourage female Muslim students, who are often the minority in class, to be more confident in voicing their concerns. As previously mentioned, demonstrating an awareness of female Muslims in the teaching content while providing a safe and inclusive space could help support female Muslim students in this respect.

This research provides a better understanding of how female Muslim students from Southeast Asia experience their tourism education at the university level. Nonetheless, there are some limitations in the study of this research. This research provides a preliminary understanding of female Muslim students' experiences and touches upon a few Islamic and cultural concepts. Due to the limited word count, in-depth discussions on other concepts are not possible to give it justice. Therefore, future research should delve deeper into this topic to identify similar or new themes and influencing concepts related to the female Muslim experience in tourism education. Areas of research could focus on the perspective of representation and cross-cultural experiences of female Muslims from different countries. Additionally, the study used structured interviews and, therefore, could not further probe the participants' answers. Future research should consider conducting in-depth semi-structured interviews to gain a deeper insight into Muslim women's tourism education experience.

References

Abdul Fatah, F. B. (2019). *Discourses of the non-veiled: Exploring discursive identity constructions among Malaysian Muslim women who do not veil* [PhD thesis, University of Warwick].

Al-deen, T. J. (2019). Agency in action: Young Muslim women and negotiating higher education in Australia. *British Journal of Sociology of Education, 40*(5), 598–613.

Anderson, D. L. (2020). Muslim international students in the United States. *Journal of International Students, 10*(2), 320–338. https://doi.org/10.32674/jis.v10i2.965

Ashcroft, B., Griffiths, G., & Tiffin, H. (1995). *The post-colonial studies reader*. Routledge.

Awan, I., & Zempi, I. (2018). *A working definition of Islamophobia: A briefing paper*, prepared for Rt. Hon Anna Soubry MP and Wes Streeting MP. https://irep.ntu.ac.uk/id/eprint/45222

Aziz, S. F. (2012). From the oppressed to the terrorist: Muslim American women caught in the crosshairs of intersectionality. *Hastings Race and Poverty Law Journal, 9*(1), 191–264.

Bontenbal, M., & Aziz, H. (2013). Oman's tourism industry: Student career perceptions and attitudes. *Journal of Arabian Studies: Arabia, the Gulf, and the Red Sea, 3*(2), 232–248. doi:10.1080/21534764.2013.864508

Brown, L. (2008). Language and Anxiety: An ethnographic study of international postgraduate students. *Evaluation and Research in Education, 21*(1), 75–95.

Brown, L. (2009). *Using an ethnographic approach to understand the adjustment journey of international students at a university in England*. In A. G. Woodside, C. M. Megehee, & A. Ogle (Eds.), *Perspectives on cross-cultural, ethnographic, brand image, story-telling, unconscious needs, and hospitality guest research* (pp. 101–142). Emerald Group.

Chapman, M. (2018). Modesty, liberty, equality: Negotiations of gendered principles of piety among Muslim women who cover. *Feminism and Psychology, 28*(4), 509–529.

Chen, B., Tabassum, H., & Saeed, M. A. (2019). International Muslim students: Challenges and practical suggestions to accommodate their needs on campus. *Journal of International Students, 9*(4), 933–953. https://doi.org/10.32674/jis.v9i3.753

Collins, F., & Bahar, E. (2000). To know shame: Malu and its uses in Malay societies. *Crossroads: An Interdisciplinary Journal of Southeast Asian Studies, 14*(1), 35–69.

Collins, P. H., & Bilge, S. (2016). *Intersectionality*. Polity Press.

Crenshaw, K. (1991). Mapping the margins: Intersectionality, identity politics, and violence against women of color. *Stanford Law Review, 43*, 1241–1299.

Dagkas, S., & Benn, T. (2006). Young Muslim women's experiences of Islam and physical education in Greece and Britain: A comparative study. *Sport, Education and Society, 11*(1), 21–38. https://doi.org/10.1080/13573320500255056

Dimandja, O. O. (2017). *"We Are Not that Different from You": A Phenomenological Study of Undergraduate Muslim International Student Campus Experiences*. [PhD thesis, University of Colorado].

Ding, X. (2016). Exploring the experiences of international students in China. *Journal of Studies in International Education, 1*(20), 1–20.

Elshinawy, M. (2023). *Ḥayāʾ: More than just modesty.* Yaqeen Institute. https://yaqeeninstitute.org. my/read/paper/haya-more-than-just-modesty (Accessed 24 April 2023).

Evolvi, G. (2019). The veil and its materiality: Muslim women's digital narratives about the burkini ban. *Journal of Contemporary Religion, 34*(3), 469–487.

Freedman, J. (2007). Women, Islam, and rights in Europe: Beyond a universalist/culturalist dichotomy. *Review of Internationalist Studies, 33*(1), 168–182.

Galloway, S. D. (2014). *The impact of Islam as a religion and Muslim women on gender equality: A phenomenological research study* [PhD thesis, Nova South-Eastern University].

Gray, D. E. (2013). *Doing research in the real world* (2nd ed.). SAGE.

Haddad, Y. Y., & Esposito J. L. (1998). *Islam, Gender, and Social Change.* Oxford University Press.

Ibnu, I. N. (2022). Education, aspiration, and everyday diplomacy: An ethnographic study of female Malaysian Muslim students in the UK. *Journal of Comparative and International Higher Education, 14*(1). https://doi.org/10.32674/jcihe.v14i1.3294

Ibnu, I. N., & Azman, N. (2022). The role of Islamic piety movements in the lives of Malaysian female Muslim students in the United Kingdom. *Journal of Studies in International Education, 26*(5), 590–605. https://doi.org/10.1177/10283153211027011

Jeffery, P., & Qureshi, K. (2022). Muslim woman/Muslim women: Lived experiences beyond religion and gender in South Asia and its diasporas. *Contemporary South Asia, 30*(1), 1–15. doi:10.1080/09584935.2021.2021859

Jones, S. H., & Unsworth, A. (2022). *The Dinner Table Prejudice: Islamophobia in Contemporary Britain.* University of Birmingham. https://discovery.ucl.ac.uk/id/eprint/10152531

Kabir, N. A. (2016). Muslim women in Australia, Britain and the United States: The role of "othering" and biculturalism in identity formation. *Journal of Muslim Minority Affairs, 36*(4), 523–539.

Karaman, N., & Christian, M. (2020). "My hijab is like my skin colour": Muslim women students, racialization, and intersectionality. *Sociology of Race and Ethnicity, 6*(4), 517–532. https://doi.org/10.1177/2332649220903740

Khokhar, F. J. (2022). Reclaiming the narrative: Gendered Islamophobia, its impacts and responses from Muslim women. *Social Identities, 28*(2), 267–281. doi:10.1080/13504630.2021.2006624

Khosrojerdi, F. (2015). *Muslim female students and their experiences of higher education* [PhD thesis, Western University]. https://ir.lib.uwo.ca/etd/2896

MacMaster, N. (2020). *Burning the veil: The Algerian war and the 'emancipation' of Muslim women, 1954–62.* Manchester University Press.

Margaretha, A. V. E. (2019). Muslim women as 'ambassadors' of Islam: Breaking stereotypes in everyday life. *Identities, 26*(4), 375–392. doi:10.1080/1070289X.2017.1346985

Mastercard-CrescentRating (2019). *Muslim women in travel 2019.* https://www.crescentrating.com/reports/muslim-women-in-travel-2019.html

McGuire, K. M., Casanova, S., & Davis III, C. H. (2016). "I'm a Black female who happens to be Muslim": Multiple marginalities of an immigrant Black Muslim woman on a predominantly white campus. *The Journal of Negro Education, 85*(3), 316–329.

Mir, R. N., & Kulibi, T. A. (2023). Tourism as an Engine for Economic Diversification: An exploratory study of Saudi Arabia's Tourism Strategy and Marketing Initiatives. *Saudi Journal of Business and Management Studies, 8*(8), 186–201.

Nagra, B. (2018). Cultural explanations of patriarchy, race, and everyday lives: Marginalizing and "Othering" Muslim women in Canada. *Journal of Muslim Minority Affairs, 38*(2), 263–279. doi:10.1080/13602004.2018.1466488

Ndimande, B. S. (2012). Decolonizing Research in Postapartheid South Africa: The Politics of Methodology. *Qualitative Inquiry, 18*(3), 215–226. doi.org/10.1177/1077800411431557

Nurein, S. A., & Iqbal, H. (2021). Identifying a space for young Black Muslim women in contemporary Britain. *Ethnicities, 21*(3), 433–453.

National Union of Students. (2018). The experience of Muslim students in 2017-18. https://www.nusconnect.org.uk/resources/the-experience-of-muslim-students-in-2017-18

Papastathopoulos, A., Kaminakis, K., & Mertzanis, C. (2020). What services do Muslim tourists want? Uncovering nonlinear relationships and unobserved heterogeneity. *Tourism Management Perspectives, 35*, 100720. https://doi.org/10.1016/j.tmp.2020.100720

Perry, B. (2014). Gendered Islamophobia: Hate crime against Muslim women. *Social Identities: Journal for the Study of Race, Nation and Culture*, 20(10), 74–89. doi:10.1080/13504630.2013.864467

Pew Research Center (2015). The Future of World Religions: Population Growth Projections, 2010-2050. https://www.pewresearch.org/religion/2015/04/02/religious-projections-2010-2050/

Rice, C., Harrison, E., & Friedman, M. (2019). Doing justice to intersectionality in research. *Cultural Studies Critical Methodologies*, 19(6), 409–420. https://doi.org/10.1177/1532708619829779

Roudi-Fahimi, F., & Moghadam, V. M. (2003). *Empowering women, developing society: Female education in the Middle East and North Africa*. Population Reference Bureau. 1–7.

Runnymede Trust (1997). Islamophobia: A Challenge for Us All. https://www.runnymedetrust.org/publications/islamophobia-a-challenge-for-us-all

Said, E. (1978). *Orientalism*. Pantheon Books.

Saunders, M. N. K., Thornhill, A., & Lewis, P. (2012). *Research methods for business students* (6th eds.). Pearson.

Seggie, F. N., & Sanford, G. (2010). Perceptions of female Muslim students who veil: Campus religious climate. *Race Ethnicity and Education*, 13(1), 59–82.

Sekerka, L. E., & Yacobian, M. M. (2018). Fostering workplace respect in an era of anti-Muslimism and Islamophobia: A proactive approach for management. Equality, Diversity and Inclusion: An *International Journal*, 37(8), 813–831.

Selod, S. (2018). Gendered racialisation: Muslim American men and Women's encounters with racialized surveillance. *Ethnic and Racial Studies*, 42(4), 552–569.

Steele, R. R., Bengali, S., Richardson, G., Disbennett, M., & Othman, Y. (2021). Muslim women negotiating their identity in the era of the Muslim ban. *Journal of Gender Studies*, 32(7), 1–12.

Stenbacka, C. (2001). Qualitative research requires quality concepts of its own. *Management Decision*, 39(7), 551–556. https://doi.org/10.1108/EUM0000000005801

Wijesinghe, S. N. R., Mura, P., & Culala, H. J. (2019). Eurocentrism, capitalism and tourism knowledge. *Tourism Management*, 70, 178–187.

Winter, T. (2009). Asian Tourism and the retreat of Anglo-Western centrism in tourism theory. *Current Issues in Tourism*, 12(1), 21–31.

Yang, E. C., & Ong, F. (2020). Redefining Asian tourism. *Tourism Management Perspectives*, 34, 100667. https://doi.org/10.1016/j.tmp.2020.100667

Zempi, I. (2016). 'It's a part of me, I feel naked without it': Choice, agency and identity for Muslim women who wear the niqab. *Ethnic and Racial Studies*, 39(10), 1738–1754.

Zempi, I. (2019). Veiled Muslim women's views on law banning the wearing of the niqab (face veil) in public. *Ethnic and Racial Studies*, 42(15), 2585–2602.

Zine, J. (2003). Dealing with September 12: Integrative antiracism and the challenge of anti-Islamophobia education. *Orbit*, 33(3), 39–41.

PART II

Researching Gender in Tourism

4

FEMINIST RESEARCH IN TOURISM

The Inclusion of Gender Non-Conforming Identities

Núria Abellan-Calvet and Marta Salvador-Almela

Abstract

Feminist research is characterised by a commitment to understand and challenge the ways in which knowledge is produced. Moreover, it looks at structural inequalities that tackle women and other marginalised groups and tries to comprehend how these structures can be changed. The aim of this chapter is to explore the potential of feminist research when applied to the study of tourism, as there has been little attention to the intersection between both approaches. To do so, the chapter presents a case study, specifically, the inclusion of gender non-conforming identities in the tourism sector, to exemplify how theoretical feminist research concepts can be landed. More concretely, it shows that in all common steps of conducting a study, feminist concepts can be applied to change the traditional ways of producing knowledge. The chapter concludes by providing some suggested questions that can lead to new paths in the study of tourism by making it more inclusive and diverse.

Keywords

Feminist research; Gender non-conforming; Intersectionality; LGBTQ+; Tourism.

Introduction

The intertwining between tourism studies and the gender perspective emerged in the 1990s (Chambers & Rakić, 2018), leading to a wide range of studies that still leave scope for development and improvement (Munar et al., 2017; Pritchard & Morgan, 2017). The gender perspective has been introduced into tourism studies to give a broader view of the industry, and its application challenges power structures by shedding light on new issues. Figueroa-Domecq et al. (2015) establish four main areas that represent the intersection between tourism and gender: (a) women as travellers and consumers, (b) gender and tourism development, (c) gendered tourism labour market, and (d) tourism research. The last of these areas is at the centre of the present chapter.

DOI: 10.4324/9781003286721-7

In this sense, tourism research needs to evolve and introduce a gender perspective to better understand the sector's complex realities (Figueroa-Domecq & Segovia-Perez, 2020). Therefore, the goal of this chapter is to broaden the scope of feminist research by exploring its potential when applied to the study of tourism. Doing so should ultimately result in the establishment of alternative paths to study tourism, as feminist research has the potential to bring light to new topics, communities, and lines of research. As explored by Letherby (2003), feminist research is a political commitment to producing knowledge that has practical impacts. The purpose of feminist research is to create social change while incorporating human diversity and fostering important relationships between those studied and the reader (Reinharz & Davidman, 1992).

We argue that feminist research allows for the emergence of new topics, challenging hegemonical and traditional research perspectives and, in this way, facilitating distinct outcomes and different ways of reaching knowledge. In this sense, we propose to integrate approaches encompassed in feminist research acknowledging intersectionality (Crenshaw, 1991), the researcher's self-reflexivity and positionality (Kobayashi, 2003), power relationships between researchers and participants (England, 1994), co-constructing feminist research (Harding, 2020; Naples & Gurr, 2013), the insider/outsider duality (Naples, 1996), and feminist standpoint theory (Harding, 1987). Thus, by changing the design, development, and evaluation of investigations, a feminist approach would result in structural transformations of knowledge production in tourism research. Some publications already highlight the potential of applying feminist research to tourism studies (Ateljevic et al., 2005; Bakas, 2017; Cohen, 2013; Simoni & McCabe, 2008).

For this reason, the chapter is divided into the following sections. First, it sets out the theoretical framework surrounding feminist research, highlighting its main characteristics and debates. Second, it introduces the research context of the case study that exemplifies how feminist research concepts can be applied. More concretely, the chapter focuses on the inclusion of non-binary, genderfluid, and genderqueer identities in Barcelona as a tourism destination. Third, it applies the main feminist research concepts to show how researchers can land abstract ideas into a specific tourism case study. To this end, results are divided into the common stages of conducting research: before, during, and after. Following this, it provides a series of questions any researcher can ask themselves to check whether they have applied feminist concepts in their research on tourism. Finally, there is reflection on the practical implications and future lines of research that can be derived from the chapter.

Conceptualising Feminist Research

The emergence of feminist research responds to the view of feminist researchers, who detected a gender bias, specifically androcentrism, in the generation of knowledge in science (Hesse-Biber, 2007). For this reason, feminist research is sometimes described as a link to women's experiences. For instance, Klein (1983, p. 90) defines it as "research that tries to take women's needs, interests and experiences into account and aims at being instrumental in improving women's lives in one way or another," while Wilkinson (1986) argues that feminist research is on women and for women, prioritising their experiences and generating knowledge based on them. Similarly, Wise (1987) states that feminist research should locate women's oppression at the centre. McHugh (2020) defines feminist research's purpose as focusing on female experiences, analysing gender oppressions and power imbalances, and advocating for women in this context to ultimately transform society.

Following this line of thought, several authors define feminist research as an approach that allows the generation of knowledge outside the methodologies developed by and for men. For instance, Payne and Payne (2004) define it as employing a specific subgroup of methods and/or selecting concrete issues, with the aim of challenging the hegemonic position of men and, therefore, the marginalisation of women in research. Thus, there is still a focus on female experiences while acknowledging that science has been based on a masculinised purpose, conception, and, consequently, results. For these reasons, some authors exclude male academics, suggesting they are unable to develop feminist research (Bernhard, 1984; Kremer, 1990; Mies, 1983), while others state that feminist research may be developed by any author regardless of their gender, as long as the research is based on a feminist approach (Harding, 1987).

Lastly, feminist research is also understood as being inscribed within feminist thought (Harding, 1987; Reinharz & Davidman, 1992), which includes the need to focus on women within a larger frame. For instance, McHugh (2020, p. 1) understands feminist research to contribute to "the transformation of science from empiricism to postmodernism," while Campbell and Bunting (1991) state that a feminist research approach is based on a particular theory of knowledge and its methodology and methods follow from this theory. For this reason, feminist research is not exclusively the study of women, and it is not enough that women undertake it (Webb, 1993), but instead is the application of a branch of theoretical knowledge. As explored by Hesse-Biber (2007), this shift in feminist thought is primarily due to critical feminist researchers who questioned the lack of integrating intersections among gender, race, and class in feminist thought and research (Anzaldúa, 1987; Butler, 1993; Collins, 1990; Crenshaw, 1991; hooks, 1984; Mohanty, 1988; Spivak, 1994). Through this lens, feminist research has also led to the exploration of new topics such as the LGBTQ+ community, where aspects such as gender identity or sexual orientation intersect.

The gender perspective as applied to research is grounded in several key concepts. Figure 4.1 summarises those taken into account in this chapter.

Feminist research also encompasses the exploration of feminist methods and methodologies. It is worth understanding the difference between these two factors: methods are the techniques available to gather evidence, while methodologies are a system of methods and postulates employed by a specific discipline. When applied to feminist research, the distinction between methods and methodologies has blurred the conceptualisation and study of each aspect (Harding, 1987).

Feminist research is unequivocally informed by feminist epistemology and, more specifically, by questioning the traditional understanding of knowledge. More specifically, it challenges the subjects of knowledge creation, the validation processes, and the intertwining with ontology, understood as the relationship between knowing and being (Landman, 2006). Through this lens, feminist research, including the methods and methodologies applied, focuses on the relationship between knowledge production with unequal global power structures and the social reality of those neglected and marginalised. In this sense, feminist research can be thought of as both theory and practice, thus requiring researchers to challenge the issues studied and the way they are explored (Letherby, 2003).

Regardless of the chosen research method, techniques must be applied that provide new insights into previously unexplored topics or aspects based on marginalisation. This new light might provide different results in terms of the questions posed, the researcher's position within the study, and the intended outcome of the research (Letherby, 2003). In this sense, rather than feminist methods, there is a growing recognition of "feminist research practice" (Kelly, 1988).

Intersectionality (Crenshaw, 1991)

- Recognises the combination and interconnection of social categorisations, such as gender, class, race, or age, among many others.

Positionality (Kobayashi, 2003)

- Defines how one is inserted in grids of power relations and how those influence methods, interpretations, and knowledge production.

Power dynamics in research (Mullings, 1999)

- Highlights how power relations between researcher and participants may alter knowledge production.

Co-construction of knowledge (Harding, 2020)

- Repositions marginalised individuals and places them at the centre of knowledge produced about (by) them.

Insider/outsider dichotomy (Dwyer & Buckle, 2009)

- Engages with the notion of space to understand how researchers flow between one position and another.

Feminist standpoint theory (Harding, 1987)

- Proposes the idea that knowledge is socially located, it is, knowledge stems from the lived experiences of individuals.

Figure 4.1   Key feminist concepts included in this chapter. (Source: Own elaboration)

For instance, there are no methods specific to feminist research, as feminist researchers employ existing techniques, such as interviews, surveys, observation, participant observation, focus groups, ethnography, life histories, and content analysis, among many others (Letherby, 2003). However, as feminist research began to emerge, qualitative methods were preferred as they were understood to be more aligned with feminist thought (Reinharz & Davidman, 1992). This tendency is still present in current feminist research, as qualitative research is perceived to achieve in-depth understanding instead of generalising and homogenising data (Naderifar et al., 2017). Despite this, feminist researchers have long employed quantitative methods (Hughes & Cohen, 2013; Miner et al., 2012; Westmarland, 2001), as these are crucial when attempting to analyse and comprehend processes and patterns within large data sets, for example, to detect gender biases and inequalities (Scott, 2012). One example of this idea is the creation of Relief Maps, a tool to study geographies by researching the interlocking of power structures, lived experiences, and places, while integrating a gender perspective in the quantitative data collection (Rodó-de-Zárate, 2014).

Research Context

The research context is based on a previous study that aimed to fill a knowledge gap encompassing tourism and gender non-conforming tourists. This study was developed by the authors of this chapter during the second half of 2021 and has not been published. The terms non-binary, genderfluid, and genderqueer refer to identities rather than physical traits, although they do not disregard intersex and trans people who may identify this way (Abrams,

> **Non-binary (Abrams, 2019)**
> - People who do not identify or do not exclusively identify as a man or as a woman. It is also an umbrella term to encapsulate all those genders that do not sit comfortably in the gender binary. Throughout this research, non-binary will be understood as both meanings.

> **Genderfluid (Ferguson, 2020)**
> - People who do not have a fixed gender throughout their life and may identify with different genders in different periods of time.

> **Genderqueer (Clements, 2018)**
> - Also known as queer, people who do not conform to binary and traditional expectations of their sexual orientation and gender. A genderqueer identity can either be outside, in between or fluctuating amongst the gender binary.

Figure 4.2   Conceptualisation of non-binary, genderfluid, and genderqueer. (Source: Own elaboration)

2019; Richards et al., 2017). Such identities are the ones that escape the rigid gender binary divided into male and female. It should be acknowledged that each identity may feel differently for those who define themselves as such, which is why it may be challenging to reach a universal definition. Throughout this chapter, "gender non-conforming" is employed to refer to all the identities included, namely: non-binary, genderfluid, and genderqueer (Figure 4.2).

More concretely, the project took a feminist approach to analyse the inclusion of non-binary, genderfluid, and genderqueer identities in Barcelona as a tourism destination. To this end, data was generated through a content analysis of LGBTQ+ tourism promotional materials for Barcelona and surveys answered by gender non-conforming people who had been to Barcelona or were willing to visit, to comprehend their experiences as tourists or future travellers to the city.

As a growing capitalist force, tourism tends to commodify places and identities to ensure its expansion (Enloe, 2000; Heller et al., 2014; Young & Markham, 2019). Consequently, tourism engages with creating hegemonic discourses that other and exoticise places and identities by prioritising Western thought (Aitchison, 2001; Santos & Caton, 2008). Therefore, promoting destinations as LGBTQ+-friendly in a global market intrinsically includes a discussion of politics, economy, and subjectivities (Waitt & Markwell, 2006). In this scenario, tourism embodies two positions simultaneously: campaigning for gender and sexual diversity and promoting visibility for the community, while at the same time making a profit from the commodification of the LGBTQ+ community (Waitt et al., 2008).

One of the main knowledge gaps found in LGBTQ+ tourism is the inclusion of non-binary identities, as there is no scholarly literature focusing on this area. Despite this, the investigation's research framework can be built based on two sources of information. Firstly, grey literature, including reports and information published by governments and specialised entities, was consulted to gain insight into the literature on gender non-conforming travellers. Secondly, blogs and open social platforms written by non-binary, genderfluid, and genderqueer tourists were included to ensure the voice of the community is represented.

Gender non-conforming identities are often considered alongside trans identities in tourism (Community Marketing & Insights [CM&I], 2019; Government of Canada, n.d.;

IGLTA, 2018). According to governments, entities, and tourism businesses, gender non-conforming travellers are more likely to face legal issues and be exposed to unsafe situations in airports, accommodation, and when in contact with the local community. Non-binary travellers are advised to research before deciding on a travel destination, considering aspects such as LGBTQ+-related laws (IGLTA, 2018; Stratton, 2019), official governmental pages (Stratton, 2019), and the social reality for members of the LGBTQ+ community (Dejarnette, 2018). Moreover, travellers should be wary of their online activity, as local authorities may have access to their accounts and interactions with local LGBTQ+ websites, which could then be used against them (Government of Canada, n.d.). Travellers are also often advised to book their trip through LGBTQ+-owned or LGTBQ+-friendly tourism operators (Dejarnette, 2018; Stratton, 2019). A report by Community Marketing & Insights (2019) revealed that 79% of respondents viewed a destination being LGBTQ+-friendly as very important.

Feminist Research Applied to Gender Non-Conforming Identities

This section aims to show how the main feminist research concepts are introduced in tourism research, concretely, in the study of the inclusion of non-binary, genderfluid, and genderqueer identities in Barcelona as a tourism destination. It offers an applied and practical vision of feminist research in tourism, divided according to the common stages of conducting research: before, during, and after.

Before Starting the Research: Choosing the Topic and the Sources

The pre-study stage usually encompasses two main aspects: choosing the topic and looking for academic literature to elaborate a theoretical framework. The decision to research non-binary, genderfluid, and genderqueer identities in tourism entails certain feminist research characteristics that can be explored. On the one hand, the decision to explore the experiences of the most marginalised identities within the LGBTQ+ community, especially in the setting of the tourism sector, involves a political element that is generally related to feminism. In this way, social and academic recognition is sought for a group severely underrepresented in academia, as is the case of gender non-conforming travellers. The academy often has the privilege of deciding what to discuss and how; feminist research, on the other hand, also encourages exploration of what is not usually named and may therefore be disregarded by many.

This idea is mirrored in the theories of Spivak (1988) and Collins (1986), who focus on the importance of the narrative and discourses created regarding marginalised identities. More concretely, Spivak (1988) argues that Global North–South power dynamics homogenise southern voices, excluding them from knowledge creation. The author states that privileged communities speak for the subalterns and questions whether oppressed groups are left with the power to speak or are stripped of it. In addition, Collins (1986) refers to the importance of a group being able to self-define, defending the importance of this act for two reasons. Firstly, it resists dehumanisation by established powers and challenges the notion of being "other." Secondly, it allows members of oppressed groups to confront internalised self-oppression. Applying Collins' conceptions to the experience of gender non-conforming travellers would facilitate a reversal of the issues they face in relation to tourism. In summary, feminist research encourages approaches in which the studied group has control over their own narrative.

Intersectionality plays an important role when choosing a research topic, since it allows us to approach realities that have not previously been studied or considered within the framework of tourism, such as non-binary, genderfluid, and genderqueer identities. Intersectionality is an analytical framework coined by Crenshaw (1991) that recognises the combination and interconnection of social categorisations such as gender, class, race, or age, among many others. Moreover, such a framework conceptualises how they overlap and the oppressions and privileges they lead to. Intersectionality has become an analytic tool employed by grassroot organisations, researchers, politicians, and other stakeholders, proving that this framework is current and is applied in different scenarios. As Collins and Bilge (2020) point out, while the concept of intersectionality is implemented worldwide, the most important question becomes what intersectionality does, rather than what intersectionality is. Every interpretation of intersectionality has in common the analysis of the intersecting power relations and the recognition of social inequalities. In this sense, several proposals have emerged in qualitative research to apply intersectionality as a tool to strengthen theoretical and methodological rigour (see Abrams et al., 2020; Al-Faham et al., 2019; Harris & Patton, 2018; Rice et al., 2019; Rodó-de-Zárate, 2021). More specifically, it is necessary to consider the three forms of intersectionality proposed by Crenshaw: structural, political, and representational.

Structural intersectionality explains the interlocking of different structures that, united, determine a unique experience for those with the included identity markers. Structural intersectionality has been used the basis of the present chapter, since it aims to disentangle diverse axes of oppression, and problematise the idea of a homogeneous LGBTQ+ community. Although there are some studies on LGBTQ+ tourism, the concept of intersectionality allows researchers to approach greater clarity on the discussed topic, since axes of oppression such as gender identity and performativity are located at the centre. In this study, intersectionality sheds light on a community that is often disregarded within the LGBTQ+ community, namely, gender non-conforming individuals.

Political intersectionality focuses on the intersection of political strategies that aim to represent subordinated identities and locate diverse inequalities. In relation to political intersectionality, it is crucial to produce knowledge with communities and identities that have long been disregarded, as gathering concrete data is the only way to foster the development of specific policies that will respond to the experiences and needs of gender non-conforming travellers.

Lastly, representational intersectionality focuses on formulating a narrative and imagery that truly depict those with the markers of oppression. Although representational intersectionality tends to refer to popular culture, in this chapter it becomes important as the voices of gender non-conforming travellers are located at the centre, thus ensuring the community is represented.

Once the research topic has been chosen, the next step generally involves the construction of the theoretical framework. For this study, another feminist research concept appeared suitable: standpoint theory. Feminist standpoint theory proposes that knowledge is socially located, that is, knowledge stems from the lived experiences of individuals. Moreover, it argues that knowledge is grounded in experience; therefore, the oppressed have a clearer vision of their context and the structures that marginalise them (Harding, 1987). More concretely, as Annas (1978) proposes, oppressed individuals have a "double vision" – awareness and understanding of the dominant world and their perspective and experiences. For this reason, Harding (2004) states that standpoint theory legitimises socially located knowledge, thus questioning the relations between power and knowledge production. Thus, through this approach the oppressed can create alternative knowledge that may destabilise hegemonic understandings. In the present case study, as previously explained, two of the most important sources of information to understand non-binary, genderfluid, and genderqueer identities' context in tourism are grey literature and the personal experiences of this specific group.

Feminist standpoint theory recognises oppressed groups' knowledge as being more valuable due to their comprehension of social structures and oppressions (Wigginton & Lafrance, 2019). For instance, feminist standpoint epistemologies allowed Collins (1990) to explore Black women's experiences of racism and sexism and, more concretely, their resistance strategies to such dominations. However, standpoint theory needs to be carefully incorporated. According to the critiques by the same author, standpoint theory must dismantle the idea that identities such as 'women' are a homogeneous group. Different groups cannot be universalised, as different social markers will change their location, experiences, and knowledge. In this sense, this study was not directed to the LGBTQ+ community as a whole, but rather to specific identities within the group, to avoid the generalisation of gender non-conforming travellers. Thus, following the works of several scholars (Anzaldúa, 1987; hooks, 1984; Mohanty, 1988), this study applies feminist standpoint theory to explore the interconnections between different categories.

During the Research: Methodology, Participants, and the Researcher

This section mainly focuses on how to develop the study, which includes establishing the methodology, obtaining results, and discussing them. This process requires the implementation of feminist research concepts to transversally integrate a gender perspective into the research. More specifically, the concepts applied at this stage of the study are co-production of knowledge, self-reflexivity and positionality, power dynamics, and the insider/outsider debate.

The research techniques used for the present study included content analysis of Barcelona's official tourism web pages to explore marketing strategies directed towards the LGBTQ+ community and identify whether gender non-conforming identities were included. In addition, surveys were conducted with non-binary, genderfluid, and genderqueer people from all over the world who had travelled to or wanted to visit Barcelona. Surveys, along with related research methods such as interviews and focus groups, elicit participants' opinions and incorporate them as part of the knowledge-creation process. For this reason, the choice of methods for this study was driven by the need to give voice to marginalised identities.

The co-construction of feminist research can be achieved by integrating participants as co-producers of knowledge and by being mindful of the power dynamics that occur during research. Participatory action research (PAR) has become a framework through which researchers scrutinise privilege in knowledge production, paying attention to their own standpoint and positionality (Langan & Morton, 2009; Naples & Gurr, 2013). For the present case study, in order to achieve a co-constructed framework, the survey's questions were extracted from a literature review which mainly explored gender non-conforming personal experiences. However, it should be noted that greater integration of participants could have been generated through the development of a focus group with gender non-conforming people determining the questions for the survey. In this sense, the co-production of knowledge is not only generated by participants' answers but also by including them at the methodological stage so as to guarantee that the research is linked to the actual needs of their community.

One of the main challenges when developing this study was to define a group of people with diverse characteristics. This was difficult because non-binary, genderfluid, and genderqueer people's experiences are not homogeneous and only a few have openly expressed their encounters within the tourism sector. This might lead us to rethink representativity in research. While standpoint theory centres on the vulnerable in its investigations, it also impedes their voices from being universalised. As feminist researchers aiming at a practical impact, establishing limits to the sample becomes problematic. Although Windsong (2016) proposes ensuring that participants have varied backgrounds in terms of gender, class, race, religion, sexuality, or citizenship status, the clashing of such metrics among the size of a snowball sampling still would not allow for its universality.

For this reason, the aim to apply intersectionality in research can meet with representational issues, which highlights the need for future studies that look into diverse social characteristics among the participants. To achieve a diverse group of participants within the gender non-conforming identity, potential respondents were approached via different means.

Social media networks such as Instagram and Twitter were the first port of call; they facilitated sharing the survey and could be used to share it among a large number of participants. However, some problems arose; for example, accidentally reaching out people who are from Barcelona who could not participate in the study, or getting a lower number of responses because it was difficult to reach the target population.

Next, we looked at various public LGBTQ+ Facebook groups that could be used to share the survey. Such groups offered a more targeted approach to finding participants who might be gender non-conforming, potentially reaching a higher number of respondents. However, many of these Facebook communities were geographically based, meaning that respondents would come from areas where non-conforming identities were widely accepted but making it harder to obtain responses from participants in non-accepting countries. There are also Facebook groups specifically by and for gender non-conforming individuals, but many of them are not open to the public and do not allow information about research studies to be posted. All of these factors hindered the creation of a diverse group of gender non-conforming respondents representative of a broad range of demographic aspects such as race, nationality, and age, among many others.

Feminist research also brings to the fore debates about the researcher's role during the investigation process, in aspects such as the prejudices they might involuntarily incorporate, their position in relation to participants, and the power dynamics that may emerge. In this sense, self-reflexivity needs to be a dynamic process that constantly reshapes the approach throughout the investigation (Sultana, 2007). England (1994) states that "reflexivity is self-critical sympathetic introspection and the self-conscious analytical scrutiny of the self as researcher" (p. 82). It is about awareness of the researcher's central role in constructing knowledge. In order to practice self-reflexivity, the researcher must reflect on their positionality by developing a critical awareness of socioeconomic, political, economic, and cultural facets of their background, education, experiences, and embodied presence in the world and how these aspects have shaped their intellectual orientation and world-view.

Positionality's importance is grounded in feminist work that claims all knowledge is shaped by the context or circumstances of its production and location (Valentine, 2002). Kobayashi (2003) states that positionality defines how one is inserted into grids of power relations and how these positions influence methods, interpretations, and knowledge production. In this way, positionality is key to shaping research objectives, the researcher's interactions during fieldwork, data analysis, and research outcomes (Peake, 2017). This is because the researcher's location within society may determine their interests and influence their perceptions and interactions.

Before and during the development of the present study, the self-reflexivity process allowed us to reflect on our positionality as researchers. Undertaking this process through an intersectional lens enables to examine different social characteristics that could have an impact on the research and the researchers. In this case, we examined our position from angles such as the roles we undertook in the study, as well as our sexuality, nationality, and race.

Firstly, we reflected on our role as researchers and what it entailed, exploring the power dynamics that might arise when working with participants. Secondly, our position as members of the LGBTQ+ community was an intersectional aspect to consider when reflecting on our positionality. Both aspects are further explored in the following sections.

Thirdly, the fact that both of us are from Barcelona and residents in the city was an aspect to reflect on as well. In this sense, it was crucial to ensure that we would approach the imaginary of Barcelona with caution, as our perceptions and those of travellers might vary greatly. It was important to design the content analysis and the survey questions objectively, with the aim of ensuring a neutral and unbiased representation of the city. Moreover, during the analysis of the results, it was crucial to be aware of our position of Barcelona's inhabitants, for instance, when understanding the feeling of safety perceived by travellers.

Finally, from a gender perspective, power dynamics such as North-South relations must be acknowledged in academia (Mostafanezhad, 2013; Salvador-Almela & Abellan-Calvet, 2021). In this sense, we developed awareness of our positionality as white researchers and how that can shape the way we approach research. In this sense, we were aware that the questions asked in the survey might not suit all experiences. We were also aware that knowledge is usually produced in white spheres, perpetuating white supremacy and the Westernisation of academia (Dupree & Boykin, 2021; Johnson, 2018). Reflecting on this, we perhaps should have broadened the definition of gender non-conforming and looked beyond the three main identities included (non-binary, genderfluid, and genderqueer), which are widely employed in the Global North. There are plenty of gender non-conforming labels, identities, and definitions in the Global South that would not fit into the white definitions of gender identity, nor should they have to.

As shown by this example, the self-reflexivity process allows researchers to understand and focus on their positionality, and in the process become aware of their location within the insider/outsider binary. Insider/outsider dichotomies have been debated extensively and claim to offer a more nuanced understanding of positionality. Researchers may occupy the position of either insider or outsider, with those based within the community they are investigating considered insiders, while those researching from outside the group are defined as outsiders. Each position offers both advantages and disadvantages (Hellawell, 2006). It has been argued that insider/outsider positions are epistemological issues, since the relation between the researcher and the subject directly affects the knowledge that is co-produced (Hayfield & Huxley, 2015).

Throughout this study, we reflected on whether our positionality would locate us as insiders or outsiders in relation to gender non-conforming participants. It was crucial to keep fluidity in mind when locating our position in this research. In both our cases, we could consider ourselves insiders, as we are both members of the LGBTQ+ community. At the same time, the LGBTQ+ community is a heterogeneous group formed of individuals who possess diverse characteristics in terms of gender orientation, gender identity, and biological sex. For this reason, we also acknowledged our position as outsiders since, as cisgender lesbians, we are not part of the gender non-conforming community.

Insider positionality comes with easier access to information, the ability to ask more meaningful questions, read non-verbal cues, and project a more candid and accurate picture of the subjects under study (Merriam et al., 2001). Thus, a researcher with inside knowledge of a group to which they belong can gain more intimate insights into the group's opinions based on their knowledge of that group (Mullings, 1999). However, an insider may also blur boundaries, producing bias by projecting their own beliefs, values, and perceptions onto others (Berger, 2015), which can influence the outcome of their investigations (Dwyer & Buckle, 2009). For example, the researcher might be deterred from asking challenging questions due to these biases (Merriam et al., 2001). Participants might also fail to disclose information, assuming that as they share the same identity as the researcher, the researcher is already aware of this information (Berger, 2015).

It is generally assumed that being an outsider is a disadvantage due to accessibility issues and the possibility of misunderstandings and misinterpretations (Merriam et al., 2001), for example, due to participants not revealing information that they take for granted (Dwyer & Buckle, 2009). Moreover, although a researcher can be located as an insider within a group, they might simultaneously be an outsider within a subgroup (Asselin, 2003). However, being an outsider has advantages, such as the ability to ask taboo questions. There is also a tendency for participants to provide more information, as the researcher is seen as someone unfamiliar with this group (Merriam et al., 2001). Furthermore, outsiders are presumed to be more objective and able to observe behaviour without distorting its meaning (Kanuha, 2000; Mullings, 1999).

The insider/outsider dichotomy can be considered from a problematised perspective by engaging with the notion of space to understand how researchers flow between one position and another (Dwyer & Buckle, 2009). In this sense, some scholars have discussed that being an outsider or insider is not a fixed or static position; they are more like permeable, ever-shifting social spaces in which experiences differ (Naples, 1996). For these reasons, defining insider and outsider status can be a complex process, and the boundaries between these positions are not always transparent, as one's positionality is influenced by a variety of intersecting factors such as race, class, gender, culture, and others (Merriam et al., 2001).

Being an insider or an outsider is not a static position, as it is modulated as the research evolves. As Naples (1996) explored, the fluidity between these two positions is constantly being adjusted in each interaction and context. Moreover, when examining their position, researchers should take into account that social groups are not homogeneous, and that they may possess identity markers that clash with those of their participants (Dwyer & Buckle, 2009).

Reflecting further on our position as lesbians, another concept arose which problematised the insider/outsider binary, namely, performativity (Butler, 1990). Performativity theory explores how gender and sexuality are not inherent principles of individuals but rather aspects of their lives that are acted out and therefore conceived of as real. These portrayals are often treated as if they have to fit the binary system of being either male or female, thus disregarding any identity that escapes this rigid logic. As Butler (1990) argues, gender is performed and acted through aspects such as speech and physical appearance, among others.

In our case, both researchers are lesbians but present very different performances. For one of us, the performance is closer to the gender performance traditionally linked to being more feminine. From this position, the researcher did not relate to the experiences of participants. The other's gender performativity can be described as far from being traditionally feminine, and rather androgyne. In this case, for the latter researcher, although as a cisgender lesbian the position with regard gender non-conforming participants was that of an outsider, many of the experiences, feelings, and emotions resonated personally.

Presenting an androgyne performance has led to various circumstances where one's gender is misread, which allows the researcher to relate to the feelings and experiences expressed by participants, such as feeling uneasy in gendered areas, being underrepresented in Barcelona's LGBTQ+ tourism campaigns, or needing more LGBTQ+ tourism spaces that are not related to nightlife, among others.

Through the process of self-reflection, researchers also gain insight into their position regarding participants. Therefore, it is imperative to reflect on the power dynamics that may emerge during the course of the research. Feminist research poses the challenge of identifying and grappling with the power dynamics in the researcher–participant relationship. Moreover, it highlights how power relations between researchers and participants may alter knowledge production (Mullings, 1999). Scholars have shown that research is driven by power dynamics, which is something to be aware of and to negotiate during the process (Merriam et al., 2001).

Reflexivity is a crucial tool for disclosing the power dynamics within research practice (Mullings, 1999) and examining how researchers' and participants' positionality contributes to the unequal power structure (Bloom, 1998; Wolf, 1996). It is worth bearing in mind that power dynamics may also emerge between researchers and participants while the research process is taking place, and they may vary following the fluid location of each actor in every step of the study.

Reflexive processes should also incorporate an awareness of intersectionality, to examine how the researcher's allegiances may affect research processes, uncover how privileges and disadvantages affect their research, and explore how power functions in a dynamic and contradictory manner (Rice et al., 2019). Several feminist works and critiques from women of colour have challenged the power of whiteness in knowledge production, including white biases and colonial assumptions, such as the false supposition that women's experiences are universal (Dinçer, 2019). Contemporary feminist scholars emphasise the need for participatory methodologies and reflexivity to understand power dynamics within research methods (Pollack & Eldridge, 2015), two aspects that are explored in depth in this chapter. It is important to bear in mind that power dynamics may refer to structural violence issues which go beyond the conducted research but are embedded in society and need to be acknowledged by researchers.

As feminist researchers, a strategy to avoid this situation is one of care for participants. Care may take various forms, especially regarding surveys. For our study, before accessing the content of the survey, participants were informed of the questions they would face, were assured that they could leave the survey at any time with no negative consequences, were offered a 24-hour helpline specialising in the LGBTQ+ community, were granted full anonymity, and were informed that they could contact the researchers at any time. Moreover, we designed the survey's content such that care for participants was transversal throughout the questions, for instance through language. Gender-neutral language was employed to ensure that all participants could feel comfortable regardless of their pronouns. Moreover, we made sure to employ language that was as stereotype-free as possible, for instance, by asking if respondents had ever been to Barcelona with a partner or partners.

After Finalising the Research: Emotions, Objectivity, and Publication

This section focuses on the last steps of a study: analysing the results and publishing. When analysing the results, it is important to consider how to deal with aspects that might challenge researchers directly, for example, taking into account the emotions and feelings that arise from the research. Several feminist scholars have integrated emotions into their re-search processes, citing the impossibility of separating feelings from aspects such as the researcher's motivations, established relations during fieldwork, self-reflexivity processes, or chosen methods (Bell, 2015; Carter, 2016; Rayaprol, 2016). One well-explored aspect is the discomfort that may arise from undertaking research and how it can be acknowledged, addressed, and managed (Ahmed, 2010; Chadwick, 2021; Falconer, 2017). Researchers also need to be aware of their own emotions and attitudes, as they may directly impact how knowledge is produced. Some authors argue that researchers who attempt to remain neutral and avoid exploring their own emotions in research participate in and reinforce power dynamics (Hale, 2008; Leacock, 1987). Feminist researchers argue that engaging with emotions is crucial to building rapport, understood as developing a trustful and safe relationship with participants (LaMarre et al., 2022; Miller, 2017; Pascoe Leahy, 2021).

In this particular case, emotions became important during the research process and, es-pecially, after finishing it. During the early stages, it was important to acknowledge several characteristics that could potentially impact the development and outcomes of the study. It was, for example, necessary to combat feelings of both researchers that we lacked compe-tence, which almost led to the investigation not being started. Moreover, it was important to reflect on our position regarding participants, which led to one of the researchers feel-ing close to the experiences extracted from the survey. As the study advanced and results from the content analysis and surveys emerged, both researchers felt disappointment and helplessness at what the data revealed about the failure to represent gender non-conforming tourists. The self-reflexive process helped us to acknowledge the frustrations and hope that the result brought – feelings that are not uncommon among researchers.

Emotions were also an important aspect after the study was complete, as the investiga-tion concluded with future lines of action and proposals. Suggestions included the need to apply intersectionality in the campaigns, providing information about further the full range of LGBTQ+ services offered in the city, to rethink gendered spaces and non-discrimination policies, to address the lack of contact between tourists and locals, and to offer information about the LGBTQ+ history in Barcelona. As Barcelonians, tourism researchers, and mem-bers of the LGBTQ+ community, acknowledging these proposals led to mixed feelings: of hope for better and more inclusive tourism in the city, as well as frustration for the long path that needs to be walked to get there. Acknowledging the feelings and emotions that emerge during a study can help the researcher understand the notion of "feminist objectiv-ity" that is widely debated within feminist studies.

Feminist research clashes directly with the notion of objectivity, which is a requisite for empirical sciences (Haraway, 1988). Thus, several feminist scholars have developed the con-cept of "feminist objectivity" based on the idea that as knowledge is located, it is linked to the individuals producing it (Bhavnani, 1993; Haraway, 1988; Harding, 1992). For these scholars, empirical objectivity is unrealistic and undesirable. Feminist objectivity is often re-lated to validating and giving importance to emotions and their role in research (Jaggar, 1997; Sprague & Zimmerman, 1993). Being aware of one's feelings leads to a more complex and global understanding of the investigation, accepting that emotions and feelings will arise, and therefore aiming to understand their potential impact rather than ignoring them.

Once the research is complete, the next step is publication. The feminist research frame-work also encourages reflection on the power relations that exist in the scholarly sphere. Researchers have the power to decide whether their scientific production remains only within the academic sphere or goes further and reaches the community involved in the study. It is necessary to find diverse ways for the results and key points of the investigation to reach as many people as possible, such as dissemination through grey literature and the translation of research into different languages. In addition, it is important to seek practical involvement for the communities that have participated in the study, either by bringing the results to them or by pressuring agents such as public administrations or private companies to embrace change where needed. As Anne Tagonist (2009) argues, researchers – and especially those engaged in feminist research – have a duty to the investigation and the participants to ensure that the results of the study have practical implications.

When applying the concept of co-producers of knowledge, it is interesting to explore how those who participated in research can become spokespeople and leaders in the dissemination of the results. In this way, feminist research enables us to understand that the research itself is not the ultimate goal; rather, the goal of research is to discover ways to change society for the better. With regard to the case study presented here, the results should be shared with participants through Facebook groups, through researchers' social media, and by encouraging the establishment of new special-interest groups so that participants who are interested in pursuing the proposals can get in touch with each other and initiate change. The results can be shared via graphs and visual summaries that can be distributed to institutions in charge of tourism in Barcelona with the hope of sparking debate.

Questions to Ask When Designing Feminist Research

The following section proposes a series of questions that any researcher can ask themselves before, during, and after carrying out feminist research in tourism in order to assess how well they have applied the various concepts of feminist research.

Before starting: choosing the topic and sources

- How is the gender perspective integrated into my research?
- How are intersectional aspects taken into account in the choice of research topic?
- What contribution does my research make in terms of gender?
- Does my research reveal structural inequalities in terms of gender?
- How will I acknowledge the knowledge of the community involved in this research?

During the research process: methodology, participants, and the researcher

- Am I applying intersectionality in the chosen methods?
- Am I applying intersectionality among participants to ensure that the group is representative?
- What aspects should I take into account regarding my positionality as a researcher?
- Which aspects of my life do I need to consider in the self-reflexivity process for this research?
- What unequal power relations might form between myself (as a researcher) and participants and how can they be avoided/minimised?
- What mechanisms can I use to ensure that research participants become co-producers of knowledge?
- Do I consider myself an insider or an outsider with respect to the community of participants and how can this position affect the research?

After finalising the research: emotions, objectivity, and publication

- How do the research results affect me?
- How can I ensure the research results reach audiences beyond the academic sphere?
- How can I involve the research participants as spokespeople in the process of disseminating the results?
- How will participants and their community benefit from the research?
- What feelings arose during the process and how did they interact with the research and the participants?
- How can this research have a positive and practical impact?

Final Thoughts

The aim of this chapter was to bring tourism research closer to feminist research. It highlighted the need to apply feminist research concepts in the various phases that an investigation entails. It was therefore been approached from two sides: the theoretical and the practical. On the theoretical side, it showed how feminist research has evolved from the recognition of women's voices to the integration of other marginal voices in society, including the LGBTQ+ community. On the practical side, it offered an example of how feminist research can be applied through a specific case study relating to tourism in Barcelona. Specifically, it showed how the main concepts within feminist research can be applied to the study of how non-binary, genderfluid, and genderqueer people perceive tourism and are represented in tourism campaigns in Barcelona.

The importance of this research resides in the need to fill a knowledge gap: to understand the relationship between gender non-conforming people and tourism. As there is limited scientific literature on the topic, this chapter aims to be the first step towards including such identities in publications on LGBTQ+ tourism. Without this, LGBTQ+ tourism's journey towards being inclusive and embracing diversity cannot be fulfilled. In addition, the choice of topic also reflects the need to recognise the growing diversity of the LGBTQ+ community and to investigate how tourism can critically embrace these realities.

To this end, the concepts of intersectionality, self-reflexivity, positionality, power dynamics, the insider/outsider binary, the co-construction of feminist research, and feminist standpoint theory were applied to specific examples and proposals. All these concepts from feminist research present opportunities to rethink who creates knowledge and how they do so. By repositioning marginalised individuals and placing them at the centre of knowledge produced about (by) them, it is possible to flatten power hierarchies typically felt in traditional research (Harding, 2020; Maynard, 1994).

Future researchers wishing to apply feminist research in a real tourism case study will be able to follow similar steps and should ask themselves the above-stated questions when developing their study. By changing the design, development, and dissemination of tourism research, the feminist approach can lead to structural change in tourism knowledge production. By approaching the sector from a feminist research perspective, new questions, and future lines of research will emerge to inform the future of the tourism studies field. Some examples include examining the intersectionality of gender and other forms of marginalisation in the tourism industry; analysing the impact of tourism on local communities, specifically looking at gendered and power dynamics; exploring the role of feminist theory and praxis in co-creating more equitable and sustainable tourism practices; or understanding other experiences of tourism researchers when applying feminist research theory. For these reasons, feminist research is a framework that should be implemented to provide a holistic approach to tourism by drawing attention to marginalised perspectives and looking for ways to make the tourism industry more inclusive and diverse.

References

Abrams, M. (2019). What does it mean to identify as nonbinary? *Healthline*. https://www.healthline.com/health/transgender/nonbinary

Abrams, J. A., Tabaac, A., Jung, S., & Else-Quest, N. M. (2020). Considerations for employing intersectionality in qualitative health research. *Social Science & Medicine, 258*, 113138. https://doi.org/10.1016/j.socscimed.2020.113138

Ahmed, S. (2010). Creating disturbance: Feminism, happiness and affective differences. In M. Liljeström & S. Paasonen (Eds.), *Working with affect in feminist readings: Disturbing differences* (pp. 31–44). Routledge.

Aitchison, C. (2001). Theorizing Other discourses of tourism, gender and culture: Can the subaltern speak (in tourism)? *Tourist Studies, 1*(2), 133–147.

Al-Faham, H., Davis, A. M., & Ernst, R. (2019). Intersectionality: From theory to practice. *Annual Review of Law and Social Science, 15*(1), 247–265. https://doi.org/10.1146/annurev-lawsocsci-101518-042942

Annas, P. J. (1978). New worlds, new words: Androgyny in feminist science fiction. *Science Fiction Studies, 5*(2), 143–156. http://www.jstor.org/stable/4239176

Anzaldúa, G. (1987). *Borderlands/la frontera: The new mestiza*. Spinsters.

Asselin, M. E. (2003). Insider research. Issues to consider when doing qualitative research in your own setting. *Journal for Nurses in Staff Development, 19*(2), 99–103. https://doi.org/10.1097/00124645-200303000-00008

Ateljevic, I., Harris, C., Wilson, E., & Collins, F. L. (2005). Getting 'entangled': Reflexivity and the 'critical turn' in tourism studies. *Tourism Recreation Research, 30*(2), 9–21. https://doi.org/10.1080/02508281.2005.11081469

Bakas, F. E. (2017). 'A beautiful mess': Reciprocity and positionality in gender and tourism research. *Journal of Hospitality and Tourism Management, 33*, 126–133. https://doi.org/10.1016/j.jhtm.2017.09.009

Bell, S. E. (2015). Bridging activism and the academy: Exposing environmental injustices through the feminist ethnographic method of photovoice. *Human Ecology Review, 21*(1), 27–58. https://doi.org/10.22459/HER.21.01.2015.02

Berger, R. (2015). Now I see it, now I don't: researcher's position and reflexivity in qualitative research. *Qualitative Research, 15*(2), 219–234. https://doi.org/10.1177/1468794112468475

Bernhard, L. A. (1984). *Feminist research in nursing research*. Poster session presented at the First International Congress on Women's Health Issues, Halifax, Nova Scotia.

Bhavnani, K.-K. (1993). Tracing the contours. *Women's Studies International Forum, 16*(2), 95–104. https://doi.org/10.1016/0277-5395(93)90001-p

Bloom, L. (1998). *Under the sign of hope: Feminist methodology and narrative interpretation*. State University of New York Press.

Butler, J. (1990). *Gender trouble: Feminism and the subversion of identity*. Routledge.

Butler, J. (1993). *Bodies that matter: On the discursive limits of "sex"*. Routledge.

Campbell, J. C., & Bunting, S. (1991). Voices and paradigms perspectives on critical and feminist theory in nursing. *Advances in Nursing Science, 13*(3), 1–15. https://doi.org/10.1097/00012272-199103000-00004

Carter, C. (2016). A way to meet queer women? Reflections on sexuality and desire in research. *Sexualities, 19*(1–2), 119–137. https://doi.org/10.1177/1363460715583594

Chadwick, R. (2021). On the politics of discomfort. *Feminist Theory, 22*(4), 556–574. https://doi.org/10.1177/1464700120987379

Chambers, D., & Rakić, T. (2018). Critical considerations on gender and tourism: An introduction. *Tourism, Culture and Communication, 18*(1), 1–8. https://doi.org/10.3727/109830418X15180180585112

Cohen, S. A. (2013). Reflections on reflexivity in leisure and tourism studies. *Leisure Studies*, *32*(3), 333–337. https://doi.org/10.1080/02614367.2012.662522

Collins, P. H. (1986). Learning from the outsider within: The sociological significance of Black feminist thought. *Social Problems*, *33*(6), 14–32. https://doi.org/10.2307/800672

Collins, P. H. (1990). *Black feminist thought: Knowledge, consciousness, and the politics of empowerment*. Routledge.

Collins, P. H., & Bilge, S. (2020). *Intersectionality*. Polity Books.

Community Marketing & Insights (CM&I). (2019). 24th annual LGBTQ tourism & hospitality survey. https://cmi.info/documents/temp/CMI_24th-LGBTQ-Travel-Study-Report2019.pdf

Crenshaw, K. (1991). Mapping the margins: Intersectionality, identity politics, and violence against women of color. *Stanford Law Review*, *43*(6), 1241. https://doi.org/10.2307/1229039

Dejarnette, K. (2018). Traveling while transgender in 2020. https://viewfinder.expedia.com/traveling-while/transgender/

Dinçer, P. (2019). Being an insider and/or outsider in feminist research: Reflexivity as a bridge between academia and activism. *Manas Sosyal Araştırmalar Dergisi*, *8*(4), 3728–3745. https://doi.org/10.33206/mjss.532325

Dupree, C. H., & Boykin, C. M. (2021). Racial inequality in academia: Systemic origins, modern challenges, and policy recommendations. *Policy Insights from the Behavioral and Brain Sciences*, *8*(1), 11–18. https://doi.org/10.1177/2372732220984183

Dwyer, S. C., & Buckle, J. L. (2009). The space between: On being an insider-outsider in qualitative research. *International Journal of Qualitative Methods*, *8*(1), 54–63. https://doi.org/10.1177/160940691878176

England, K. V. L. (1994). Getting personal: Reflexivity, positionality, and feminist research. *Professional Geographer*, *46*(1), 80–89. https://doi.org/10.1111/j.0033-0124.1994.00080.x

Enloe, C. (2000). *Maneuvers: The international politics of militarizing women's lives*. University of California Press.

Falconer, E. (2017). Moments of collusion? Close readings of affective, hidden moments within feminist research. *Women's Studies International Forum*, *61*, 75–80. https://doi.org/10.1016/j.wsif.2016.10.001

Figueroa-Domecq, C., Pritchard, A., Segovia-Pérez, M., Morgan, N., & Villacé-Molinero, T. (2015). Tourism gender research: A critical accounting. *Annals of Tourism Research*, *52*, 87–103. https://doi.org/10.1016/j.annals.2015.02.001

Figueroa-Domecq, C., & Segovia-Perez, M. (2020). Application of a gender perspective in tourism research: A theoretical and practical approach. *Journal of Tourism Analysis*, *27*(2), 251–270. https://doi.org/10.1108/JTA-02-2019-0009

Government of Canada. (n.d.). Travel and your sexual orientation, gender identity, gender expression and sex characteristics. https://travel.gc.ca/travelling/health-safety/lgbt-travel

Hale, C. (2008). *Engaging contradictions: Theory, politics, and methods of activist scholarship*. University of California Press.

Haraway, D. (1988). Situated knowledge: The science question in feminism and the privilege of partial perspective. *Feminist Studies*, *14*(3), 575–599. https://doi.org/10.2307/3178066

Harding, N. A. (2020). Co-constructing feminist research: Ensuring meaningful participation while researching the experiences of criminalised women. *Methodological Innovations*, *13*(2). https://doi.org/10.1177/2059799120925262

Harding, S. (1987). *Feminism and methodology*. Indiana University Press.

Harding, S. (1992). Rethinking standpoint epistemology: What is "strong objectivity?" *The Centennial Review*, *36*(3), 437–470. https://www.jstor.org/stable/23739232

Harding, S. (2004). *The feminist standpoint theory reader: Intellectual and political controversies*. Routledge.

Harris, J. C., & Patton, L. D. (2018). Un/Doing intersectionality through higher education research. *The Journal of Higher Education*, *90*(3), 347–372. https://doi.org/10.1080/00221546.2018.1536936

Hayfield, N., & Huxley, C. (2015). Insider and outsider perspectives: Reflections on researcher identities in research with lesbian and bisexual women. *Qualitative Research in Psychology*, *12*(2), 91–106. https://doi.org/10.1080/14780887.2014.918224

Hellawell, D. (2006). Inside-out: Analysis of the insider-outsider concept as a heuristic device to develop reflexivity in students doing qualitative research. *Teaching in Higher Education, 11*(4), 483–494. https://doi.org/10.1080/13562510600874292

Heller, M., Pujolar, J., & Duchêne, A. (2014). Linguistic commodification in tourism, *Sociolinguistics, 18,* 539–566. https://doi.org/10.1111/josl.12082

Hesse-Biber, S. N. (2007). Putting it together: Feminist research praxis. In S. N. Hesse-Biber & P. L. Leavy (Eds.), *Feminist research practice* (pp. 330–349). SAGE Publications. https://dx.doi.org/10.4135/9781412984270.n11

hooks, B. (1984). *Feminist theory: From margin to center.* South End Press.

Hughes, C., & Cohen, R. (2013, May 31). *Feminism counts: Quantitative methods and researching gender* (1st ed.). Routledge.

IGLTA. (2018). Travel tips for transgender, genderqueer and non-binary wanderlusters. https://www.iglta.org/Blog/Travel-Blog/ArtMID/9209/ArticleID/603/Travel-Tips-for-Transgender-Genderqueer-and-Non-Binary-Wanderlusters

Jaggar, A. M. (1997). Contemporary Western feminist perspectives on prostitution. *Asian Journal of Women's Studies, 3*(2), 8–29. https://doi.org/10.1080/12259276.1997.11665794

Johnson, A. (2018). An academic witness: White supremacy within and beyond academia. In A. Johnson, J-S. Remi, & B. Kamunge (Eds.), *The fire now: Anti-racist scholarship in times of explicit racial violence.* Zed Books.

Kanuha, V. K. (2000). "Being" native versus "Going Native": Conducting social work research as an insider. *Social Work, 45*(5), 439–447. https://doi.org/10.1093/sw/45.5.439

Kelly, L. (1988). How women define their experiences of violence. In K. Yllö & M. Bograd (Eds.), *Feminist perspectives on wife abuse* (pp. 114–132). SAGE Publications.

Klein, R. (1983). How to do what we want to do: Thoughts about feminist methodology. In G. Bowles & R. Klein (Eds.), *Theories of women's studies.* Routledge.

Kobayashi, A. (2003). GPC ten years on: Is self-reflexivity enough? *Gender, Place and Culture, 10*(4), 345–349. https://doi.org/10.1080/0966369032000153313

Kremer, B. (1990). Learning to say no. Keeping feminist research for ourselves. *Women's Studies International Forum, 13*(5), 463–467. https://doi.org/10.1016/0277-5395(90)90098-I

LaMarre, A., Rice, C., Friedman, M., & Fowlie, H. (2022). Carrying stories: Digital storytelling and the complexities of intimacy, relationality, and home spaces. *Qualitative Research in Psychology, 19*(4), 1143–1168. https://doi.org/10.1080/14780887.2022.2047246

Landman, M. (2006). Getting quality in qualitative research: A short introduction to feminist methodology and methods. *Proceedings of the Nutrition Society, 65*(04), 429–433. https://doi.org/10.1079/pns2006518

Langan, D., & Morton, M. (2009). Reflecting on community/academic "collaboration": The challenge of "doing" feminist participatory action research. *Action Research, 7*(2), 165–184. https://doi.org/10.1177/1476750309103261

Leacock, E. (1987). Theory and ethics in applied urban anthropology. In L. Mullings (Ed.), *Cities of the United States* (pp. 317–336). Columbia University Press.

Letherby, G. (2003). *Feminist research in theory and practice.* Open University Press.

Maynard, M. (1994). Methods, practise and epistemology: The debate about feminism and research. In M. Maynard & J. Purvis (Eds.), *Researching women's lives from a feminist perspective* (pp. 10–26). Taylor & Francis.

McHugh, M. C. (2020). Feminist qualitative research: Working toward transforming science and social justice. In P. Leavy (Ed.), *The Oxford handbook of qualitative research* (2nd ed.). https://doi.org/10.1093/oxfordhb/9780190847388.013.16

Merriam, S. B., Johnson-Bailey, J., Lee, M. Y., Kee, Y., Ntseane, G., & Muhamad, M. (2001). Power and positionality: Negotiating insider/outsider status within and across cultures. *International Journal of Lifelong Education, 20*(5), 405–416. https://doi.org/10.1080/02601370120490

Mies, M. (1983). Towards a methodology for feminist research. In G. Bowles & R. D. Klein (Eds.), *Theories of Women's studies.* Routledge.

Miller, T. (2017). Telling the difficult things: Creating spaces for disclosure, rapport and 'collusion' in qualitative interviews. *Women's Studies International Forum, 61,* 81–86. https://doi.org/10.1016/j.wsif.2016.07.005

Miner, K., Jayaratne, T., Pesonen, A., & Zurbrügg, L. (2012). Using survey research as a quantitative method for feminist social change. In S. N. Hesse-Biber (Ed.), *Handbook of feminist research: Theory and praxis* (pp. 237–263). SAGE Publications. https://dx.doi.org/10.4135/9781483384740.n12

Mohanty, C. T. (1988). Under western eyes: Feminist scholarship and colonial discourses. *Feminist Review, 30*, 61–88. https://doi.org/10.4324/9781003135593-13

Mostafanezhad, M. (2013). 'Getting in touch with your inner Angelina': Celebrity humanitarianism and the cultural politics of gendered generosity in volunteer tourism. *Third World Quarterly, 34*(3), 485–499. https://doi.org/10.1080/01436597.2013.785343

Mullings, B. (1999). Insider or outsider, both or neither: Some dilemmas of interviewing in a cross-cultural setting. *Geoforum, 30*(4), 337–350. https://doi.org/10.1016/s0016-7185(99)00025-1

Munar, A. M., Khoo-Lattimore, C., Chambers, D., & Biran, A. (2017). The academia we have and the one we want: On the centrality of gender equality. *Anatolia, 28*(4), 582–591. https://doi.org/10.1080/13032917.2017.1370786

Naderifar, M., Goli, H., & Ghaljaie, F. (2017). Snowball sampling: A purposeful method of sampling in qualitative research. *Strides in Development of Medical Education, 14*(3). https://doi.org/10.5812/sdme.67670

Naples, N. A. (1996). A feminist revisiting of the insider/outsider debate: The "outsider phenomenon" in rural Iowa. *Qualitative Sociology, 19*(1), 83–106. https://doi.org/10.1007/BF02393249

Naples, N., & Gurr, B. (2013). Feminist empiricism and standpoint theory: Approaches to understanding the social world. In S. N. Hesse-Biber (Ed.), *Feminist research practice. A primer* (pp. 14–41). SAGE Publications.

Pascoe Leahy, C. (2021). The afterlife of interviews: Explicit ethics and subtle ethics in sensitive or distressing qualitative research. *Qualitative Research, 2013*. https://doi.org/10.1177/14687941211012924

Payne, G., & Payne, J. (2004). *Key concepts in social research*. SAGE Publications.

Peake, L. (2017). Feminist methodologies. In D. Richardson, N. Castree, M. Goodchild, A. Kobayashi, W. Liu, & R. Marston (Eds.), *The AAG international encyclopedia of geography* (pp. 2331–2340). John Wiley and Sons.

Pollack, S., & Eldridge, T. (2015). Complicity and Redemption: Beyond the Insider/outsider research dichotomy. *Social Justice, 42*(2), 132–145.

Pritchard, A., & Morgan, N. (2017). Tourism's lost leaders: Analysing gender and performance. *Annals of Tourism Research, 63*, 34–47. https://doi.org/10.1016/j.annals.2016.12.011

Rayaprol, A. (2016). Feminist research: Redefining methodology in the social sciences. *Contributions to Indian Sociology, 50*(3), 368–388. https://doi.org/10.1177/0069966716657460

Reinharz, S., & Davidman, L. (1992). *Feminist methods in social research*. Oxford University Press.

Rice, C., Harrison, E., & Friedman, M. (2019). Doing justice to intersectionality in research. *Cultural Studies ↔ Critical Methodologies*, (6), 409–420. https://doi.org/10.1177/1532708619829779

Richards, C., Bouman, W. P., & Barker, M.-J. (2017). Genderqueer and Non-Binary Genders, in C. Richards, W. P. Bouman, & Barker, M.-J. (Eds.), *Genderqueer and Non-Binary Genders* (pp. 1–8). Palgrave Macmillan UK. https://doi.org/10.1057/978-1-137-51053-2

Rodó-de-Zárate, M. (2014). Developing geographies of intersectionality with relief maps: Reflections from youth research in Manresa, Catalonia. *Gender, Place and Culture, 21*(8), 925–944. https://doi.org/10.1080/0966369X.2013.817974

Rodó-de-Zárate, M. (2021). *Interseccionalitat. Desigualtats, llocs i emocions*. Tigre de Paper.

Salvador-Almela, M., & Abellan-Calvet, N. (2021). Volunteer tourism and gender: A feminist research agenda. *Tourism and Hospitality Research, 21*(4), 461–472. https://doi.org/10.1177/14673584211018497

Santos, C., & Caton, K. (2008). Reimagining Chinatown: An analysis of tourism discourse. *Tourism Management, 29*, 1002–1012. https://doi.org/10.1016/j.tourman.2008.01.002

Scott, J. (2012). Quantitative methods and gender inequalities. In C. Hughes & R. Cohen (Eds), *Feminism Counts* (pp. 4–14). Routledge.

Simoni, V., & McCabe, S. (2008). From ethnographers to tourists and back again: On positioning issues in the anthropology of tourism. *Civilisations, 57*(1), 173–189. https://doi.org/10.4000/civilisations.1276

Spivak, G. C. (1988). Can the subaltern speak? In C. Nelson & L. Grossberg (Eds.), *Marxism and the interpretation of culture* (pp. 271–313). Macmillan Education.

Spivak, G. C. (1994). Can the subaltern speak? In P. Williams & L. Chrismen (Eds.), *Colonial and post-colonial theory: A reader*. Columbia University Press.

Sprague, J., & Zimmerman, M. K. (1993). Overcoming dualism: A feminist agenda for sociological methodology. In P. England (Ed.), *Theory on gender: Feminism on theory* (pp. 255–280). Aldine De Gruyter.

Stratton, A. (2019). #TravelingWhileTrans: How to stay safe while seeing the world. https://matador-network.com/read/travelingwhiletrans-stay-safe-seeing-world/

Sultana, F. (2007). Reflexivity, positionality and participatory ethics: Negotiating fieldwork dilemmas in international research. *Acme, 6*(3), 374–385.

Tagonist, A. (2009). Fuck you and fuck your fucking thesis: Why I will not participate in trans studies. *Livejournal*. https://tagonist.livejournal.com/199563.html

Valentine, G. (2002). People like us: Negotiating sameness and difference in the research process. In P. Moss (Ed.), *Feminist geography in practice: Research and methods* (pp. 116–117). Blackwell Publishers.

Waitt, G., & Markwell, K. (2006). Gay tourism. *Culture and context*. The Haworth Hospitality Press.

Waitt, G., Markwell, K., & Gorman-Murray, A. (2008). Challenging heteronormativity in tourism studies: Locating progress. *Progress in Human Geography, 32*(6), 781–800. https://doi.org/10.1177/0309132508089827.

Webb, C. (1993). Feminist research: Definitions, methodology, methods and evaluation. *Journal of Advanced Nursing, 18*(3), 416–423. https://doi.org/10.1046/j.1365-2648.1993.18030416.x

Westmarland, N. (2001). The Quantitative/Qualitative debate and feminist research: A subjective view of objectivity. *FQS Forum: Qualitative Social Research, 2*(1), 1–10. https://doi.org/10.17169/fqs-2.1.974

Wigginton, B., & Lafrance, M. N. (2019). Learning critical feminist research: A brief introduction to feminist epistemologies and methodologies. *Feminism and Psychology*, 1–17. https://doi.org/10.1177/0959353519866058

Wilkinson, S. (1986). *Feminist social psychology: Developing theory and practice*. Open University Press.

Windsong, E. A. (2016). Incorporating intersectionality into research design: An example using qualitative interviews. *International Journal of Social Research Methodology, 21*(2), 135–147. https://doi.org/10.1080/13645579.2016.1268361

Wise, S. (1987). A framework for discussing ethical issues in feminist research a review of the literature. In V. Griffiths, M. Humm, R. O'Rourke, J. Barsleer, J. Poland, & S. Wise (Eds.), *Writing feminist biography II: Using life histories*. Studies in Sexual Politics, No. 19. Department of Sociology, University of Manchester.

Wolf, D. (1996). Situating feminist dilemmas in fieldwork. In D. L. Wolf (Ed.), *Feminist dilemmas in fieldwork* (pp. 1–55). Westview.

Young, M., & Markham, F. (2019). Tourism, capital, and the commodification of place. *Progress in Human Geography, 44*(2), 276–296. https://doi.org/10.1177/0309132519826679

5

METHODOLOGICAL REFLECTION ON DATA COLLECTION WITHIN LGBT+ SOCIAL INCLUSION RESEARCH

Vizak Gagrat

Abstract

Recently, events research on lesbian, gay, bisexual, transgender, and other queer (LGBT+) communities has increased as a segment of interest. However, despite this proliferation, the process of research into these communities explores experiences that are closely intertwined with negative experiences and trauma. Additionally, the pleasant experiences of tourism and events – have not received attention. In particular, autoethnographic research is less prominent in tourism and events studies, especially within the LGBT+ research space.

Despite having a background in LGBT+ event volunteering, while conducting research within the LGBT+ space, the author encountered various hurdles and challenges within their methodology. This research involved participants from diverse sexualities and gender preferences. Furthermore, the author's volunteering experience has now provided substantial insight into how to overcome these challenges while maintaining the integrity of the research. This chapter aims to provide insight into this research experience through autoethnography.

This chapter uses self-reflection to discuss the author's experience researching within the LGBT+ community – mainly focused on the bisexual and transgender communities. Thus, in keeping with the sentiment of the United Nations Sustainable Development Goals, "No one is left behind," this chapter aims to pave the way for further research into the LGBT+ space.

Keywords

LGBT+, Reflection, Research methods, Ethics, Bisexual, Transgender.

Introduction

The lesbian, gay, bisexual, transgender, and other queer (LGBT+) communities-based research has been increasing and entered mainstream discussions within the area of tourism and events research space (Ong et al., 2020; Ram et al., 2019). However, despite the abundance of research around these communities, the methodological processes that may

DOI: 10.4324/9781003286721-8

be closely intertwined with negative experiences and trauma have not received attention. Especially, self-reflection on practices and research methods and auto-ethnography are less prominent in the tourism and events studies, especially within the LGBT+ research.

I had the fortunate experience of conducting research in 2021 to better understand LGBT+ events volunteers' perceptions of the social inclusion of bisexual and transgender volunteers. While conducting this research within the LGBT+ space, the methodology had numerous challenges and hurdles. As a cisgender gay male, despite having a background in the LGBT+ volunteering space in Brisbane, Queensland, the following challenges were observed/experienced. This research involved participants from diverse sexualities (including gay, lesbian, bisexual, pansexual, and queer) and gender identities (including cisgender, transgender, non-binary, non-conforming, and gender-diverse). However, numerous challenges and hurdles were encountered during that time of research, which will be discussed thoroughly in this chapter. Overcoming these challenges and hurdles was a steep learning curve, especially since professional boundaries of the ethics of research practice had to be maintained, and it was important to maintain sensitivity towards the participants. Nevertheless, this experience provided me with substantial insight into ways to overcome these challenges while maintaining the integrity of the research.

This chapter uses self-reflection to discuss the experience of researching within the LGBT+ community focusing on the bisexual and transgender communities. Especially how important it is for researchers to be sensitive and inclusive of diversity while conducting research within the LGBT+ community (Lewis & Reynolds, 2021) while also keeping the focus on the respect given to the LGBT+ community and maintaining ethical research practices.

Thus, this chapter has two research objectives: first, understanding the challenges of conducting research within the bisexual and transgender community, and second, exploring strategies to overcome those challenges. And in keeping with the sentiment and philosophy of the United Nations Sustainable Development Goals, "No one is left behind" (United Nations, 2014, p. 9), this chapter aims to pave the way for further research into the LGBT+ space.

Literature Review

Social Inclusion

To understand social inclusion, it is first essential to understand marginalisation. The American Psychological Association (APA) (American Psychological Association [APA], 2020, Marginalization) defines marginalisation as "a reciprocal process through which an individual or group with relatively distinctive qualities, such as idiosyncratic values or customs, becomes identified as one that is not accepted fully into the larger group." There are many forms of marginalisation, such as racism, prejudice, discrimination, oppression, and even segregation or social exclusion (Causadias & Umaña-Taylor, 2018). While it is common for numerous groups or individuals to be marginalised because they are underrepresented or in the minority, specific groups may be more disadvantaged and discriminated against than others (Ong et al., 2020). For example, within the LGBT+ community, the transgender community is often more discriminated against socially and economically than the bisexual, gay, and lesbian community members (UNWTO, 2017). Furthermore, an example of social inclusion in practice was demonstrated at the London 2012 Olympics, wherein the key focus message was "Unity in diversity." Thus, promoting sexual diversity was essential to London's reputation as a gay-friendly destination and to further promote the marginalised non-normative queer sexualities (Hubbard & Wilkinson, 2015).

Social inclusion is studied heavily to counter marginalisation within various underrepresented and minority groups. Social inclusion has been significantly studied in various areas with each bringing its unique perspectives such as mental health, disabilities, and education. For example, in the space of mental health and well-being, social inclusion has been regarded as a "fundamental" human need to belong in a social world, which thus affects their need for belonging and self-esteem (Begen & Turner Cobb, 2015; Evans & Repper, 2000). Furthermore, in the space of disability, social inclusion is often researching issues such as lack of engagement, relationships and participation, thus causing social exclusion of people living with disabilities (Gannon & Nolan, 2007; Koller et al., 2018).

In addition, within the field of tourism, there are parts of societies, especially those who are living with disabilities or those belonging to marginalised groups are often excluded from general leisure and tourism activities (Kastenholz et al., 2015), whereas tourism and leisure activities can potentially enrich one's personal development and deepen one's social relations. Thus, Kastenholz et al. (2015) suggest more inclusive means of tourism such as accessible leisure tourism. Furthermore, events activities can provide opportunities in a public space for people to interact and engage with each other, and these encounters play a vital role in social inclusion (Duffy & Mair, 2018).

The Australian Government (2010, Social inclusion) defines social inclusion as "Social inclusion is recognised as the provision of opportunities that will allow all individuals to feel valued and to participate fully as members of society."

Based on the earlier definition of social inclusion, the research aims to provide opportunities for bisexual and transgender volunteers to participate and a platform to have a say in research that could make a difference in their volunteering life. Furthermore, this research provided a platform for bisexual and transgender volunteers to have a space wherein they could discuss their issues and challenges on social inclusion.

Just as studying social inclusion was necessary for that research, it was also imperative to demonstrate social inclusion within the method applied in conducting the research. Therefore, this chapter explores how social inclusion was practised throughout the research process, including recruitment, interviewing, collecting, and analysing data through autoethnography.

Method

Autoethnographic and Reflective Narrative

Autoethnography is not common and is an unorthodox method of research (Shepherd et al., 2020), especially within the field of tourism and events (Noy, 2008). Autoethnography allows authors to reflect on their experiences during research to further understand a particular concept or discipline they are studying (Holt, 2003) and explore their personal experiences in relation to the wider phenomenon (Mair & Frew, 2018; Zazkis & Koichu, 2015). Despite this chapter not being an autoethnographic study, I use my experience researching within the LGBT+ community to reflect back on the challenges incurred during the data collection phase of the project and present it in an autoethnographic and reflective narrative.

While studying bisexual and transgender volunteers' perceptions of social inclusion in LGBT+ events, it was also imperative to consider the social inclusion of bisexual and transgender volunteers within my research data set. How ironic would it be to have a study on social inclusion without being inclusive of the population being studied?

In this chapter, autoethnographic approach is used by reflecting on my experience while conducting research; and using the notes and transcripts from the data collection phase. These notes and transcripts help bring back those memories of the challenges I faced while conducting my research. Furthermore, as someone who constantly puts words on paper about my feelings, emotions, and experiences, it provided an excellent tool for me to go back to those notes I had written while conducting the interviews. For example, during one interview, I wrote in my notes, "misogyny and lateral violence destroying the vigour of transgender volunteers." This note immediately got my memory back to the discussion and story shared by a transgender female who had suffered from misogyny from gay males and lateral violence from the lesbian female population of the LGBT+ community. As a researcher, I could use this reflection to better understand why these challenges and hurdles occurred during the data collection. Furthermore, it helped me work on negotiating strategies to help overcome these hurdles and challenges.

The use of autoethnography and reflective narrations as a research method provides two advantages. Firstly, as discussed above, it is an uncommon research method within tourism and event studies (Noy, 2008), thus providing a novel way to address a research problem, in this case the challenges and hurdles faced during the data collection of this research. This chapter could also be beneficial for future tourism and events researchers, especially those studying sensitive concepts or phenomena, to be able to articulate their findings by bringing in their own lived experiences. Secondly, it helps bring back the emotive reflection process, which often gets lost within empirical studies. This is especially true of stories and emotions vital to the nature of the study like this one. The impact that lived experiences can bring using this emotive reflection process is often missing within empirical research.

Results

Positives and Challenges in Researching Bisexual and Transgender Participants

For many reasons, research within the LGBT+ community can be a challenging experience (Ivory, 2005). For example, sexual minorities often have anxiety and reluctance to identify themselves due to the social stigma associated with their sexuality (Leider, 2000). Similarly, this research had challenges, including difficulty recruiting bisexual and transgender participants, sensitivity around vulnerable community members, and fears of biphobia and transphobia.

However, before moving into the challenges, first, let's understand some positive experiences that emerged from this research. During this research, two significant positives were noted in the following:

The first positive benefit observed was that this research provided a platform for bisexual and transgender volunteers to have their say. While a growing number of studies within the tourism and events areas are conducted within the LGBT+ events (Hubbard & Wilkinson, 2015; Ivory, 2005; Ong et al., 2021), there is an apparent lack of research on the bisexual, transgender and gender-diverse population within the LGBT+ community (Ong et al., 2020). This means there is an evident lack of platform for these sexual and gender minorities to express their opinions. This research on social inclusion focused on the minorities of bisexual and transgender community members. Numerous participants said that research like this provides opportunities to marginalised groups like transgender and bisexual communities. As a researcher, this was a very rewarding process to find out that the participants appreciate such research projects, and thus passionately want to be a part of it.

I remember numerous bisexual and transgender participants appreciated the research where they felt included. It provided them with a platform to share their experiences as members of the LGBT+ community. For example, one participant, Frankie (name changed), said, *"Thank you for attempting to include the bisexual and transgender community in this research; it sounds really important."* Frankie was very emotional throughout the interview because she felt that the wider LGBT+ group had forgotten about her and always felt marginalised. As discussed above, this research stemmed from the sentiment of the United Nations Sustainable Development Goals, *"No one is left behind."* However, as a researcher it was eye-opening to see that the bisexual and transgender community were feeling the opposite of this sentiment. It was vital as a researcher for me to make this research on "social inclusion" as inclusive as possible to the LGBT+ community.

The second positive benefit was that this research could get access to raw and authentic first-hand lived experiences of bisexual and transgender volunteers. Most qualitative research, especially within the social sciences, end up with stories that help deepen our understanding of the concept or phenomenon being studied (Lewis, 2011). Similarly, within this research, I had the fortunate experience of learning numerous stories and lived experiences, both positive and negative, from transgender and bisexual participants. One such story was so emotive that I remember sitting in front of my camera on Zoom, conducting the interview and trying hard not to cry. In fact, despite it being a long narrative experience, I remember persuading my supervisors to keep this story in the final draft of my thesis, irrespective of the word count.

This story was narrated by Kai (name changed), who said, *"When I fly the flags up... I fly all of them... but I know for a fact that one of the young trans community members was going for a walk, was only 13 (years old), was walking past and stopped and cried because they could see the trans flag flying in the park...and it was just one of those moments that little person has gone on to be a really strong advocate for the young trans community...just by just seeing a visual, a visual that says, I am welcomed here."* This story was so important as well as moving for two reasons. Firstly, as a researcher it made me aware that young LGBT+ community members are often overlooked when conducting research within these spaces. However, it is these research topics that can help create platforms for further research, by opening up a dialogue for discussion. Secondly, as a researcher, we often tend to forget those populations that live in regional and rural areas. While there are a few LGBT+ focused research papers that have been conducted in the rural and regional parts of Australia (Lewis & Mehmet, 2021; Lewis et al., 2023), there is still an untapped population of LGBT+ community members who have not been involved in these research topics.

While these positive experiences were rewarding and uplifting, the research also came with challenges and obstacles. The three significant challenges were as follows:

The first challenge noted was the recruitment of bisexual and transgender participants. While researching the LGBT+ community is complex and challenging (Ivory, 2005), conducting research within the bisexual and transgender community proved to be even more challenging. Upon planning for data collection, I was slightly optimistic about my bisexual and transgender population recruitment; however, the reality set in as soon as the recruitment process actively commenced.

I was actively contacting referrals from the transgender and bisexual community through multiple modes of communication. For example, I was provided referrals to transgender and gender-diverse volunteers connected to the same organisation wherein I volunteer. But, despite sending them multiple emails and Facebook messages, most of them did not respond. This was when I started panicking and realised this would be a complicated process.

While most referrals did not respond, some that did were very reluctant to participate in the research. Some of them showcased their reluctance by not providing any availability or cancelling the interviews a day or two before the actual date.

As a researcher, this process was by far the most challenging part of the research. Not because I did not have enough participants to reach saturation, but because these numbers were not representative of the bisexual, transgender, or gender-diverse population I had hoped for. Furthermore, at one point, due to these constant rejections, I was at risk of doing social inclusion research for the bisexual and transgender community without anyone participating from the bisexual and transgender community.

The second challenge was the fear of biphobia and transphobia. This particular challenge is closely linked as a significant reason for participants' unwillingness to participate in this research. Biphobia and transphobia contribute to enormous trauma within the bisexual and transgender communities (Marzetti et al., 2022). Biphobia is often associated with negative attitudes towards stereotypes and discrimination towards bisexual individuals (Eliason, 1997). Similarly, transphobia is traditionally expressed as fear, emotional disgust, or hatred towards individuals who do not conform to society's gender expectations (Hill & Willoughby, 2005; Morrison et al., 2017). Furthermore, the terms biphobia and transphobia are often critiqued similarly to homophobia due to the cultural and structural oppression of heterosexuals towards the LGBT+ community (Eliason, 2001). These forms of discrimination often lead to social marginalisation, which creates separation within the LGBT+ community.

Throughout the interviews, numerous transgender, gender-diverse, and bisexual participants mentioned that they had experienced biphobia or transphobia in some way or the other. However, it was shocking that most of their experiences were from within the wider LGBT+ community, thus resulting in internalised phobias. For example, Xander (name changed) stated, *"there is a tendency for people to be dismissive"*; therefore, noting that this internalised transphobia ranges from subtle discrimination to outright aggression. This explained why bisexual and transgender participants are so apprehensive about participating in research due to these fears of internalised phobias from the LGBT+ community.

Numerous participants also discussed how these negative experiences could be traumatising and take away their sense of confidence and belief. For example, Kai (name changed) stated, *"they [bisexual and transgender] don't feel worthy and they're not part of it, then they don't include themselves, because they feel like they're not wanted."* Participants added that these bisexual and transphobic assumptions and comments could also adversely impact the community members and their mental well-being. As a researcher, it was vital for me to understand this particular challenge, especially as a cisgender gay male to not come across as biphobic or transphobic, instead to try and promote inclusivity.

The third challenge was the sensitivity around vulnerable participants. As evidently seen above and based on numerous types of research that have shown that bisexual and transgender community members live with extensive trauma from negative stereotypes, discrimination, and marginalisation, thus, making them very vulnerable (Lewis & Reynolds, 2021). Therefore, as a researcher, it was vital for me to ensure that I respected the sensitivity of these participants.

Ethical research often protects the researcher and the participants by minimising harm while increasing the sum of good (Israel & Hay, 2006; Lewis & Reynolds, 2021). Furthermore, the researcher must consider the participant's potential risks may experience due to the questions asked and be sensitive towards their welfare (Lewis & Reynolds, 2021).

This sensitivity was especially challenging within this research because the research questions were focused on the social inclusion of themselves and fellow bisexual and transgender volunteers within the LGBT+ event volunteering space. This meant they all had to re-live those negative experiences and traumatic moments they lived through as transgender and bisexual individuals. I was always sensitive and aware that re-living adverse or traumatic events could have severe implications on their mental well-being. The recommendations highlight numerous ways in which I had to ensure that the participants were taken care of, including providing helpline numbers, breaks during interviews, and other such measures for their well-being.

Furthermore, the vulnerabilities experienced by LGBT+ people differ based on the specific LGBT+ community that they identify with. And while the wider LGBT+ community experiences similar stigma, i.e., diverging from heteronormativity, each of these LGBT+ community experiences different trauma and experiences (Lewis & Reynolds, 2021). This made it difficult within this research because, as a cisgender gay male, my experiences are very different to bisexual, transgender, and gender-diverse people. However, I had to maintain sensitivity towards their experiences without being tokenistic. Therefore, it was necessary as a researcher to respect their vulnerability and be mindful of their sensitivity at all times, from the recruitment process to the end of the data collection.

Discussion

Negotiation Strategy

As a new researcher, while the challenges mentioned previously were very all-consuming, I had to be resourceful and creative in finding ways to overcome those challenges, namely lack of bisexual and transgender participants, fear of biphobia and transphobia, and sensitivity around vulnerable participants. It is crucial for a researcher to be inclusive of the participants when researching because it adds significant value to the outcome, offers novel perspectives, and access to hard-to-reach groups, and empowers people by participating (Grant & Ramcharan, 2009; Walmsley et al., 2018). Also, Slaley (2009) argues that inclusivity in research helps strengthen the evidence base for assessing the impact and finding more nuanced ways to capture it. Thus, the following three strategies were used to maintain the ethical boundaries of this research; and to be socially inclusive and respectful of the participants within this research.

The first recommendation is the inclusivity of marginalised groups. As discussed above in the challenges, it was a difficult process to recruit bisexual, transgender, and gender-diverse people for the research. It would have been ironic if research on social inclusion did not strive to be inclusive in its recruitment. One of the ways to overcome this hurdle was the use of snowball sampling. Snowball sampling is a research technique whereby initial participants are identified based on the closeness to the population criteria needed for this research (Ritchie et al., 2005), which was that they need to be members of the LGBT+ community and volunteer at LGBT+ festivals and events. Furthermore, snowball sampling enables the researcher to better access individuals and groups for the study that may otherwise remain inaccessible, such as marginalised and underrepresented communities (Woodley & Lockard, 2016). Thus, within this research, snowball sampling provided me with a strong initial contact, and because of that relationship, I could access other community members for the research (Woodley & Lockard, 2016). For example, in this research, my initial contact, Kris (name changed), a transgender male, provided me with leads and references to other members of the transgender and gender-diverse communities. While almost half of the leads did not respond, the other half were successfully recruited because of the initial reference provided by Kris. Also, the fact that Kris was a participant in this research provided the other transgender participants with a sense of relatability and that they were not the only ones from the community. Overall, while I thought snowball sampling was a viable method for this type of research, I was amazed to see the power of how quickly the word spread within the community, when the recruitment was done.

Another way to overcome this recruitment hurdle was by visiting the Trans Fair Day in Brisbane. Kris was aware of my struggles with the recruitment of transgender participants for this study and invited me to visit the Trans Fair Day as a volunteer. Being there as a volunteer helped me form connections with other transgender and gender-diverse participants to be able to recruit them. This opportunity was vital for me to contribute to their community and use these events as a platform to establish trust and relationships with the community members (Hom et al., 2011). Establishing this trust was a positive step for dual reasons- first, to showcase my genuine support towards the transgender community, and second, to establish contact with potential participants for the research. Thus, a strong recommendation for showcasing inclusivity and forming positive relationships is attending events and contributing to the community.

Overall this was an essential part of demonstrating inclusivity within my research. Practising inclusivity within my research was vital because the research was about inclusivity.

The second recommendation is to form a relationship of trust and respect. As discussed earlier in the first recommendation, establishing trust is essential as a starting point while recruiting participants. However, this trust also has to be sustained throughout the research project. As researchers, building and sustaining trust can be done by developing positive relationships and ensuring accountability, shared interests, and a concern for the best interests of others (Hom et al., 2011). For me forming that trust and relationship with the participants was a positive experience. This helped me significantly in collecting data and established a sense of comfort wherein they were willing to share deeply rooted personal stories and experiences.

Furthermore, as a cisgender gay male, I had to ensure the transgender and bisexual participants that I respected their gender and sexual identities. Respect is an essential part of the research process, especially if your participants are vulnerable or have been marginalised (Lewis & Reynolds, 2021). Especially since many/several participants opened up about how they suffer from internalised biphobia and transphobia, as stated earlier in the chapter. For example, Xander (name changed), a transgender female, said they suffer from *"lateral antagonisms,"* especially from the cis gay male population of the LGBT+ community. Thus, I had to demonstrate the utmost respect for the transgender and bisexual community for them to form that trust with me as a researcher. An example of demonstrating respect is by ensuring that I refer to the participants by the appropriate pronouns. Another example was to constantly check in if they needed breaks and if they were happy to continue with the interviews. Furthermore, I also ensured that I respectfully and actively listened to the participants during the interviews and ensured that at no point did I come across as dismissive or try to cut the interview short.

Moreover, I constantly provided positive affirmations to the participants. Providing positive affirmations can be a fantastic way to make someone feel recognised and rewarded, which is a vital part of social inclusion (Cobigo et al., 2012). Conversely, in the absence of recognition and positive affirmation, people may see themselves as devalued and thus value others less (McConkey, 2007). Therefore, I needed to constantly provide positive affirmations to volunteers by using quotes such as, *"Thank you for your time"* *"Your voice is valuable to this research"*, and *"Really appreciate your time and the valuable insight."* These constant affirmations and recognition were a great way to maintain trust with the participants, make them feel rewarded, and try to make them feel included.

The third recommendation is to be mindful of the vulnerability and be sensitive towards the participants. As discussed above, bisexual and transgender people often are vulnerable due to the extensive trauma caused by negative stereotypes, discrimination and marginalisation (Lewis & Reynolds, 2021). Therefore, as a researcher, I needed to be mindful of this vulnerability. Furthermore, I also had to ensure that the data collection was conducted while maintaining the boundaries of ethical research practices.

The mindfulness of these issues was reflected in the ethics application drafted for this project. There were some key points emphasised to the interviewees, including the awareness that *"there is a risk that these reflections may invoke negative feelings and memories... If these negative feelings or memories occur during the interview, we encourage you to inform the interviewer if you wish to take a break from or stop the interview. If these persist beyond the interview, we encourage you to contact a counselling helpline."* And the numbers from various LGBT+ helplines and counselling services were provided to the participants. This strategy provided me with a better assurance that the participants are aware of the potential risks within this research and are equipped with solutions if the risk arises. Furthermore, participants were also informed that the research is voluntary, and they can end the interview at any given time if they feel any sense of discomfort during this process. These steps were added to the typical ethics application template as I recognised the particular vulnerabilities of the population I was interviewing.

In addition to ethics documentation, I had to ensure that during the interview, I constantly checked in with the participants to ensure they were okay. Thus, it was vital for me to continually check in on the verbal and non-verbal cues from the participants during the interviews. Numerous studies state that in qualitative data collection methods, it is vital for the researcher to be mindful of the verbal and non-verbal cues as a part of their data collection (Castelli et al., 2012; Jones & LeBaron, 2002). This is especially important because it helps us observe the attitude of the participants to help deepen our understanding of the phenomena being studied (Jones & LeBaron, 2002). But in this study, it was also vital to gauge their non-verbal cues and constantly check in on the participants to ensure they were comfortable during the interview process. I often used phrases like *"are you happy for me to move ahead?"* to ensure there was constant checking in on the participants verbally. And in addition to that, I was fortunate to have all the participants keep their cameras on, which helped me check on their non-verbal cues. For example, if I observed a participant getting too emotional or visibly angry while discussing their negative experiences, I would first provide empathy by using phrases like *"that must have been very challenging to experience"* or *"I am very sorry that you had to experience something like this,"* to calm them down, and then ask if they were happy to continue.

Overall, these were some strategies used to overcome the challenges and hurdles experienced within the research on social inclusion. As discussed earlier, it was vital to ensure that all participants felt included and to maintain the ethical integrity of this research.

Reflection and Lessons Learned

As a naïve gay male, I have always assumed that most members of the LGBT+ community share the same set of experiences, positive and negative. Furthermore, as an active volunteer within LGBT+ events in Brisbane, Australia, on the surface, it always felt like each community within the wider LGBT+ framework had a similar range of trauma, attitudes, and feeling of being marginalised. I couldn't have been more wrong!

This research was possibly a first little under-the-surface dive into a very deep ocean of understanding social inclusion within marginalised communities. And the deeper I looked and tried to understand the lived experiences of the bisexual, gender-diverse, and transgender community members, the more confronting it got for me. While some stories of inclusivity were positive, most were very difficult to process, especially from a Western society. The stories, only a few of which I have explored within the research, had so much depth, trauma, and discrimination lying within them – which made me often wonder how much privilege I possess as a gay male.

I remember going to the Trans Fair Day and talking to some transgender activists with a positive voice to create change within the community. Just during our talks, numerous sensitive topics arose, such as the high number of transgender people in Brisbane who had become homeless during the COVID-19 pandemic due to a lack of jobs and the downfall of the economy. In particular, one community member discussed how transgender people sometimes have to start living in their cars because of how expensive hormones are in Australia. It was a very eye-opening experience for me as a researcher.

Based on the findings of this chapter and upon reflection as a researcher working with marginalised groups I do present some takeaway pointers for the readers. Firstly, it is important to not underestimate the power of snowball sampling, because it helps form a network within a community that is not easily accessible. The snowball sampling and recruitment of participants for research can also be done by immersing oneself in the events, meetings, and forums that occur within these communities. Furthermore, it is also vital to form a trusting relationship because it does help make the participants feel more comfortable.

Also, as stated above in the discussion, another important takeaway is to be mindful of the vulnerability of the participants and be sensitive to their issues. In addition, provide positive affirmations and be genuine about them rather than tokenistic. And the final takeaway is to demonstrate respect at all given times.

To sum up, numerous researchers have worked in this area of LGBT+ events and social inclusion, yet there is a lot more that could be done within this area from a tourism and events perspective. So many such stories are still untold and need to be said to help us understand how to make the LGBT+ community more inclusive.

Conclusion

This chapter, through autoethnography, used self-reflection to discuss the experience I had researching social inclusion within the bisexual, transgender, and gender-diverse communities. There were positives in this experience, such as providing these communities with a platform to hear their voice and to get first-hand lived experiences from members of these communities. However, there were also challenges which created unforeseen obstacles within this research. These challenges included the recruitment of transgender and bisexual participants, fear of biphobia and transphobia, and sensitivity of the vulnerable participants.

Thus, this chapter helped with two objectives: firstly, it helped deepen the understanding of the challenges of conducting research within the bisexual and transgender communities, and second, exploring strategies to overcome these challenges.

Furthermore, this chapter has highlighted through reflective practices what I learned as a cis gay male about the experiences of transgender and bisexual volunteers within LGBT+ events. Overall this chapter highlights how important it is to be inclusive and respectful of all the marginalised communities within the LGBT+ framework.

Overall my research philosophy comes from the United Nations Sustainable Development Goals sentiment of "No one is left behind" (United Nations, 2014, p. 9). Keeping this sentiment in the forefront, I hope this chapter paves the way for future researchers to keep looking for those untold stories within the unexplored or marginalised communities within the LGBT+ events and tourism space.

References

American Psychological Association (APA). (2020, April 6). American Psychological Association. Retrieved from APA Dictionary of Psychology: https://dictionary.apa.org/marginalization

Australian Government. (2010). Canberra: Australian social inclusion board. Retrieved April 9, 2014.

Begen, F. M., & Turner-Cobb, J. M. (2015). Benefits of belonging: Experimental manipulation of social inclusion to enhance psychological and physiological health parameters. *Psychology & Health*, *30*(5), 568–582. https://doi.org/10.1080/08870446.2014.991734

Castelli, L., Carraro, L., Pavan, G., Murelli, E., & Carraro, A. (2012). The power of the unsaid: The influence of nonverbal cues on implicit attitudes. *Journal of Applied Social Psychology*, *42*(6), 1376–1393. https://doi.org/10.1111/j.1559-1816.2012.00903.x

Causadias, J. M., & Umaña-Taylor, A. J. (2018). Reframing marginalization and youth development: Introduction to the special issue. *The American Psychologist*, *73*(6), 707–712. https://doi.org/10.1037/amp0000336

Cobigo, V., Ouellette-Kuntz, H., Lysaght, R., & Martin, L. (2012). Shifting our conceptualization of social inclusion. *Stigma Research and Action*, *2*(2), 75–84.

Duffy, M., & Mair, J. (2018). Social inclusion, social exclusion and encounter. In *Festival encounters* (1st ed., pp. 83–93). Routledge. https://doi.org/10.4324/9781315644097-8

Eliason, M. (2001). Bi-negativity. *Journal of Bisexuality*, *1*(2–3), 137–154. https://doi.org/10.1300/J159v01n02_05

Eliason, M. J. (1997). The prevalence and nature of biphobia in heterosexual undergraduate students. *Archives of Sexual Behavior*, *26*(3), 317–326. https://doi.org/10.1023/A:1024527032040

Evans, J., & Repper, J. (2000). Employment, social inclusion and mental health. *Journal of Psychiatric and Mental Health Nursing*, *7*(1), 15–24. https://doi.org/10.1046/j.1365-2850.2000.00260.x

Gannon, B., & Nolan, B. (2007). The impact of disability transitions on social inclusion. *Social Science & Medicine (1982)*, *64*(7), 1425–1437. https://doi.org/10.1016/j.socscimed.2006.11.021

Grant, G., & Ramcharan, P. (2009). Valuing people and research: Outcomes of the learning disability research initiative. *Tizard Learning Disability Review*, *14*(2), 25–34. https://doi.org/10.1108/13595474200900016

Hill, D. B., & Willoughby, B. L. B. (2005). The development and validation of the genderism and transphobia scale. *Sex Roles*, *53*(7–8), 531–544. https://doi.org/10.1007/s11199-005-7140-x

Holt, N. (2003). Representation, legitimation, and autoethnography: An autoethnographic writing story. *International Journal of Qualitative Methods*, *2*(1), 18–28. https://doi.org/10.1177/160940690300200102

Hom, E. J., Edwards, K., & Terry, S. F. (2011). Engaging research participants and building trust. *Genetic Testing and Molecular Biomarkers*, *15*(12), 839–840.

Hubbard, P., & Wilkinson, E. (2015). Welcoming the world? Hospitality, homonationalism, and the London 2012 Olympics. *Antipode*, *47*(3), 598–615. https://doi.org/10.1111/anti.12082

Israel, M., & Hay, I. (2006). *Research ethics for social scientists*. SAGE.

Ivory, B. T. (2005). LGBT students in community college: Characteristics, challenges, and recommendations. *New Directions for Student Services*, *2005*(111), 61–69. https://doi.org/10.1002/ss.174

Jones, S. E., & LeBaron, C. D. (2002). Research on the relationship between verbal and nonverbal communication: Emerging integrations. *Journal of Communication*, *52*(3), 499–521. https://doi.org/10.1111/j.1460-2466.2002.tb02559.x

Kastenholz, E., Eusébio, C., & Figueiredo, E. (2015). Contributions of tourism to social inclusion of persons with disability. *Disability & Society*, *30*(8), 1259–1281. https://doi.org/10.1080/09687599.2015.1075868

Koller, D., Pouesard, M. L., & Rummens, J. A. (2018). Defining social inclusion for children with disabilities: A critical literature review. *Children & Society*, *32*(1), 1–13. https://doi.org/10.1111/chso.12223

Leider, S. (2000). *Sexual minorities on community college campuses*. ERIC Digest.

Lewis, P. J. (2011). Storytelling as research/research as storytelling. *Qualitative Inquiry, 17*(6), 505–510. https://doi.org/10.1177/1077800411409883

Lewis, C., & Mehmet, M. (2021). When a pandemic cancels pride: An exploration of how stakeholders respond to the cancellation of a rural Australian pride event. *Event Management, 26*(5), 949–966. https://doi.org/10.3727/152599521X16367300695807

Lewis, C., Mehmet, M., & McLaren, S. (2023). 'A lot of gay energy in the city': An identity-based exploration of leisure travel to domestic cities for rural queer people in Australia. *Journal of Hospitality and Tourism Management, 54*, 22–31. https://doi.org/10.1016/j.jhtm.2022.12.001

Lewis, C., & Reynolds, N. (2021). Considerations for conducting sensitive research with the LGBTQIA+ communities. *International Journal of Market Research, 63*(5), 544–551. https://doi.org/10.1177/14707853211030488

Mair, J., & Frew, E. (2018). Academic conferences: A female duo-ethnography. *Current Issues in Tourism, 21*(18), 2152–2172. https://doi.org/10.1080/13683500.2016.1248909

Marzetti, H., McDaid, L., & O'Connor, R. (2022). "Am I really alive?": Understanding the role of homophobia, biphobia and transphobia in young LGBT+ people's suicidal distress, *Social Science & Medicine (1982), 298*, 114860–114860. https://doi.org/10.1016/j.socscimed.2022.114860

McConkey, R. (2007). Variations in the social inclusion of people with intellectual disabilities in supported living schemes and residential settings. *Journal of Intellectual Disability Research, 51*(3), 207–217. https://doi.org/10.1111/j.1365-2788.2006.00858.x

Morrison, M. A., Bishop, C., Gazzola, S. B., McCutcheon, J. M., Parker, K., & Morrison, T. G. (2017). Systematic review of the psychometric properties of transphobia scales. *The International Journal of Transgenderism, 18*(4), 395–410. https://doi.org/10.1080/15532739.2017.1332535

Noy, C. (2008). The poetics of tourist experience: An autoethnography of a family trip to Eilat 1. *Journal of Tourism and Cultural Change, 5*(3), 141–157. https://doi.org/10.2167/jtcc085.0

Ong, F., Lewis, C., & Vorobjovas-Pinta, O. (2021). Questioning the inclusivity of events: The queer perspective. *Journal of Sustainable Tourism, 29*(11–12), 2044–2061. https://doi.org/10.1080/09669582.2020.1860072

Ong, F., Vorobjovas-Pinta, O., & Lewis, C. (2020). LGBTIQ + identities in tourism and leisure research: A systematic qualitative literature review. *Journal of Sustainable Tourism*, 1–24. https://doi.org/10.1080/09669582.2020.1828430

Ram, Y., Kama, A., Mizrachi, I., & Hall, C. M. (2019). The benefits of an LGBT-inclusive tourist destination. *Journal of Destination Marketing & Management, 14*, 100374.

Ritchie, B. W., Burns, P., & Palmer, C. (2005). Tourism research methods: Integrating theory with practice. In *Tourism research methods: Integrating theory with practice*. CABI Publishing. https://doi.org/10.1079/9780851999968.0000

Shepherd, J., Laven, D., & Shamma, L. (2020). Autoethnographic journeys through contested spaces. *Annals of Tourism Research, 84*, 103004. https://doi.org/10.1016/j.annals.2020.103004

Slaley, K. (2009). *Exploring impact: Public involvement in NHS, social care and public health research*. INVOLVE.

United Nations. (2014). Report of the open working group of the general assembly on sustainable development goals: A/68/970.

UNWTO. (2017). Second global report on LGBT tourism. https://www.unwto.org/archive/global/publication/affiliate-members-global-reports-volume-fifteen-second-global-report-lgbt-tourism

Walmsley, J., Strnadová, I., & Johnson, K. (2018). The added value of inclusive research. *Journal of Applied Research in Intellectual Disabilities, 31*(5), 751–759. https://doi.org/10.1111/jar.12431

Woodley, X., & Lockard, M. (2016). Womanism and snowball sampling: Engaging marginalized populations in holistic research. *Qualitative Report, 21*(2), 321–329. https://doi.org/10.46743/2160-3715/2016.2198

Zazkis, R., & Koichu, B. (2015). A fictional dialogue on infinitude of primes: Introducing virtual duoethnography. *Educational Studies in Mathematics, 88*(2), 163–181. https://doi.org/10.1007/s10649-014-9580-0

PART III

Practising Gender in Tourism I

Tourism Development

6

GENDER AND SUSTAINABLE DEVELOPMENT

A Post-1994 South African Responsible Tourism Policy Perspective

Shireen van Zyl and Hugh Bartis

Abstract

The United Nations Tourism World Organisation supports tourism that contributes to gender equality and the empowerment of women. The chapter aims to discuss how gender is perceived in South Africa within the context of its Responsible Tourism policy. Responsible Tourism is seen to be a means to achieve the goals of Sustainable Development by the post-1994 South African Government. Sustainable Development Goal 5, of Agenda 2030 for Sustainable Development, addresses issues pertaining to gender equality and the empowerment of women and girls. The equality and empowerment of women in post-1994 South Africa is a prerogative for the South African Government, based on the marginalisation of women in South Africa prior to 1994. For purposes of this chapter, content analysis was undertaken, utilising the Responsible Tourism policy in South Africa. Findings point to the potential socio-economic benefits for women, through Responsible Tourism practice. Gender equality and empowerment in the South African Responsible Tourism workplace require further development in terms of interpretation and implementation of policy. Furthermore, terminology needs to be aligned, bearing cognisance of South Africa's historical context, with reference to the term, women, in policy, in relation to terminology such as gender, age, sex, race, and the disabled.

Keywords

Sustainable development, Responsible Tourism, Equality, Empowerment, Women

Introduction

The issue of gender in terms of Responsible Tourism falls within the ambit of Sustainable Development and has significant implications for women. In South Africa, the National Tourism Sector Strategy (NTSS) recognises Responsible Tourism as a guiding principle and value (National Department of Tourism, 2016, p. 16). Responsible Tourism is geared towards

DOI: 10.4324/9781003286721-10

achieving the goals of Sustainable Development (National Department of Tourism, 2011, p. 1). Millennium Development Goal 3 addressed the issue of gender equality and the empowerment of women. The United Nations World Tourism Organisation recognises that the 2030 Agenda for Sustainable Development, which was embraced in 2016, replaced The Millennium Development Goals and aims to build upon what was not achieved by The Millennium Development Goals. Sustainable Development Goal (SDG) 5 specifically addresses the issue of gender. In 2017, the United Nations General Assembly's theme – International Year of Sustainable Tourism for Development – acknowledged the potential of tourism to contribute to the achievement of the 2030 Agenda for Sustainable Development and the SDGs.

In a post-1994 democratic South Africa, Responsible Tourism was first addressed in The White Paper on the Promotion and Development of Tourism in South Africa (1996). In this policy document, Responsible Tourism was proposed as the underlying philosophy for tourism development by the African National Congress (ANC) Government in the new democratic South Africa. The Responsible Tourism Manual (2002) and the Responsible Tourism Handbook (2003) presented guidelines for tourism businesses to conduct business in a more responsible and sustainable manner. These three documents were published by the erstwhile Department of Environment Affairs and Tourism (DEAT). The last policy addressing Responsible Tourism is the South African National Minimum Standard (2011) that presents the National Minimum Standards for Responsible Tourism and attempts to align the different criteria being used, by providing a baseline standard for the implementation of Responsible Tourism by businesses to attest the sustainability of tourism businesses (National Department of Tourism, 2011, p. 1). This document was published by the National Department of Tourism (NDT). The NDT came about after DEAT was split into two national government departments in 2009. The split of the two national government departments can be attributed to two reasons, namely, tourism being one of the main contributors to the National Gross Domestic Product (GDP) of South Africa and tourism accounting for more than 1 million employment opportunities at the time. The creation of a separate government department led to a more focused approach to tourism, as opposed to being part of another department and possibly treated as an addition (Department of Environmental Affairs and Tourism, 2009).

These documents are thus very significant in terms of understanding the South African Government's approach regarding the way Responsible Tourism was, and is, to be interpreted and conducted by tourism businesses in South Africa. There are various definitions for the term Responsible Tourism in the different policy documents mentioned in the previous paragraph; as the concept of Responsible Tourism developed, the definitions evolved. In the last document, the South African National Minimum Standard (2011), Responsible Tourism is defined as a "tourism management strategy in which the tourism sector and tourists take responsibility to protect and conserve the natural environment, respect and conserve local cultures and ways of life, and contribute to stronger local economies and a better quality of life for local people" (National Department of Tourism, 2011, p. 5). In this regard, it allows for tourism to be consumed in a more responsible manner and allows the tourists, the tourism product owners and host communities to benefit through the adoption of Responsible Tourism management strategies. The function of Government, in terms of tourism, is to create an enabling environment, so that all stakeholders are able to benefit optimally. Thus, it is purported that Responsible Tourism is to be driven by the private sector, with the Government creating the environment for the private sector to succeed, as well as to empower communities, including previously marginalised communities and women (Department of Environmental Affairs and Tourism, 1996).

This chapter aims to illustrate how gender is perceived within Responsible Tourism policy, bearing in mind the contested history of South Africa, and the construct of gender. Documents are written for various purposes and timeframes differ. Consequently, it is understandable that there could be significant anomalies in South Africa's Responsible Tourism policy, when interpreted with a view to achieve sustainable development and applying the 2030 Agenda for Sustainable Development document. This chapter, whilst addressing the issue of gender, takes cognisance of South Africa's historical past and its diverse population in terms of race, ethnicity, etc. In South Africa's endeavour to achieve transformation in the tourism industry, it is understandable that South Africa's approach to gender differs from other countries.

This chapter introduces the issue of gender within the Responsible Tourism policy in terms of SDG 5 of the 2030 Agenda for Sustainable Development, with a focus on gender, specifically women. The methodology used in this research study involves the analysis of documents pertaining to Responsible Tourism policy in relation to SDG 5 of the 2030 Agenda for Sustainable Development and is further explained in the following section of this chapter. Thereafter the references to gender in Responsible Tourism policy and SDG 5 of the 2030 Agenda for Sustainable Development are outlined, followed by a discussion of the inferences drawn about gender. Recommendations for government and industry, as well as further research opportunities on the topic of gender in Responsible Tourism, are presented. Key messages are listed, followed by the conclusion.

Methodology

The content analysis of documents (also known as document analysis) is recognised as a form of qualitative data analysis. The sample of documents for this research is the Responsible Tourism policy for South Africa. According to Altinay et al. (2016) content analysis as a data collection technique involves the analysis of the meanings and relationships of words, phrases, or concepts in the text, to draw conclusions and make inferences about the messages within the texts. Relational analysis, also known as semantic analysis, is a category of content analysis that involves identifying the words, phrases, or concepts in the text, analysing the text for the relationships between them and thereafter evaluating these relationships according to their strength (considering the occurrences and co-occurrences of words, phrases, or concepts in a sentence and/or paragraph), sign (considering whether words, phrases, or concepts are positively or negatively related), and direction (indicating how words, phrases or concepts are positioned). When making inferences about the messages in texts, cognisance must and was taken of the writer/s of the text, the intended audience as well as the time when the documents were written; and not merely that there was a presence of word/s, phrases, or concepts. Some words, phrases, and concepts are explicit (e.g. when the word women is used, or a synonym like the word female), whilst others are implicit (where, by implication, women form part of a broader context and category).

Overview of Gender within the Context of Sustainable Development and Responsible Tourism

SDG 5 consists of 9 Targets and 14 Indicators, none of which mention the word tourism. Nevertheless, in terms of tourism, Sustainable Development has been interpreted significantly since the definition in 1987 (with reference to the Brundtland Report). Sustainable Development is defined as development that "meets the needs of the present generation without compromising the ability of the future generations to meet their own needs" (World Commission of the Environment and Development, 1987, p. 16). The concept of Sustainable Development cuts across various disciplines and is particularly applicable to tourism because tourism encompasses diverse sectors directly and indirectly. Furthermore, tourism as a concept requires a plethora of issues to be addressed in its quest to achieve sustainability. Consequently, Sustainable Tourism Development involves issues pertaining to gender.

The term "gender" is defined as "... the array of socially constructed roles and relationships, personality traits, attitudes, behaviors, values, relative power and influence that society ascribes to the two sexes on a differential basis. Whereas biological sex is determined by genetic and anatomical characteristics, gender is an acquired identity that is learned, changes over time, and varies widely within and across cultures" (United Nations World Tourism Organisation, 2011, p. 2).

The fifth goal of the 2030 Agenda for Sustainable Development is to "Achieve gender equality and empower all women and girls." The choice of terminology specifically refers to the female gender with reference to women and girls. The reader is left having to interpret the difference between women and girls, as well as speculate why this goal omits the male gender. The 2030 Agenda for Sustainable Development document iterates, with reference to the Universal Declaration of Human Rights document, that respect for human rights is crucial for all; it recognises that the human rights of women and girls have been and still are denied: "Women and girls must enjoy equal access to quality education, economic resources and political participation as well as equal opportunities with men and boys for employment, leadership and decision-making at all levels...All forms of discrimination and violence against women and girls will be eliminated, including through the engagement of men and boys" (United Nations World Tourism Organisation, n.d., p. 10). For purposes of this chapter, the focus is the equality and empowerment of women with specific reference to Target 5.5 and Indicator 5.5.2:

Target 5.5: "Ensure women's full and effective participation and equal opportunities for leadership at all levels of decision-making in political, economic and public life.";

Indicator 5.5.2: "Proportion of women in managerial positions"

SDG 5 contains two keywords: "equality" and "empowerment." Gender equality "describes the concept that all human beings, both women and men, are free to develop their personal abilities and make choices without the limitations set by stereotypes, rigid gender roles, or prejudices" (United Nations World Tourism Organisation, 2011, p. 2). Empowerment "means that people – both men and women – can take control over their lives: set their own agendas, gain skills (or have their own skills and knowledge recognised), increase self-confidence, solve problems, and develop self-reliance" (United Nations World Tourism Organisation, 2011, p. 2).

In terms of South Africa, the Bill of Rights forms "the cornerstone of democracy in South Africa...enshrines the rights of all people in our country and affirms the democratic values of human dignity, equality and freedom" (Constitution of the Republic of South Africa, 1996, p. 5). In terms of equality, the document states that "Equality includes the full and equal enjoyment of all rights and freedoms...measures are designed to promote the achievement of equality...to protect or advance persons, or categories of persons, disadvantaged by unfair discrimination may be taken." As such, the state or anyone may not "unfairly discriminate directly or indirectly against anyone on...race, gender, sex, pregnancy, marital status, ethnic or social origin, colour, sexual orientation, age, disability, religion, conscience, belief, culture, language and birth." The term women empowerment is defined, in the Women Empowerment and Gender Bill, as "... the advancement of women as contemplated by section 9(2) of the Constitution" (Republic of South Africa, 2013, p. 4). This document proposes strategies to empower women (with reference to economic empowerment, socio-economic empowerment of rural women and socio-economic empowerment of women with disabilities). The addition of the term "disability" demonstrates further inclusivity in respect of Responsible Tourism.

The international definitions for the terms, equality and empowerment, with reference to gender, differ significantly within the South African context. The constructs of gender and tourism cannot be critiqued without taking cognisance of South Africa's historical past. As such, issues of race and ethnicity have bearing upon women in the tourism industry. A brief synopsis is provided in an attempt to contextualise why. Women were discriminated against as a result of their ethnicity or skin colour (the term skin colour can also be synonymously interpreted as race, although it also reflects a manner of speaking in South Africa). In the apartheid context, white women were also discriminated against, because society was largely patriarchal, based on political, social, and economic reasons. Black women (terminology referring to racial groups in South Africa is contentious in terms of references to Black, African, and Coloured; however, for purposes of this chapter, the word Black includes African, Coloured, and Indian women) bore the greatest brunt of this discrimination, as these women were also discriminated against based on their gender and the colour of their skin. In other words, because the women were also Black, the political dispensation at the time was extremely repressive. For example, Black women could not acquire property in their name; and they could not hold down permanent employment if they were married. With the dawn of democracy in South Africa, political, social, and economic repressive policies were abolished and made way for policies which were based on equality, as enshrined in the South African Constitution.

Whilst the focus of this chapter is women, it is important to note that SDG 5 refers to women and girls and Responsible Tourism policy refers to the women and the youth. Until what age is a child not an adult and when does a girl become a woman? Does this differ from country to country? In South Africa, the Bill of Rights defines a child to be under the age of 18 (Constitution of the Republic of South Africa, 1996, p. 12). However, in South Africa, the youth is categorised to be between the ages 14–35, according to the National Youth Policy (National Youth Commission, 1997). Technically there is a contradiction between being regarded as part of the youth in South Africa as opposed to being an adult when one has reached 18 years.

Table 6.1 reflects the use of the word gender, utilised in the Responsible Tourism policy.

The reference to the Responsible Tourism Handbook (2003) and the South African National Standard (2011) in Table 6.1 provides examples from the Responsible Tourism documents that illustrate the proximity of the word gender to other demographic variables such as race, ethnicity, age, and the disabled; this is a significant aspect when understanding the concept of the previously disadvantaged or marginalised groups in pre-1994 South Africa. To do justice to the concept of gender within the context of Responsible Tourism, it must be contextualised in terms of the implications for women as marginalised in terms of gender, race, and ethnicity, etc. The White Paper on the Promotion and Development of Tourism in South Africa (1996) addresses gender in Section 6.7 titled the "Role of Women." This is the only Responsible Tourism document that dedicates a separate section to discuss the role of women. The White Paper on the Promotion and Development of Tourism in South Africa (1996), in Section 6.5, also recognises women as part of labour and states: "ensure equitable pay and working conditions as well as special conditions for female employees."

Table 6.1 Proximity analysis of the word gender to other demographic variables in Responsible Tourism documents

Policy document	Quotation
The White Paper (1996, p. 26)	"To use tourism as a catalyst for human development, focusing on gender equality, career development and the implementation of national labour standards"
The Responsible Tourism Manual (2002, p. 35)	"Tourism can lead to new domestic arrangements and gender roles that create new opportunities for women and young people"
	"Tourism can lead to new domestic arrangements and gender roles that create social tension" referring as an example to, "reduced esteem for elders and/or men"
The Responsible Tourism Handbook (2003, p. 8)	"Recruit and employ staff transparently, aiming to create a diverse workforce in terms of gender, ethnicity, age and disability"
The South African National Standard (2011, Economic criteria 5.3.1)	"The organisation shall use fair and equitable processes for recruitment and advancement in relation to race, gender and disability"

Source: own construction

In this case, the word "female" is used as a synonym for the word woman. In this reference, the onus is on the tourism businesses to create the particular conditions required for women in the workplace.

Inferences Drawn from the Responsible Tourism Documents

This section will draw upon the similarities and limitations in the various documents, as applicable to women within the context of Responsible Tourism in South Africa.

The Role of Women in Rural Areas

The White Paper on the Promotion and Development of Tourism in South Africa (1996) recognises that women, more so in rural areas, have a key role to play in the development of Responsible Tourism. The employment of women is reflected in terms of the developmental needs of industry. Reference is made to a survey conducted among female farm workers in the eastern part of the Mpumalanga Province (previously referred to as the "Lowveld") that concludes that there is a strong association between salaries and household welfare amongst women who are employed. Furthermore, it is recognised that employment in the tourism industry improves family life for both men and women in rural areas. A reference is made to "urban drift," referring specifically to men who seek employment in cities and the mines; by implication women are left behind to deal with the laborious work in the rural fields. These rural areas often have poor access to infrastructure and basic resources; furthermore these areas are often AIDS-stricken. The White Paper on the Promotion and Development of Tourism in South Africa (1996) recognises and regards the role of women in communities to be mothers, teachers, mentors; to stimulate community growth and development through tourism; to lead and implement community projects; to lobby support from developers and local authorities and to broaden the skills base of rural women, e.g. craft training.

The multiple roles attributed to women are significant. By implication, the role of the woman in the family as a mother suggests the absence of the male figure (spouse, husband, father) in terms of the scenario of the men who left the rural areas to access employment in urban areas, positioning women as breadwinners in the family in the rural community. This paragraph also is suggestive of the networks that women form in communities and that the role that they play extends beyond their immediate family. The influence of the woman is also recognised in that she has a role to play in developing tourism in the area.

Although the woman's role as a resident in a community has been emphasised, whether working in the tourism industry or not, women also engage in tourism as an activity and in so doing expand their worldview. They can be regarded as part of emerging markets, which are defined in The White Paper on the Promotion and Development of Tourism in South Africa (1996) as "population groups entering the market in increasing numbers as domestic tourists, especially those previously neglected"; or traditional domestic markets, which are defined as "previously advantaged domestic leisure groups" (Department of Environmental Affairs and Tourism, 1996).

The Role of Women in the Tourism Industry

The White Paper on the Promotion and Development of Tourism in South Africa (1996) also recognises the role of women as "policy-makers, entrepreneurs, entertainers, travel agents, tourist guides, restauranteurs, workers, managers and guest house operators and other leading roles in the tourism business environment." Furthermore, to ensure gender equality in terms of the conditions of employment of women; as well as respect and dignity for women in the promotion, marketing, and development of women in tourism. It is also expected that women also provide safety and security to women tourists.

The diverse voices of women, as employees or managers in the tourism workplace are thus essential in terms of influencing the conditions of employment, the marketing messages portraying women in a dignified and respectful manner, and in making inputs in policy. However, the extent to which an employee versus a manager is given a voice is often a contentious issue in the workplace.

The Previously Marginalised Woman as the Entrepreneur/SMME Owner, Manager and/or Employee

Various projects and initiatives are available to empower women, such as the Women in Tourism (WiT) programme, for example, that provides training and networking opportunities for women entrepreneurs, professionals, and leaders. Such initiatives often involve government and industry working together with other stakeholders such as higher education institutions. Women should make use of these opportunities to overcome challenges in the tourism workplace that hampers their career development.

Tourism businesses are required to comply with regulation and legislation. It is identified that many historically disadvantaged individuals (HDIs) and entrepreneurs are often unaware of some of the legalities required to operate as a tourism business and do not always understand the different operational spheres of the local, provincial, and national government (Department of Environmental Affairs and Tourism, 2002, p. 17).

The Responsible Tourism Manual for South Africa (2002, p. 42) proposes how to encourage locals to partake in the tourism industry at an enterprise level. Tourism enterprises are encouraged to set targets to raise the number of HDIs and local people employed and to report on the progress achieved, as well as to monitor and report on increasing the salary (wage bill) accruing to HDIs and residents (Department of Environmental Affairs and Tourism, 2002, p. 49).

The need to provide preferential support to locally owned, locally managed, and locally staffed tourism products is qualified with the clause "with an at least equal gender ratio of staff, and HDI complement…support will help to maximise the percentage of revenue that is retained in the local economy…this retained revenue tends to circulate (through secondary purchasing) into the poorer, more marginalised sectors of the community" (Department of Environmental Affairs and Tourism, 2002, p. 25).

Women as Part of Local Communities (Including Previously Marginalised Communities)

The term "local communities" is defined in the South African National Standard (2011) document as all the people who reside in the local area of the tourism organisation; the term "local area" is defined as being where tourism organisations are located within urban areas and the word "local" is defined as referring to the local area of the organisation (Department of Environmental Affairs and Tourism, 2011, p. 5).

The Responsible Tourism Manual of South Africa (2002) defines Responsible Tourism as "…enabling local communities to enjoy a better quality of life through increased socio-economic benefits…" (Department of Environmental Affairs and Tourism, 2002, p. 8). Whilst it is recognised in the Responsible Tourism Handbook (2003) that tourism can improve the living standards of local communities, it is cautioned that uncontrolled development can have a destabilising effect, presenting a range of negative impacts; thus, it is encouraged that staff and local communities are involved in planning and decision-making (Department of Environmental Affairs and Tourism, 2003, p. 13). In this respect, it proposes that planning and decision-making should involve all stakeholders, with the aim of uplifting local communities.

A "Responsible Tourism Self-Evaluation Form" for Tourism Operators is provided and the first question under the "Economic Impact" section asks: "What are you doing to recruit, employ and train local and previously disadvantaged people?" (Department of Environmental Affairs and Tourism, 2003, p. 38). Upskilling of local people, particularly the previously disadvantaged, is thus important for the advancement of women.

Cognitive Map

Figure 6.1 provides a visual illustration to show the link between Responsible Tourism and Sustainable Development in this chapter. Whilst the gender focus is on women for both sets of documents, the Responsible Tourism documents are indicative of the synonymous and often ambiguous positioning of women in various categories that are reflective of South Africa's context. These terms run the risk of coming across as amorphous concepts; however, they need to be understood within the context of South Africa's policy development. A synopsis is outlined in the following section.

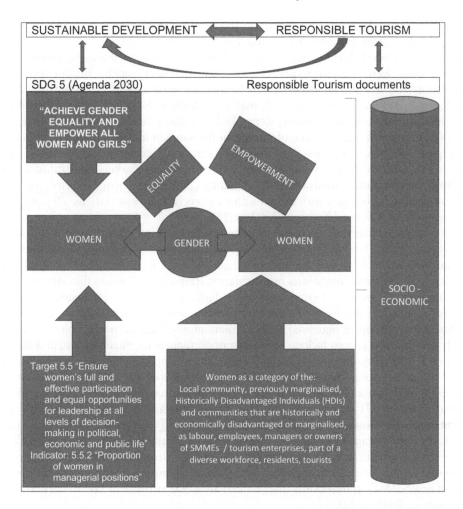

Figure 6.1   A visual representation to illustrate the relationship between the word/s, phrases, and concepts as relevant to this research.

Policy Context for Responsible Tourism in Post-1994 South Africa

The White Paper on the Promotion and Development of Tourism in South Africa (1996) defines the term, previously neglected communities, as "population groups that were largely excluded from mainstream tourism activities." A key constraint identified for tourism in South Africa is the limited integration of local communities and previously neglected or marginalised groups, largely due to the pre-1994 government policy. It is further recognised that the past apartheid policies constrained tourism development. Concerns voiced by previously neglected groups regarding factors that they perceived to have inhibited them from accessing a successful tourism industry are "Tourism is a white man's thing and not for us," lack of meaningful exposure and involvement in the tourism industry, suspicion and mistrust of not benefiting from tourism, not understanding the wider opportunities that tourism presents, lack of market access, language barriers, barriers to entry, inequalities, and negative attitudes that exist in the industry about community products being inferior (Department of Environmental Affairs and Tourism, 1996, p. 12).

In post-1994 South Africa, communities and previously neglected groups, particularly in the rural areas, were expected to play an important role as drivers for the development of Responsible Tourism and Sustainable Development (Department of Environmental Affairs and Tourism, 1996). Factors identified, that limited the satisfactory involvement of local communities in the tourism industry, were a lack of information and awareness in training and finance, coupled with the lack of interest in the tourism industry to build partnerships and capacity with local communities through job creation, incentives, and rewards (Department of Environmental Affairs and Tourism, 1996). Access to markets is identified as a key constraint for historically disadvantaged communities and entrepreneurs; therefore, they need to form meaningful linkages with formal sector enterprises and associations (Department of Environmental Affairs and Tourism, 2002, p. 28).

Government envisaged that a new way of conducting tourism could sustain South Africa's Reconstruction and Development Programme (RDP) in terms of contributing to transformation in South Africa. The RDP was the democratic government's first comprehensive approach to remedying the structural inequalities of the country. Responsible Tourism thus emerged as the new paradigm, allowing the Government and industry to include the previously neglected in the tourism industry. Involvement extends to planning and decision-making so that potential benefits of tourism can be optimised by local communities. Socio-economic objectives for previously marginalised groups include optimising opportunities for Small, Micro, and Medium Enterprises (SMMEs), creating sustainable employment opportunities, eliminating discrimination and encouraging mutual respect whilst developing tourism with dignity, encouraging community participation, empowering community structures, etc. The Responsible Tourism Manual (2002, p. 17) identifies one of the economic responsibilities of a Responsible Tourism enterprise as: "Creating and promoting employment and entrepreneurial opportunities for historically disadvantaged individuals...Putting to work locally available labour and skills." It further recognises that, in line with South Africa's Tourism Growth and Development Strategy, the diversification of the sector can be achieved through product development and the development of new SMMEs in historically marginalised areas (Department of Environmental Affairs and Tourism, 2002, p. 22).

Key Messages

Sustainability in tourism is important. It is essential that tourism is resilient when confronted with external challenges. The United Nations World Tourism Organisation 2022 World Tourism Day theme addressed the topic of having to Rethink Tourism to achieve sustainability. The coronavirus disease of 2019 (COVID-19) health crisis presented challenges for many tourism businesses in their endeavour to remain operational during the pandemic. On a global level, the tourism industry is reinventing itself to become more resilient and sustainable. In South Africa, COVID-19 impacted very forcibly on the sustainability of tourism businesses. Many of these impacts were not within the ambit of the entrepreneur/ business owner/corporate business control. In terms of the traditional business model, one has very little control over the external factors that impact the business and therefore the entrepreneur must adapt the elements that they have control over, to accommodate the external impacts upon the business. Did practising Responsible Tourism play a role in the survival of businesses during COVID-19? Did tourism businesses continue practising

Responsible Tourism during the pandemic or did they focus more on sustainability? What was its impact on the issue of gender and women? An opportunity exists within the post-COVID-19 Tourism environment to readdress the issues pertaining to gender and women. It is thus crucial to consider, taking due cognisance of race, ethnicity, age, and the disabled:

- The extent to which the tourism sector employs more women than other sectors;
- The number of women versus men in South Africa who are employed in the formal tourism workforce;
- The number of women versus men employed at professional levels, involved in decision-making;
- The number of women versus men in South Africa who occupy top positions in the tourism workplace;
- The extent to which self-employed women in South Africa are engaged in unpaid work whilst engaged in community development;
- The number of women in family tourism businesses in South Africa and issues of ownership;
- The number of women versus men who are employed by women employers in the tourism industry;
- The extent to which women in the tourism workforce in South Africa are earning less than their male counterparts; and
- The number of women versus men in leadership positions in tourism.

Recommendations to Government

Even though a new Tourism Act was legislated by the South African Government in 2014 which promoted the practice of Responsible Tourism, The White Paper on the Promotion and Development of Tourism in South Africa (1996) was not promulgated by Parliament as official legislation.

Despite Responsible Tourism being identified as the underlying philosophy for the country (Department of Environmental Affairs and Tourism, 1996). South Africa, as a country, is not a leading Responsible Tourism destination. However, the City of Cape Town is aligned with Responsible Tourism practices and in 2009 won an international best practice award for Responsible Tourism. Tourism destinations that implement Responsible Tourism practices have a market advantage (Department of Environmental Affairs and Tourism, 2003, p. 6), as they can use this approach to attract a distinctive type of tourist to the country. In this regard, the distinctive type of tourist may be those who are keen to embrace and buy into the philosophy of Responsible Tourism. Cape Town, located in the Western Cape Province of South Africa, is governed by the Democratic Alliance (DA) and is presently considered to be the official opposition party in South Africa. It should be noted that the DA governs the City of Cape Town and is supportive of promoting Cape Town as a Responsible Tourism destination. The DA is seemingly aware of the potential benefits that Responsible Tourism can bring about to all stakeholders. Cape Town, like the rest of South Africa, is a city of deep-seated inequalities, brought about by the historic past. These inequalities, specifically the systematic marginalisation of women and Blacks, require specific policies to address those gender and racial imbalances, which have been perpetuated over decades. Whilst these policies have been developed, there is a need to determine whether the policies have yielded the desired outcomes.

Recommendations to Industry

The Responsible Tourism Manual (2002) and the Responsible Tourism Handbook (2003) provide practical guidelines for tourism businesses to utilise Responsible Tourism to further the advancement of women in the post-apartheid South African context. Industry potentially has a crucial role to play in the empowerment of women in South Africa. As a management strategy, Responsible Tourism allows for a conduit to create a diverse workforce and has the potential to bring about large-scale changes to local communities. Thus, planning and decision-making should involve all stakeholders, with the aim of achieving gender equality and empowerment in the tourism workplace and in local communities. Special conditions may be required to create the desired working conditions, and as such, managers and business owners should be conscious of the historical context of women in terms of race, ethnicity, age, and the disabled in South Africa.

Further Research Opportunities

It is recommended that researchers be conscious of terminology. Currently many anomalies exist within the terminology utilised throughout the Responsible Tourism and Sustainable Tourism Development documents. The diction encompassing gender in tourism policy requires further attention. Many of the categories utilised to describe women within the tourism realm vary significantly and consequently the different descriptions vary in terms of the emphasis placed on gender, e.g. women as a category of race, women as a category of the disabled, and women as a category of being historically disadvantaged. Consequently, it is difficult for the tourism business owner/entrepreneur to contextualise the significance of the diction utilised to describe women within the various categories. Policy documents need to be clear in terms of diction utilised to describe the specific elements relating to gender and women. New documentation should be aligned accordingly, so that there is one voice in terms of diction and terminology. This will enable the tourism industry to align itself with future government policy documents. There is a lack of alignment in terms of terminology and diction between Responsible Tourism and the SDG 5 document. The issue of terminology utilised in Responsible Tourism versus Sustainable Development is to be addressed and merged, where possible and should be applied within the industry.

Conclusion

This chapter presents a snapshot of South Africa's Responsible Tourism policy framework in terms of its inferences to gender, particularly regarding women and their envisaged role in Responsible Tourism. Due to South Africa's political past, it is cautioned against merely categorising gender as a demographic term next to other terms such as race and ethnicity. Furthermore, the synonymous and implicit use of the word gender with other phrases (e.g. women as part of HDIs, etc.) needs to be addressed as this highlights the incongruity of the role of women across racial and ethnic lines within South Africa's historical context. It is recommended that the anomalies identified within the policy framework regarding gender in Responsible Tourism, particularly in terms of terminology, be rectified to achieve alignment between Responsible Tourism practice and Sustainable Development within the South African context. It is also important to review Responsible Tourism documentation against SDG 5 to better position the concept of gender taking due cognisance of South Africa's historical context and gender implications thereof for tourism.

References

Altinay, L., Paraskevas, A., & Jang, S. (2016). *Planning research in hospitality and tourism* (2nd ed.). Routledge.

Department of Environmental Affairs and Tourism. (1996). *White paper on the development and promotion of tourism in South Africa.* https://www.gov.za/sites/default/files/gcis_document/201411/tourism-white-paper.pdf

Department of Environmental Affairs and Tourism. (2002). *The responsible tourism manual for South Africa.* https://www.tourism.gov.za/CurrentProjects/ResponsibleTourism/Responsible%20Tourism/Responsible%20Tourism%20Manual.pdf

Department of Environmental Affairs and Tourism. (2003). *The responsible tourism handbook.* https://www.tourism.gov.za/CurrentProjects/ResponsibleTourism/Responsible%20Tourism/Responsible%20Tourism%20Handbook.pdf

Department of Environmental Affairs and Tourism. (2009). *Ministerial and department environmental affairs and tourism briefings: Strategic plans 2009 to 2012.* Parliamentary Monitoring Group. https://pmg.org.za/committee-meeting/10432/

National Department of Tourism. (2011). *South African national standard (SANS 1162).* https://www.tourism.gov.za/CurrentProjects/ResponsibleTourism/Responsible%20Tourism/Responsible%20Tourism%20Publications.pdf

National Department of Tourism. (2016). *National tourism sector strategy (NTSS) 2016-2026.* https://www.gov.za/sites/default/files/gcis_document/201712/national-tourism-sector-strategy-ntss-2016-2026a.pdf

National Youth Commission. (1997). *National youth policy.* https://www.westerncape.gov.za/text/2004/11/ny_policy_2000_compressed.pdf

Republic of South Africa, 1996. *Constitution of the Republic of South Africa108 of 1996.* https://www.gov.za/sites/default/files/images/a108-96.pdf

Republic of South Africa. 2013. *Women empowerment and gender equality bill.* https://www.gov.za/documents/women-empowerment-and-gender-equality-bill

United Nations World Tourism Organisation. (n.d.). *Transforming our world: The 2030 agenda for sustainable development (A/RES/70/1).* https://sustainabledevelopment.un.org/content/documents/21252030%20Agenda%20for%20Sustainable%20Development%20web.pdf

United Nations World Tourism Organisation and United Nations Entity for Gender Equality and the Empowerment of Women (UN Women). (2011). *Global report on women in tourism 2010.* UNWTO. https://www.e-unwto.org/doi/pdf/10.18111/9789284413737

World Commission of the Environment and Development. (1987). *Report of the world commission on environment and development: Our common future.* https://sustainabledevelopment.un.org/content/documents/5987our-common-future.pdf

7

WOMEN PARTICIPATION, EMPOWERMENT, AND COMMUNITY DEVELOPMENT IN SELECT TOURISM DESTINATIONS IN THE PHILIPPINES

Eylla Laire M. Gutierrez

Abstract

The role of women in tourism industries has long been part of existing discourses. For years now, the industry is lauded for the economic opportunities it offers to women compared to other sectors. As a result, women make up the majority of tourism workforce. Despite these, issues relating to their participation in the industry persist. Arguably, they remain marginalised and not empowered. The existing debate suggests that women are involved in lower paid, unskilled jobs, with low representation in key decision-making positions. The aim of this chapter is to examine whether tourism participation contributes to the empowerment of women by looking at their experiences within their own communities. In doing so, suggestions are made on how women can act as active agents facilitating community development in tourism communities in the Philippines. Positioned as an exploratory study, primary data was obtained by conducting interviews with women in select tourism communities in the Philippines. The findings of this study contribute to the enhancement of understanding about women in the Philippine tourism industry.

Keywords

Women, Empowerment, Participation, Community Development, The Philippines

Introduction

Determining the role of women in tourism has long been part of the development discourse due to their presence in tourism industries across the globe. As emphasised by Momsen and Nakata (2010), tourism creates gendered impacts that vary depending on an individual's role

DOI: 10.4324/9781003286721-11

in the industry. On the one end of the spectrum, tourism is perceived as a sector allowing women to progress and become empowered given its flexible and dynamic nature (Ateljevic, 2008). Arguably, the industry continues to provide better opportunities for women's participation in the workforce, entrepreneurship, and leadership than other sectors of the economy (Fruman & Twining-Ward, 2017) – which enhance both women's empowerment and gender equality in societies (Boley & McGehee, 2014). This is evidenced by the fact that women make up the majority of tourism workforce where globally 59% of people employed in tourism are women (United Nations World Tourism Organization [UNWTO], 2020).

As pointed out by Cole (2006), however, participation does not always lead to empowerment. Beyond providing economic opportunities for women through participation, empowerment extends to psychological, social, political aspects, as Scheyvens (1999, 2000) suggested, and entails the satisfaction of three As of agency, autonomy, and authority identified by Cole (2018). Thus, it is crucial to look at the extent to which women's participation in tourism activities leads to their empowerment. Well beyond these conceptualisations, this study acknowledges that the pursuit of empowerment through participation does not come in a "one-size-fits-all recipe" (Cole, 2018, p. 3). As several studies have noted, it is important to contextualise gender analysis in a specific locality to better understand the gender dynamics involved in tourism activities (Feng, 2013). Thus, it is important to look at the context of women's participation in a specific setting, which is in this case the Philippines, the top Asian country known for its efforts in closing the gender gap (Mathkar, 2019).

In the Philippine tourism sector, little attention has been paid to gender studies despite women's significance in tourism and its impact on the Philippine economy. Tourism remains to be a development pillar in the country. Prior to the pandemic, the industry contributed to 12.7 of the country's gross domestic product (GDP), employing 5.7 million Filipinos (Philippine Statistics Authority [PSA], 2019). During the outbreak of the coronavirus (COVID-19) pandemic, however, the industry suffered from a substantial decline in tourism activities and receipts – decreased by 61% in domestic travels, 83% in international travel, and 17.5% in employment (Bengzon, 2021). As Basuil et al. (2020) noted, the impact of the pandemic is exacerbated for disadvantaged communities and individuals, including women.

Against this backdrop, this chapter aims to examine whether tourism participation contributes to the empowerment of women by looking at their experiences within their own communities. More specifically, the study aims to determine how women can act as active agents facilitating community development in tourism destinations. In doing so, the chapter contributes to better understanding of the situation of women in the Philippine tourism industry. Suggestions are made on how women can act as active agents facilitating community development in tourism communities in the Philippines.

Women, Communities, and Tourism in the Philippines

Women in Communities

For years now, the Philippines is recognised as one of the best performing countries in the world in terms of achieving gender parity. According to the recent World Economic Forum's (WEF's) 2022 Global Gender Gap Report, the Philippines retains its position as the best performing country in Asia in terms of bridging gender gaps. Globally, the country ranks 19th out of the 146 countries surveyed in terms of its progress in achieving gender parity along the lines of economic participation and opportunity, educational attainment, health and survival, and political empowerment. The same observations were made in World Bank's Women, Business, and Law report (2022) suggesting that the country's score of 78.8 out of 100 remains higher than the regional average score of 71.9. The country received perfect scores in the following indicators: workplace, pay, and entrepreneurship. Moreover, the Philippines' gender wage gap is considered small when compared to many advanced economies (Albert & Vizmanos, 2017).

Despite these, the situation of women in the Philippines proves paradoxical – despite women's advancement in professional areas, their experience at the community and individual levels varied (David et al., 2017). For instance, despite its recognition for bridging gender gap, the country remains to have the lowest female labour force participation in ASEAN (Cabegin & Gaddi, 2019). In fact, 6.6 million Filipino women are estimated to work in the informal sector (Global Network of Women Peacebuilders [GNWP], 2022), often characterised by income volatility, lack of social protection, poor working conditions, among others (Pascual, 2008). As noted by the World Bank (2022), the country needs to work on women's equality in the Philippine society, specifically in terms of mobility, marriage, parenthood, assets, and pension.

Beyond economic perspectives, numerous factors entrenched in the Philippine society, including gender roles (i.e., where women are assigned to reproductive, domestic, and care work), religious and cultural restrictions, lack of access to skills training, occupational gender segregation, among others, continue to influence the opportunities available to them (Cabegin & Gaddi, 2019; Pascual, 2008). Several studies have also established a link between existing societal norms and women's economic opportunities. Cabegin and Gaddi (2019), for example, found that Filipino women have a higher likelihood to withdraw from their careers due to marriage and childbirth compared to men. That is, society still expects women to take a *back seat* in their careers upon marriage. In some parts of the country, religious practices and beliefs remain influential to women's occupation (e.g., Muslim religion restricting women's choice of occupations) (Santos, 2022). In terms of governance and leadership positions, David et al. (2017) noted that Filipino women remain underrepresented in decision-making positions both in the public and private sectors of the country. Institutional factors, pertaining to sexuality and reproductive health, remain restrictive and regressive as access to birth control remains limited, and the promotion of sex education, abortion, and divorce is still considered a taboo (Santos, 2022). Despite women's advancement in societies, Garcia (2020) still found that violence against women (VAW) remains an issue that is prevalent in the country where cultural factors continue to influence women's behaviour and response towards violence.

Despite these, women's movement towards attaining empowerment has advanced. Over the years, Filipino women have gained increased control and influence over themselves, their communities, and societies to which they belong in. In fact, several studies suggest the crucial role played by women in various developmental initiatives that benefit society. At the community level, women's contribution to environmental protection, conservation, and resource management has been noted by several authors (Dasig, 2020; Gabriel et al., 2020; Jumawan-Dadang, 2015; Lontoc, 2020). As suggested by Jumawan-Dadang (2015), Filipino women are hailed to be natural stewards of environmental protection across the country. The same case is observed in the indigenous tribes of Kalanguya and Aetas where women undertake substantial work to ensure that resources are protected for the survival of their own communities (Gabriel et al., 2020; Lontoc, 2020). To an extent, Filipino women were also found to play a unique role in climate change adaptation initiatives (Graziano et al., 2018) and in efforts to implement disaster risk reduction mechanisms (Ramalho, 2019). The same critical roles are played by Filipino women in cultural preservation (Pelayo et al., 2020; Tang et al., 2022; Yankowski, 2019) and in agricultural and coastal communities (Angeles & Hill, 2009; Maligalig et al., 2021; Ramirez et al., 2020).

Through Filipino women's involvement in these crucial aspects of community's survival and development, they were found to have bigger social networks, to an extent social capital, than men (Quetulio-Navarra et al., 2017). Studies have also noted how their participation in such activities assists in the achievement of their empowerment. Women were found to have the ability to negotiate their roles in highly gendered communities. Dasig (2020) and Lontoc (2020), for example, found that women were able to negotiate and sustain their active participation in traditionally patriarchal systems. In some other cases, women were found to lead cultural and economic activities, whereas men were found to provide them assistance as is the case of pottery making in Bohol (Yankowski, 2019). As a result, they were also able to influence the community's sense of pride and self-worth. Moreover, the increased involvement of women in various economic activities resulted in their augmented power over their own households (Maligalig et al., 2021). As Eder (2006) noted, wives have gained more control on household expenditures and budgeting, as husbands entrust a huge portion, if not all their income to their wives (Pajaron, 2016).

The situation and experience of Filipino women prove to be more complex, however, when further examination is made. For instance, despite women's increased participation and voice in social, political, and economic realms, their experiences in their households differ. Caring work, as Prieto-Carolino and Mamauag (2019) noted, remains to be the primary responsibility of women and proves to be undervalued by Philippine society. Even at present, Filipino women remain solely responsible for undertaking care work. While women have grown flexible and are now able to share the burden of economic work, men appear to be less willing to take on the responsibility of sharing domestic work (Angeles & Hill, 2009). In family ventures, women's labour is usually not paid for their work is perceived as extensions of their assigned domestic work (Ramirez et al., 2020). The gender dynamics in Filipino communities becomes more complex when the behaviours of men have been accounted for. In the case of agrarian households for example, Angeles and Hill (2009) found that men preferred being unemployed instead of taking lower paid "women's jobs."

Women in Tourism

Discourses examining the role of women in tourism are becoming more pronounced across the globe. Unfortunately, this may not be the case for the Philippines. To date, only a few studies were conducted in this field. Despite this, several authors have made significant contributions to enhancing the understanding of the situation of women in the Philippine tourism industry.

The seminal work of Chant (1996), for example, earlier noted that the type of work women are involved with in tourism reflects the cultural and historical treatment of women in the country. Arguably, women in the industry were bound in labour niches which were considered subordinate positions following societal gender constructions – women as secondary income earners and primary persons in charge of domestic work. Historically, this resulted in women being commodified as observed in the prevalence of sex tourism in the country in the early 1970s (Chant, 1996). Several years after, Dulnuan and Mondiguing (2000), also found that tourism development in Sagada reinforced existing gender division of labour where women were involved in work that were seen as an extension of their reproductive work (e.g., managing lodging houses and restaurants). While men were observed to engage in tourism activities, women were found to be situated in worse situations as they took in additional agricultural work and domestic work. Pleno (2006), on the other hand, found that through women's engagement in ecotourism activities in Bohol, they lived happier lives due to their newly acquired skills. Yet, issues in relation to income sufficiency and negative attitude of their husbands towards their work persisted. In examining the entrepreneurial engagements of women in tourism areas in the Philippines, Gumba (2020) found that women were engaged in small enterprises (e.g., variety stores, eateries, and souvenir shops) which were highly dependent on tourism activities within their province. While women were provided with extra income, they still faced issues with access to credit as they were discriminated against as "secondary income earners" (Gumba, 2020, p. 358). Further aggravating their situation, reproductive work remains their sole responsibility.

Following an overview on the discourses surrounding Filipino women in tourism and in communities across the country, it is notable that not all forms of economic activities contribute to women's empowerment (Gutierrez & Vafadari, 2022). In fact, nuances in their situation and experiences can be observed. Thus, generalisations do not accurately depict their varied realities across geographical locations and socioeconomic groups in the country (Ramalho, 2019; Santos, 2022). To ensure the accuracy of information, contextual and cultural differences must be considered in analysing the situation of women in the industry (Gutierrez & Vafadari, 2022).

Defining Empowerment

Empowerment in this study is understood from multi-dimensional perspective as suggested by Boley and McGehee (2014), Elshaer et al. (2021), and Scheyvens (2000) capturing its psychological, economic, social, and political dimensions. As Cole (2018) posits, this entails the satisfaction of three As of agency, autonomy, and authority that underpin the ability and capacity of women to make choices for themselves and to take control of their lives, decisions, and resources to affect their own lives (Briedenhann & Ramchander, 2006). In the long run, Floro (1995) suggested that women empowerment assists in the realisation of gender equality. That is, women empowerment exponentially results in increased women engagement. In this chapter, the augmented model by Gutierrez and Vafadari (2022), establishing the relationship between women empowerment and community development, will be adopted (see Figure 7.1).

These dimensions of empowerment are not interdependent from each other, and can be achieved independently (Moswete & Lacey, 2015), and may even be overlapping at times (Scheyvens, 1999). This model therefore suggests that empowerment refers to a long and tedious process that facilitates the elevation of women's psychological, economic, social, and political status in societies (Lenao & Basupi, 2016). Empowerment is therefore a "negotiated process" (Movono & Dahles, 2017, p. 10) that requires interactions, consultations, and adjustments among members of community. For women empowerment to accrue, stakeholders must understand "how and why women need to be empowered" (Vukovic et al., 2021, p. 4) – where women are recognised as "critical constituency within the local communities needing empowerment" (Lenao & Basupi, 2016, p. 56).

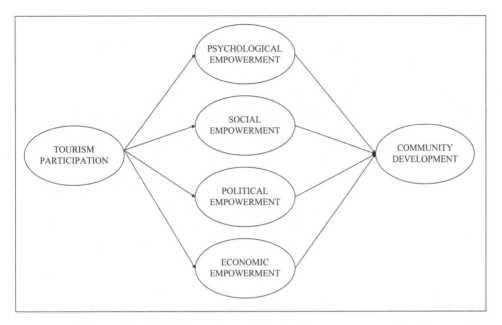

Figure 7.1 Women's participation, empowerment, community development model. (Source: Author's own, adapted from Gutierrez & Vafadari, 2022)

Study Context and Methods

For the purposes of this study, the provinces of Palawan and Cebu served as the research focus of this study. These two provinces are considered tourism hubs in the Philippines – annually, Palawan receives over a million tourists (Fabro, 2020), meanwhile Cebu receives 2 million tourists (Erram, 2022). Palawan is dubbed as the last ecological frontier, while Cebu is known as the premier heritage and culinary destination in the country. With diverse product offerings (i.e., ecotourism in Palawan and cultural tourism in Cebu) and contrasting historical representation of women, the two research locales offer a comprehensive image about the situation of women in select tourism communities in the Philippines.

Women in Cebu

Cebu is known as the "Queen City of the South," the oldest city and first capital in the Philippines receiving over 2 million tourists annually before the pandemic (Arnaldo, 2021). Known for its rich culture and history, Cebu became the first Spanish settlement in the country and the Spanish colonial capital. Thus, the Spanish culture readily influenced the dynamics and social relations of women in the province. In general, women are respected in the Cebuano culture (Quisumbing, 1963). In terms of economic and social aspects, Cebuano households establish a sharing culture – shared work in domestic and care work, earning income, among others. While women are expected to manage household finances, they are also contributing to family income. Because of this, women in Cebu can engage in economic activities belonging to the cottage industry such as pottery making and weaving.

Women in Palawan

Palawan is known as the largest province in the country situated in the western part of the Philippines receiving over a million visitors annually before the pandemic (Formoso, 2022). Culturally, Palawan is considered a patrilineal society where men are seen to have a better social status than women (Pondorfer et al., 2017). As an archipelago home to various indigenous ethnic groups, its culture and social relations have been directly influenced by the indigenous cultures of various tribes. Within households, women are assigned to take domestic and care work while managing household finances (Alcantara, 1994). Compared to the Cebuano culture, women are not expected to contribute to the income of the household or to participate in economic activities. Communal activities such as leading ceremonial and religious activities are considered a man's job (Pondorfer et al., 2017). In this sense, community leadership follows a royal blood line where men are the only eligible leaders.

Due to the dearth of data monitoring and describing women's participation in tourism activities in the Philippines, this study employs an exploratory qualitative research approach. This empirical investigation provides a nuanced perspective on the experiences of women in the Philippine tourism industry. The approach employed in this study was guided by the methods used by several studies examining women's experiences in tourism (Gutierrez & Vafadari, 2022; McCall & Mearns, 2021). To examine whether tourism participation contributes to the empowerment of women, two questions were asked: first, do women's work contribute to their psychological, economic, social, and political empowerment? Second, how do women's participation in tourism contribute to community development?

The findings presented are based on a fieldwork conducted in July to August 2022 in tourism-rich destinations, Palawan and Cebu. A mix of purposive and snowball sampling techniques was employed in selecting the participants for this study. The authors primarily communicated with the respective tourism offices of Palawan and Cebu to identify potential participants who are directly engaged in tourism activities. From this, other women were also recommended to be interviewed during the field visit. All participants expressed their interest to participate in the study. The primary means of data collection was face-to-face semi-structured interviews exploring the individual opinions and experiences of the participants (Caparrós, 2018; Hesse-Biber, 2013). There was a total of ten women, five in each research locale. All are employed in a tourism-related enterprise. Table 7.1 presents the profile of participants. The interviews were conducted in both English and Filipino

Table 7.1 Participant profile

Name	Age	Marital status	Education	Employment status
Joyce	48	Married	Bachelor	Self-employed: restaurant owner
Neris	68	Married (widowed)	High school	Self-employed: food vendor
Jona	32	Married	Bachelor	Food and beverage staff
Janine	52	Married	Bachelor	Accommodation attendant
Mica	34	Single	Bachelor	Self-employed: media business owner
Elaine	24	Single	Bachelor	Tourism officer
Maria	54	Married	Bachelor	Self-employed: souvenir shop owner
Jamela	35	Married	Bachelor	President of tourism association
Mae	51	Married (widowed)	High school	Tour guide
Tere	48	Married	Bachelor	President of tourism association

which meant that translation was necessary. Audio-recorded transcriptions were subjected to deductive coding to identify patterns from which themes were deciphered.

The Experiences of Women in Philippine Tourism

Women's Participation in Tourism

For both destinations, the economic participation of women was not ideal and, to an extent, prohibited in the past few years. At present, however, perceptions of women and their involvement in economic activities are gradually changing as suggested by the participants from both Palawan and Cebu. As Neris (Palawan) suggested:

in the past, women like me were prohibited by our husbands to work because they want us to focus on taking care of our kids. In our community, we are not encouraged to work outside. They were not happy to see me going out to work.

Similarly, Maria (Cebu) noted that:

hiring women was difficult in the past as their husbands did not allow them to work. People were expecting them to just stay at home. I needed to convince them at first that this work will add income to their households. But now, it is acceptable for women to go to work.

This supposed transition in the perception of women may be attributed to the growing financial needs in Filipino households as Mae (Cebu) noted *"...because his earnings were not enough, I needed to step in."* As Joyce (Palawan) argued: *"nowadays, we need to work. We are expected to work to help our own families earn income."* To some extent, women's participation in economic activities has become an expectation rather than an exception (Duffy et al., 2015).

For tourism destinations such as Palawan and Cebu, women's involvement in the industry became inevitable: *"...because our place (El Nido, Palawan) gets almost all our income from tourism, it was natural to join the industry,"* said Joyce (Palawan). For Cebu, however, the case was different because there were existing stereotypes towards women working in the industry dubbed to commodify them. As Elaine (Cebu) mentioned:

they had stereotypes about tourism before since Cebu was a sex tourism hotspot. Women were sold in the past and there are still very few who does it. It took some effort and time to convince them (her family) that I work as a professional.

To an extent, there were still gender-designated work in the industry: *"women are mostly involved in guest relations, cooking, housekeeping before, because naturally, men would take up jobs that might be difficult and dangerous for women to do such as tour guiding which requires carrying heavy luggage, etc."* (Joyce, Palawan). In fact, tourism was seen as a feminine job where women are naturally drawn into. But as Jona (Palawan) noted: *"it's changing since I see some women becoming tour guides, dive masters, etc. while it was mostly men now who in charge of housekeeping."* The same observation is noted by Jamela (Cebu) in suggesting that:

I don't think gender is an issue for work as long as you are qualified. We have boatmen and tour guides who are women. We also have male performers. For us, as long as you are skilled and willing to be trained, you can take the job.

Women's Empowerment in Tourism

All the women interviewed agreed that they were happy and proud working for tourism. As emphasised by Mica (Palawan): "*I am happy to know that I am doing my part to promote the culture here. I am proud to showcase the beauty of Palawan through my work.*" Even Neris (Palawan) suggested that "*even when I was younger and now, I am old, I have always been proud of the work I do here. This is how I remain happy and contented just talking to visitors and showing them what we have.*" For others, they even aspired to work for the industry because of the benefits it brings to their communities. As Jona (Palawan) mentioned: "*it was really my dream to work for tourism here. I always wanted to be where I am now that is why I even took this degree. This was my dream workplace as well.*" Elaine (Cebu) also echoed this by saying that "*even if I am not a tourism graduate, when I decided to work for the government, I wanted to know how I can make tourism develop because it helped a lot of families in our province.*" Most of these women are thankful that their involvement in tourism allowed them to develop and improve their skills as well as their self-confidence. As Joyce (Palawan) noted:

> before, I was really shy and I did not want to talk to strangers. But now, I learned how to communicate and socialize. I also had to learn some basic managerial skills so I can help my husband in managing our business.

For others like Neris (Palawan): "*because of tourism, I learned how to speak English. I was also driven to finish my studies even when my classmates were younger than me. I became more confident that I can face anyone now.*" Across all women interviewed, they were positive about the psychological impacts that their participation in tourism brought them. The same positive psychological impacts of women's involvement in tourism activities were noted by several studies. Kunjuraman and Hussin (2016), Moswete and Lacey (2015), and Pleno (2006), for example, found that women's involvement in ecotourism developed their self-esteem and confidence. The development of their skills (e.g., communication, cooking, managerial) enabled them to achieve self-fulfilment and happiness (Lenao & Basupi, 2016; Miettinen, 2006; Tucker, 2007). Through their engagement with visitors, women were also found to develop a sense of pride for the work that they do, along with their ability to showcase their culture (Elshaer et al., 2021; Kunjuraman & Hussin, 2016) and conserve their environment (Jumawan-Dadang, 2015; Nutsugbodo & Mensah, 2020).

In terms of their satisfaction with the economic contributions brought by their participation in tourism, mixed responses were noted. For one, the income they received through tourism acted as the turning point for their work to be accepted by their families and communities. As Tere (Cebu) stated: "*it was only when I was able to bring home money for our meals and for my children's education when my husband finally agreed that I continue working.*" For Jona (Palawan), "*because I am earning more than my husband, we decided that he will be staying at home to take care of our newborn baby while I work.*" Through her participation in tourism, Neris (Palawan) emphasised that "*I can now buy things for myself without asking money from my husband.*" As observed by McCall and Mearns (2021) and Moswete and Lacey (2015), through the income they receive from tourism, they were able to improve the economic conditions of their own families. To an extent, the additional income allowed them to gradually negotiate their roles within their own households and communities (Foley et al., 2018; Movono & Dahles, 2017). However, some also noted that they still cannot fully depend on their income from tourism as some of them are considered seasonal. This was the sentiment of Janine (Palawan) in saying that *"because visitors come and go, during lean season, we really cannot depend on it yet. Without the assistance from the government, I don't think it will be enough."* Joyce (Palawan) on the one hand noted that "since the pandemic, we are still trying to recover. My income (their business) is now very dependent on whether tourists will start coming again." In some other cases, due to the seasonality of tourism activities (i.e., natural calamities and disasters, pandemic), some women still struggle to achieve economic independence (Bras & Dahles, 1998).

Beyond the individual impact of tourism on women, it was also worth noting how their participation influenced their status within their households and communities. Again, their responses revealed the complexity of their situation. Within their community, Neris (Palawan) stated that *"our status improved. Now they see that women are not just housewives but can also bring in money."* The same sentiments were shared by Jamela (Cebu) who noted that *"because I am the president and tourism officer is also a woman, they see us differently. Now they know that women can lead and help the community."* However, their situation within their households remains diverse. As Neris (Palawan) also mentioned: *"my husband was not happy that I was earning more. He started drinking while I was out working. He just stayed at home while I tried to balance both my work and taking care of the house."* As Ferguson (2011) and Chant (1996) noted, in some cases, tourism exacerbates the burden (i.e., domestic and economic work) carried by women. The same case is observed by Tere (Cebu) in suggesting that *"even if I was working, I am expected to take care of the kids and the house. I still do the cleaning and cooking."* Because domestic work is believed to be women's work, men are often inclined to refuse this responsibility (Feng, 2013). Others, like Joyce (Palawan), Jona (Palawan), and Maria (Cebu), were happy to say that their husbands were very supportive in taking charge of domestic work while they were out working. The experiences of women in this aspect suggest that women tend to have increased external empowerment (i.e., outside household), but limited internal empowerment (i.e., within their household) (Caparrós, 2018; Dunn, 2007). Despite women's improved status in their respective communities, gender-assigned roles remained within their households. Men still refused to take up and share household and care work (Caparrós, 2018; Feng, 2013), while resorting to antagonising actions (e.g., drinking, violence) (Ishii, 2012; Tran & Walter, 2014).

In terms of their political interactions and involvement, several women noted that women's occupation of key leadership positions across the industry proved beneficial to uplifting their status. For one, Mica (Palawan) noted that *"because I was exposed to women leaders both in the government and in the private sector, I became inspired to be participate as well."* The same sentiments were echoed by Elaine (Cebu) who said that *"I decided to work for the government because of governor who was a woman. Through her, I realized that women can actually lead."* The impact of women's leadership on influencing other women was first noted by Cole (2018) arguing for increased avenues for women's voices to be heard. For others, their political involvement became heightened because of their participation in tourism. Jamela (Cebu) noted that as the president of the association, she *"(I) became more active in working with the government to talk about our concerns and our suggestions. I even organize the members before elections to talk about who is the best candidate to support the aims of their association."* The same observations were made by Elshaer et al. (2021), Pleno (2006), Tran and Walter (2014), and Vizcaino-Suárez (2018) in suggesting that through tourism, women were given opportunities to take up leadership positions and be involved in decision-making processes concerning tourism development.

That is, their involvement signifies that they are "critical constituency within the local communities needing empowerment" (Lenao & Basupi, 2016, p. 56)

Women's Contribution to Community Development

Women's participation in tourism also affected the community surrounding them. As suggested by the women interviewed, they are glad that their involvement in tourism results in positive spillovers to their respective communities. As Mica (Palawan) mentioned: "*I am happy that through my business, I am also able to employ other talented women since we are a woman dominated-company.*" The same case is observed by Marie (Cebu) in saying that "*women just needed to hone their skills to work… key to my business' success is all the women who patiently created each of my products carefully.*" Others, such as Neris (Palawan), who became one of the first members of the community who provided products and services to the nearby private tourism estate, mentioned:

> *I am happy that my work in tourism allowed other women and other members of the community to earn extra income. When they needed someone to entertain the guests, I asked women to join me and form a singing choir. I also recruited some youth members when I needed help in producing kakanin (rice cake) for the guests.*

As observed by Caparrós (2018), women-owned and -led enterprises tend to employ other women as well. As suggested by Bakas et al. (2018) and Vizcaino-Suárez (2018), increased women representation in leadership and top-level management positions leads to greater gender equality measures to be implemented. Elaine (Cebu) further went on to say that "*through my work, I realized that if more women are able to work, the better is for our industry and for the economy,*" suggesting that encouraging more women to join economic activities would be beneficial to their community. Meanwhile, Janine (Palawan) emphasised the importance of government intervention to "*…hear out our concerns and to help develop our skills so we can participate.*" According to her, it is difficult not just for women, but other local community members to participate in tourism without the right knowledge and skills. For Jamela (Cebu), the success of their association meant that "*women can lead a group…that we can help our community*" in talking about the scholarship program, cash assistance initiative, among other programs that their association funds through their community tourism enterprise. The same case is shared by Tere (Cebu) who talked about how her leadership does not only capacitate women but specifically senior citizens, fishermen, and artisans in their community who want to work and to receive extra income, "*I am happy to serve the senior members, fishermen, artisans, and other service providers in our community. When we have a visitor, we also need their help.*" In this sense, women's participation in tourism allows them not only to help themselves but their respective communities as well. The contribution of women in uplifting the quality of lives in their own communities has also been noted by existing studies (Caparrós, 2018; Yankowski, 2019).

Conclusion and Recommendations

This chapter contributes to the enhancement of understanding about the situation and experiences of women in the Philippine tourism industry. The accounts presented in this chapter, while should not be generalised, should be accounted for as they reveal the situation of women in the select tourism destinations. As suggested, there seems to be a gradual transition in the perception of women – from the patrilineal society of Palawan and the well-revered status of women in Cebu to the present treatment and expectations of women. Due to the financial challenges facing Filipino households, women were forced to take part in economic activities, including tourism. Based on the accounts, women's participation in tourism opened several opportunities that contribute to their psychological, economic, and political empowerment. The accounts illustrate how women were able to develop their skills and capacities, self-confidence, and positive self-perception through their involvement in tourism. Women's exposure to other people paved the way for their work to be recognised, their increased self-esteem, and the improvement of their economic conditions. Similarly, the accounts exemplified women's capacity to participate and influence decision-making processes within their communities. However, the same accounts also revealed the continuous struggle women face in enhancing their voice and status within their own households. As earlier pointed out by Vizcaino-Suárez (2018), women's interaction within their household is a crucial aspect of their ability to negotiate their roles – redistributing and sharing domestic work with other members of the household, and acceptance of their participation in economic activities.

The ability of women to contribute to community development has also been proven through these accounts. Empowered women can empower other women to take up space in communities and societies. The accounts revealed that part and parcel of women's motivation to participate in the industry is their ability to facilitate development within their communities and in the industry. If and when women are given the chance to take the lead, they will take this as an opportunity to improve themselves and the members of their community. In this same perspective, women empowerment is redefined to describe the ability of women to affect change that contributes to community development.

The movement towards women empowerment in the Philippines has come a long way, yet still offers some rooms for improvement. For women to act as agents facilitating community development in tourism destinations, the following suggestions are recommended: first, women's capacities and skills need to be developed by recognising that in doing so, they can better contribute to economic activities; second, women's ability to lead in both the public and private sectors should be recognised and promoted – this is a way to ensure that women's voices are brought to the forefront of decision-making processes; and finally, men should be encouraged and acknowledged as partners in pursuing gender equality and women empowerment – their participation could fast track women's empowerment.

Declaration

This chapter is culled from the author's working research study titled: "Women Participation, Empowerment, and Community Development in Tourism Areas in the Philippines."

This work was supported by the Ritsumeikan Asia Pacific University, Beppu, Oita, Japan and the Ryoichi Sasakawa Young Leaders Fellowship Fund (SYLFF) Association.

References

Albert, J. R. G., & Vizmanos, J. F. V. (2017). Do men and women in the Philippines have equal economic opportunities? In *PIDS policy notes no. 2017–09*. Philippine Institute for Development Studies.

Alcantara, A. N. (1994). Gender roles, fertility, and the status of married Filipino men and women. *Philippine Sociological Review*, 42(1/4), 94–109. https://www.jstor.org/stable/41853665

Angeles, L. C., & Hill, K. (2009). The gender dimension of the agrarian transition: Women, men and livelihood diversification in two peri-urban farming communities in the Philippines. *Gender, Place and Culture, 16*(5), 609–629. http://dx.doi.org/10.1080/09663690903148465

Arnaldo, M. S. F. (2021, March). Cebu's come-on for visitors: 70% cut in rates. *BusinessMirror.* https://businessmirror.com.ph/2021/03/17/cebus-come-on-for-visitors-70-cut-in-rates/

Ateljevic, I. (2008). *Women empowerment through tourism.* Wageningen University.

Bakas, F. E., Costa, C., Durão, M., Carvalho, I., & Breda, Z. (2018). 'An uneasy truth?': Female tourism managers and organizational gender equality measures in Portugal. In S. Cole (Ed.), *Gender equality and tourism: Beyond empowerment* (pp. 34–43). CABI United Kingdom.

Basuil, D. A., Lobo, K., & Faustino, C. M. (2020, June 29). The Philippines' gender-blind COVID-19 response. *East Asia Forum.* https://www.eastasiaforum.org/2020/06/29/the-philippines-gender-blind-covid-19-response/

Bengzon, B. C. (2021, June 09). *Philippine tourism updates* [PowerPoint slides]. Department of Tourism. https://drive.google.com/file/d/1QcZVWpq-eWZZzRYAghF2M5pQ0Ips0QLW/view

Boley, B. B., & McGehee, N. G. (2014). Measuring empowerment: Developing and validating the resident empowerment through tourism scale (RETS). *Tourism Management, 45,* 85–94. https://doi.org/10.1016/j.tourman.2014.04.003

Bras, K., & Dahles, H. (1998). Women entrepreneurs and beach tourism in Sanur, Bali: Gender, employment opportunities, and government policy. *Pacific Tourism Review, 1*(3), 243–256. https://www.ingentaconnect.com/contentone/cog/ptr/1998/00000001/00000003/art00007

Briedenhann, J., & Ramchander, P. (2006). Township tourism: Blessing or blight? The case of Soweto in South Africa. In E. Smith & M. Robinson (Eds.), *Cultural tourism in a changing world: Politics, participation and (re) presentation* (pp. 99–102). Channel View.

Cabegin, E. C. A., & Gaddi, R. S. (2019). Determinants of female labor force participation in the Philippines. *National Economic and Development Authority.* https://neda.gov.ph/wp-content/uploads/2021/09/Determinants-of-Female-Labor-Force-Participation-in-the-Philippines.pdf

Caparrós, B. M. (2018). Trekking to women's empowerment: A case study of a female-operated travel company in Ladakh. In S. Cole (Ed.), *Gender equality and tourism: Beyond empowerment* (pp. 57–66). CABI United Kingdom.

Chant, S. (1996). Women's roles in recession and economic restructuring in Mexico and the Philippines. In T. Sinclair (Ed.), *Gender, work and tourism* (pp. 297–327). Routledge.

Cole, S. (2006). Cultural tourism, community participation and empowerment. In K. Smith & M. Robinson (Eds.), *Cultural tourism in a changing world.* Channel View Publications.

Cole, S. (2018). *Gender equality and tourism: Beyond empowerment.* CABI.

Dasig, S. M. M. (2020). Difficult but fulfilling: Women's lived experiences as leaders in fisherfolk organizations in Bolinao, Philippines. *Gender, Technology and Development, 24*(1), 10–27. https://doi.org/10.1080/09718524.2020.1728158

David, C. C., Albert, J. R. G., & Vizmanos, J. F. V. (2017). Filipino women in leadership: Government and industry. In *PIDS policy notes no. 2017–22.* Philippine Institute for Development Studies.

Duffy, L. N., Kline, C. S., Mowatt, R. A., & Chancellor, H. C. (2015). Women in tourism: Shifting gender ideology in the DR. *Annals of Tourism Research, 52,* 72–86. https://doi.org/10.1016/j.annals.2015.02.017

Dulnuan, J. R., & Mondiguing, R. P. (2000). *Gender and Tourism in the Cordillera: A study on how tourism affects the women and men of Banaue, Ifugao* [Master's thesis, University of the Philippines].

Dunn, S. (2007). *Toward empowerment: Women and community-based tourism in Thailand.* University of Oregon.

Eder, J. F. (2006). Gender relations and household economic planning in the rural Philippines. *Journal of Southeast Asian Studies, 37*(3), 397–413. https://doi.org/10.1017/S0022463406000701

Elshaer, I., Moustafa, M., Sobaih, A. E., Aliedan, M., & Azazz, A. M. (2021). The impact of women's empowerment on sustainable tourism development: Mediating role of tourism involvement. *Tourism Management Perspectives, 38,* 100815. https://doi.org/10.1016/j.tmp.2021.100815

Erram, M. M. B. (2022, December 07). South Korea remains Cebu's top market for foreign arrivals. *Cebu Daily News.* https://cebudailynews.inquirer.net/478836/south-korea-remains-cebus-top-market-for-foreign-arrivals

Fabro, K. A. (2020, April 30). No tourism income, but this Philippine community still guards its environment. *Mongabay*. https://news.mongabay.com/2020/04/no-tourism-income-but-this-philippine-community-still-guards-its-environment/

Feng, X. (2013). Women's work, men's work: Gender and tourism among the Miao in rural China. *Anthropology of Work Review, 34*(1), 2–14. https://doi.org/10.1111/awr.12002

Ferguson, L. (2011). Promoting gender equality and empowering women? Tourism and the third millennium development goal. *Current Issues in Tourism, 14*(3), 235–249. https://doi.org/10.1080/13683500.2011.555522

Floro, M. S. (1995). Economic restructuring, gender and the allocation of time. *World Development, 23*(11), 1913–1929. https://doi.org/10.1016/0305-750X(95)00092-Q

Foley, C., Grabowski, S., Small, J., & Wearing, S. (2018). Women of the Kokoda: From poverty to empowerment in sustainable tourism development. *Tourism Culture & Communication, 18*(1), 21–34. https://doi.org/10.3727/109830418X15180180585158

Formoso, C. A. (2022, April). Palawan records over 40k tourist arrivals from January to April. *Palawan News*. https://palawan-news.com/palawan-records-over-40k-tourist-arrivals-from-januaryapril/#:~:text=Palawan%20has%20registered%20a%20total,of%20travelers%20provided%20by%20Atty

Fruman, C., & Twining-Ward, L. (2017, October 23). Empowering women through tourism. *World Bank Blogs*. https://blogs.worldbank.org/psd/empowering-women-through-tourism-0

Gabriel, A. G., De Vera, M., & B. Antonio, M. A. (2020). Roles of indigenous women in forest conservation: A comparative analysis of two indigenous communities in the Philippines. *Cogent Social Sciences, 6*(1), 1720564. https://doi.org/10.1080/23311886.2020.1720564

Garcia, T. M. R. (2020). *Violence against women in the Philippines* [Master's thesis, Norwegian University of Life Sciences]. https://nmbu.brage.unit.no/nmbu-xmlui/handle/11250/2678663

Global Network of Women Peacebuilders (GNWP). (2022). *COVID-19 and women, peace and security database*. https://gnwp.org/resources/covid-19-wps-database/

Graziano, K., Pollnac, R., & Christie, P. (2018). Wading past assumptions: Gender dimensions of climate change adaptation in coastal communities of the Philippines. *Ocean & Coastal Management, 162*, 24–33. https://doi.org/10.1016/j.ocecoaman.2018.01.029

Gumba, B. G. (2020). Entrepreneurial engagement of women in selected tourism areas in the Philippines. *International Journal of Scientific Research in Science and Technology, 7*(4), 349–359. https://doi.org/10.32628/IJSRST207455

Gutierrez, E. L. M., & Vafadari, K. (2022). Exploring the relationship between women's participation, empowerment, and community development in tourism: A literature review. *International Marketing Journal of Culture and Tourism, 2*, 39–68. https://doi.org/10.33001/18355/IMJCT0105

Hesse-Biber, S. (2013). *Feminist research practice: A primer* [online] (2nd ed.). SAGE.

Ishii, K. (2012). The impact of ethnic tourism on hill tribes in Thailand. *Annals of Tourism Research, 39*(1), 290–310. https://doi.org/10.1016/j.annals.2011.05.004

Jumawan-Dadang, R. (2015). Saving marine life: An empirical assessment of ecofeminist thought in coastal communities. *Philippine Sociological Review*, 61–83. https://www.jstor.org/stable/24717160

Kunjuraman, V., & Hussin, R. (2016). Women participation in ecotourism development: Are they empowered. *World Applied Sciences Journal, 34*(12), 1652–1658. DOI: 10.5829/idosi.wasj.2016.1652.1658

Lenao, M., & Basupi, B. (2016). Ecotourism development and female empowerment in Botswana: A review. *Tourism Management Perspectives, 18*, 51–58. DOI:10.1016/j.tmp.2015.12.021

Lontoc, G. (2020). Negotiating indigenous identities within mainstream community livelihoods: Stories of Aeta women in the Philippines. *Studies in the Education of Adults, 52*(2), 157–174. https://doi.org/10.1080/02660830.2020.1763099

Maligalig, R., Demont, M., Umberger, W. J., & Peralta, A. (2021). Understanding Filipino rice farmer preference heterogeneity for varietal trait improvements: A latent class analysis. *Journal of Agricultural Economics, 72*(1), 134–157. https://doi.org/10.1111/1477-9552.12392

Mathkar, M. (2019). Philippines leads Asia in gender equality. *The Hindu*. https://www.thehindu.com/news/international/philippines-leads-asia-in-gender-equality/article26351544.ece

McCall, C. E., & Mearns, K. F. (2021). Empowering women through community-based tourism in the Western cape, South Africa. *Tourism Review International, 25*(2–3), 157–171. https://doi.org/10.3727/154427221X16098837279967

Momsen, J., & Nakata, M. (2010). Gender and tourism: Gender, age and mountain tourism in Japan. In J. Mosedale (Ed.), *Political economy of tourism* (pp. 151–162). Routledge.

Moswete, N., & Lacey, G. (2015). "Women cannot lead": Empowering women through cultural tourism in Botswana. *Journal of Sustainable Tourism, 23*(4), 600–617. https://doi.org/10.1080/09669582.2014.986488

Movono, A., & Dahles, H. (2017). Female empowerment and tourism: A focus on businesses in a Fijian village. *Asia Pacific Journal of Tourism Research, 22*(6), 681–692. http://dx.doi.org/10.1080/10941665.2017.1308397

Nutsugbodo, R. Y., & Mensah, C. A. (2020). Benefits and barriers to women's participation in eco-tourism development within the Kakum Conservation Area (Ghana): Implications for community planning. *Community Development, 51*(5), 685–702. https://doi.org/10.1080/15575330.2020.1825977

Pajaron, M. (2016). Heterogeneity in the intrahousehold allocation of international remittances: Evidence from Philippine households. *Journal of Development Studies, 52*(6), 854–875. https://doi.org/10.1080/00220388.2015.1113261

Pascual, C. G. (2008). *Social and economic empowerment of women in the informal economy: Impact case study of Sikap Buhay.* ILO, Subregional Office for South-East Asia and the Pacific.

Pelayo, E. T. C., Zerrudo, E. B., & Ancheta, A. A. (2020). Framing vernacular memories of the women fisherfolks: A vanishing cultural heritage in Namayan Island, Philippines. *Journal of Nature Studies, 19*(1), 66–80. https://www.journalofnaturestudies.org/files/JNS19-1/66-80_Pelayo_Framing%20Vernacular%20Memories.pdf

Philippine Statistics Authority (PSA). (2019). *Philippine Tourism Satellite Accounts (PTSA) Report.* https://psa.gov.ph/sites/default/files/2019%20Philippine%20Tourism%20Satellite%20Accounts%20%28PTSA%29%20Report_1_0.pdf

Pleno, M. J. L. (2006). Ecotourism projects and women's empowerment: A case study in the province of Bohol, Philippines. *Forum of International Development Studies, 32,* 137–155. https://dl.ndl.go.jp/view/download/digidepo_8380218_po_08.pdf?contentNo=1&alternativeNo=

Pondorfer, A., Barsbai, T., & Schmidt, U. (2017). Gender differences in stereotypes of risk preferences: Experimental evidence from a matrilineal and a patrilineal society. *Management Science, 63*(10), 3268–3284. http://dx.doi.org/10.1287/mnsc.2016.2505

Prieto-Carolino, A., & Mamauag, B. L. (2019). Pagdipara: Caring work by poor elderly women in coastal communities in Iloilo, Philippines. *Asian Journal of Women's Studies, 25*(3), 375–395. https://doi.org/10.1080/12259276.2019.1646493

Quetulio-Navarra, M., Znidarsic, A., & Niehof, A. (2017). Gender perspective on the social networks of household heads and community leaders after involuntary resettlement. *Gender, Place & Culture, 24*(2), 225–246. https://doi.org/10.1080/0966369X.2016.1277185

Quisumbing, L. R. (1963). Characteristic features of Cebuano family life amidst a changing society. *Philippine Sociological Review, 11*(1/2), 135–141. https://www.jstor.org/stable/43596654

Ramalho, J. (2019). Empowerment in the era of resilience-building: Gendered participation in community-based (Disaster) risk management in the Philippines. *International Development Planning Review, 41*(2), 129–149. https://doi.org/10.3828/idpr.2018.25

Ramirez, P. J., Narvaez, T. A., & Santos-Ramirez, E. J. (2020). Gender-inclusive value chains: The case of seaweed farming in Zamboanga Peninsula, Philippines. *Gender, Technology and Development, 24*(1), 110–130. https://doi.org/10.1080/09718524.2020.1728810

Santos, A. P. (2022, August). [Dash of SAS] The Philippines' consistent top performance in the Global Gender Gap is ironic. *Rappler.* https://www.rappler.com/voices/thought-leaders/dash-of-sas-philippines-consistent-top-performance-global-gender-gapironic/#:~:text=Once%20again%2C%20the%20Philippines%20is,Global%20Gender%20Gap%20since%202006.

Scheyvens, R. (1999). Ecotourism and the empowerment of local communities. *Tourism Management, 20*(2), 245–249. https://doi.org/10.1016/S0261-5177(98)00069-7

Scheyvens, R. (2000). Promoting women's empowerment through involvement in ecotourism: Experiences from the Third World. *Journal of Sustainable Tourism, 8*(3), 232–249. https://doi.org/10.1080/09669580008667360

Tang, A. P. R., Ac-ac, M. A., Aragon, L. C., Mercado, J. M. T., & Sembrano, E. A. M. (2022). *Women of Laguna: Continuing old traditions.* DFA Foreign Service Institute.

The World Bank. (2022). *Women, Business, and the Law 2022*. https://openknowledge.worldbank. org/handle/10986/36945

Tran, L., & Walter, P. (2014). Ecotourism, gender and development in northern Vietnam. *Annals of Tourism Research, 44*, 116–130. https://doi.org/10.1016/j.annals.2013.09.005

Tucker, H. (2007). Undoing shame: Tourism and women's work in Turkey. *Journal of Tourism and Cultural Change, 5*(2), 87–105. https://doi.org/10.2167/jtcc089.0

United Nations World Tourism Organization (UNWTO). (2020). *Women's empowerment and tourism*. https://www.unwto.org/gender-and-tourism

Vizcaino-Suárez, P. V. (2018). Tourism as empowerment: Women artisan's experiences in Central Mexico. In S. Cole (Ed.), *Gender equality and tourism: Beyond empowerment* (p. 46). CABI.

Vukovic, D. B., Petrovic, M., Maiti, M., & Vujko, A. (2021). Tourism development, entrepreneurship and women's empowerment–Focus on Serbian countryside. *Journal of Tourism Futures*. https:// doi.org/10.1108/JTF-10-2020-0167

Yankowski, A. (2019). Salt making and pottery production: Community craft specialization in Albuquerque, Bohol, Philippines. *Ethnoarchaeology, 11*(2), 134–154. https://doi.org/10.1080/19442890. 2019.1642570

8

GENDERED UBUNTU

Exploring the Intersection of Ubuntu, Gender Equity, and Tourism Development in Africa

Ogechi Adeola and Albert Nsom Kimbu

Abstract

This chapter examines the impact of patriarchy and Ubuntu on leadership and tourism development in Africa. It underscores the importance of challenging patriarchal power structures and promoting gender equity in tourism development and broader contexts. The study also discusses the potential of Ubuntu philosophy in fostering gender equity and education in tourism. It posits that Ubuntu can provide a foundation for socially and environmentally sustainable tourism development that prioritises the well-being of local communities and the preservation of natural and cultural resources. Furthermore, the chapter explores how Ubuntu contributes to inclusivity and revitalising African traditions to advance gender equality and tourism development. It proposes that Ubuntu can promote inclusive tourism development by embracing African traditions and values that prioritise collective well-being and community progress.

Keywords

Gendered Ubuntu, gender equity, gender-based violence, tourism development, Africa

Introduction

In Africa, widespread cultural beliefs often limit women to domestic roles and discriminate against them in the areas of education, property ownership, and employment (Kimbu et al., 2021; Ribeiro et al., 2021). This patriarchal mentality perpetuates gender inequality and hinders the full participation of women in society, allowing for male dominance. The irrelevance often connected with women's social position has restrained many women from reaching their full potential. Ncube (2010) asserts that many women in Africa continue to experience conditions similar to those prior to independence. Even though there have been considerable legal reforms, their lifestyles have not significantly changed. Restrictions placed by colonialism, religion, culture, postwar regimes, and socio-economic factors persist, severely limiting women's rights in many African nations

DOI: 10.4324/9781003286721-12

(Ncube, 2010). The impact of colonialism on the African culture is still evident today, as it introduced Western capitalist ideals that promoted gender inequality, a concept that was not generally widespread in traditional African society (Ntshangase & Tlhakodisho, 2022). Thus, the colonial era altered the trajectory that post-colonialism is attempting to restore through Ubuntu. In clear terms, colonialism promoted social inequality through the introduction of new orientations that contradicted the ethos of equality which characterised Africa (Jaiyeola, 2020; Ncube, 2010). Ubuntu, on the other hand, promotes equality, togetherness, and oneness. The adoption of Ubuntu values is therefore suggested in this chapter as a mechanism that can drive equality in Africa and the development of the tourism sector.

Ubuntu expresses the African idea of what it means to be human, which is mirrored in collective personhood and ethics (Khoza, 2012; Mbigi, 1996; Ndlovu, 2016). Karsten and Illa (2005) state that Ubuntu is an African worldview that is established and embedded in people's everyday lives. It is best summed up by the adage that highlights that our individual identities are inseparable from our shared existence (Kim, 2007). African societies are recognised for their communitarian principle of Ubuntu, yet it is distressing to note that many women continue to suffer gender discrimination. Violations and socio-economic differences against girls and women reflect a society's patriarchal attitude, and closing the gender gap requires the united effort of everyone.

The Ubuntu concept is a guiding light in this direction. It recognises that each person has both legal and moral obligations to advance their social rights and promote gender justice (Magadla & Chitando, 2014). A person who practices Ubuntu is affirming of others and open to them. He or she does not perceive others' ability and goodness as a threat. A person's sense of belonging to a larger total gives them a healthy sense of self-assurance. The foundation of Ubuntu is the idea that everyone is important, irrespective of gender (Abraham & Prabha, 2022).

Additionally, Ubuntu philosophy offers a promising approach to addressing the gender inequalities that persist in many African societies and contribute to reducing violence and exploitation of women and girls in tourism destinations. As a communitarian principle, Ubuntu emphasises the importance of collective well-being and recognises the moral obligations of each person to advance the welfare of the community (Karsten & Illa, 2005; Ncube, 2010). This can provide a basis for community-led tourism development initiatives that empower and prioritise the well-being of women and girls.

Cultural beliefs and patriarchal attitudes continue to limit the potential of women, thereby perpetuating gender inequality in many African societies (Abraham & Prabha, 2022; Magadla & Chitando, 2014; Ndlovu, 2016), including tourism destinations. This hinders the full participation of women in society and makes them vulnerable to violence and exploitation. The Ubuntu philosophy can, therefore, offer a way to challenge these power structures and promote gender equity in tourism development and beyond.

The African worldview of Ubuntu offers an alternative framework for tourism development that prioritises community-led initiatives, promotes gender equity and education, and challenges patriarchal power structures. This chapter explores the potential of the Ubuntu philosophy for promoting gender equality and inclusivity in the tourism industry in Africa. Specifically, it examines the role of tourism in combating violence against women, the influence of patriarchy in tourism, and the potential of Ubuntu for promoting gender equity and education. Furthermore, the chapter considers Ubuntu as a means of reviving African traditions to promote inclusivity and gender equality in tourism.

Methodology

This chapter examines the potential of Ubuntu as a tool for addressing gender-based violence and promoting gender equity through tourism in Africa, drawing on evidence and examples from the existing literature published in Scopus indexed management journals. To provide a detailed examination, the chapter considers not only the gender equity debate but also other aspects of education and violence against women separately. A series of research review approaches are employed, incorporating keywords such as "Ubuntu philosophy," "tourism and ubuntu," "gender equity," "violence against women," "women and education," "Ubuntu and tourism," "women and Ubuntu," and "gender equity and tourism." Most of the research articles were examined in this chapter sourced from major publishers such as Elsevier, Emerald, Springer, Sage, Wiley, Taylor & Francis, among others. Additionally, we evaluated both country-level, regional and global (annual), reports examining the gender equity debate, such as those from the Mo Ibrahim Foundation, UN Women, World Tourism Organisation, and World Health Organization.

The Ubuntu Philosophy

The word "Ubuntu" originates from the Bantu Nguni languages spoken by the Zulu, Xhosa, Swati, and Ndebele (Ncube, 2010). Many traditional African societies have foundational principles similar to Ubuntu. It promotes a mindset oriented towards fostering compassion, nurturing community spirit, cultivating harmonious relationships and embracing reverence and receptivity (Mangaliso, 2001). Further attributes include the ability for empathy, reciprocity, and decency (Bekker, 2008). As described by Karsten and Illa (2005, p. 613), "Ubuntu expresses an African view of the life world anchored in its own person, culture and society, which is difficult to define in a Western context."

Ubuntu advocates for the value of humanity, promoting unwavering respect for others. It embodies a leadership concept that harmoniously balances the past, present, and future. By drawing lessons from the past, addressing urgent issues in the present, and providing a visionary outlook for the future, Ubuntu holds immense potential, as noted by Boele van Hensbroek (2001), to foster a more inclusive and accepting discourse towards non-Western cultures that have often been misunderstood and undervalued. Additionally, the principles of Ubuntu encompass a leadership ideology that prioritises connections and solidarity over material possessions. This ideology, highlighted by Malunga (2009), emphasises the significance of shared opportunities, responsibilities, and challenges. By embracing Ubuntu, leaders can forge stronger bonds and create a sense of collective purpose among individuals and communities. Consideration for all persons, regardless of background, collaboration and participative decision-making are central to Ubuntu. It expresses Africans' notion of what it is to be human, which is mirrored in the sense of "collective personhood and morality" (Khoza, 2012; Mbigi, 1996; Ncube, 2010; Ndlovu, 2016; Venter, 2004). As Tutu observed (1999), a person who embodies Ubuntu is congenial, giving, loving, empathetic, sensitive to the pain of others, as well as approving of the humanity of others. Therefore, with these qualities, particularly the emphasis on humanity, Ubuntu is a concept that should be incorporated in to leadership orientation and values across sectors in Africa.

Ubuntu and the Role of Tourism in Combating
Violence Against Women in Africa

In many African countries, cultural discrimination exists towards girls and women. This is exacerbated by governance and socio-economic challenges, which have been the bane of the African continent (Zengenene & Susanti, 2019). The term violence originated from the Latin word "violare," which means to violate (Bosha, 2018). The African Charter on Human and Peoples' Rights characterises violence against women as "any actions done against women that cause or might cause them bodily, sexual, mental, and economic harm, including the menace to commit such acts; or to execute the installation of unjustified limitations on or denial of basic rights in private or general life in peacetime and during times of armed conflict" (Deutsche Welle, 2022).

Gender-based violence encompasses a wide range of maltreatment, from sexual and female genital mutilation to abusive behaviour and exploitation (Manyonganise, 2017; Prescott & Madsen, 2011; Walby & Towers, 2017). Internationally, an estimated 736 million women aged 15 and older, or almost one-third (30%), have been victims of gender-based violence at some point in their lives, either by an abusive partner or by someone with whom they have not had an emotional bond (WHO, 2022). In Africa, where 36% of women have experienced violence, the issue is particularly severe (WHO, 2022). A 2020 survey indicated that over 44% of African women, or more than two in five, have experienced gender-based violence (Deutsche Welle, 2022). According to Manyonganise (2017), patriarchal views of women's struggles have also contributed to violence against women in Africa. Without doubt, violence and discrimination against women breach the ideals of equal opportunity and basic human rights (Zengenene & Susanti, 2019). At the national and international levels, laws and conventions are necessary.

In pre-colonial African communities, morality was seen as influencing every facet of a person's life. Ubuntu represents the African tendency to exhibit empathy, equality, honesty, peaceful cooperation, and humanism in order to create and uphold a just and caring community. According to Mbiti (1969), Ubuntu stresses connectedness, common humanity, and communal understanding. This emphasizes a focus on community.

Hailey (2008) discusses Archbishop's Tutu's interpretation of Ubuntu, noting that an individual embodying this philosophy is approachable, tolerant, and accessible to others, validating others' ability and goodness. The significance of Ubuntu may be observed in the way African communities, in principle, view humanity as being practised via the ethics of caring. Krog (2008) described this as being "interconnected-toward-wholeness."

Individualism has not only tainted the essence of Ubuntu but also eroded African ideals of recognising the other in the spirit of community life, and this death of community life, where the "we" has transformed into the "I" (Metz, 2014). In an Ubuntu setting, "the identity or subjectivity of the individual and the society are mutually constitutive and, consequently, none is superior" (Eze, 2008, p. 388). Hence, Ubuntu's value serves two purposes: first, it may be used in one-on-one relationships; and second, it can be implemented on an individual vs. collective level (Eze, 2008). However, the Ubuntu maxim "I am because we are; since we are, therefore I am" (Mbiti, 1969) runs the risk of an unbalanced togetherness that weakens the "I" as a result of imposed repressive cultures. Moreover, in view of the issue of gender and gender-based violence in Africa, the "I" (i.e. women) risks being annihilated and silenced by the oppressive "we" (perpetrators of violence, as well as social structures that encourage male dominance) (Metz, 2014). Going forward, it is critical to reconsider how Ubuntu is conceptualised to account for women's positions and address theoretical flaws (Sanni & Ofana, 2021).

Aspiration 6 of Agenda 2063 envisions "an Africa whose growth is people-driven, capitalising on the potential afforded by the African people, particularly its women and youth, and caring for children" (African Union, n.d.). One of the three aims of this aspiration is full gender equality in all aspects of life, with combating violence and discrimination against women and girls recognised as a priority. Another component of Agenda 2063's signature initiative, "Silencing the Guns," is the abolition of gender-based violence. Tourist destinations often create environments which expose women and girls to different forms of harassment, assault, and exploitation, making them vulnerable (Eger, 2021; Yasegnal, 2023). The commodification of women's bodies and sexuality in the tourism industry reinforces gender-based violence and perpetuates patriarchal power structures. However, in the Ubuntu philosophy, violence against women and girls is considered a violation of the community's interconnectedness and harmony. It emphasises the importance of recognising and valuing the contributions of all members of the community, including women and girls, and protecting their rights to safety and security. Ubuntu philosophy can provide an alternative framework for tourism as it prioritises the well-being of local communities and the prevention of violence against women and girls.

Community-led tourism development initiatives that prioritise the empowerment of women and girls can play a critical role in preventing violence against women and girls. These initiatives can include training and empowerment programmes that provide women and girls with the requisite skills and knowledge (Afenyo-Agbe & Adeola, 2020; World Tourism Organization, 2020). A recent World Tourism Organization (2023) study on gender in national tourism strategies of 77 countries revealed the absence of clearly defined and targeted gender policies in many of them. The report highlighted the need for governments to start designing national tourism policies and strategies that move away from being gender blind or simply being gender aware to policies that mainstream gender and are gender transformative. This is relevant to Africa. One of the means of doing this could be through drawing on the Ubuntu philosophy to develop tourism initiatives, involving local communities at all stages of the decision-making processes from the planning through to delivery. This will create a sense of ownership and accountability within the community and promote sustainable tourism development that prioritises the well-being of all members of the community.

Patriarchy and Ubuntu: Exploring Their Influence on Leadership and Tourism in Africa

Globally, women only hold 25.7% of the possible legislative roles, 7.2% of head-of-state seats, 6.2% of head-of-government roles, and 21.3% of cabinet posts (Brookings, 2022). About 24% of parliamentary seats worldwide are held by women, and this percentage is very similar in Africa (Brookings, 2022).

Patriarchy is a well-established feature in African society, despite countries in the continent having some of the world's most progressive constitutions. The term patriarchy refers to a societal structure where men and women are socially designated and treated unequally in political, social, and economic interactions and institutions (Nash, 2009). According to Krishnan et al. (2020), a patriarchal society is one in which men hold and maintain dominance in all significant aspects of life. Studies (e.g., DeKeseredy, 2020; Dobash & Dobash,

1979) have shown that patriarchy has two components. First, patriarchy, as a hierarchical system of social structures and interpersonal connections enables males to hold positions of privilege, authority, and dominance in society. Second, patriarchy justifies itself as a system of thought, implying that it offers strategies for fostering acceptance of subordination among those who gain from such behaviours, as well as among those who are positioned by society (DeKeseredy, 2021). According to Nash (2009), in a patriarchal society, the qualities considered to be "feminine" or associated with women are underestimated. Women are largely denied full involvement in political and economic life, and their rights are often violated as a result of this male-dominant ideology.

In the last two decades, there has been an increase in the number of occurrences that may be attributed to patriarchy in Africa. The state and the myriad of social institutions are male-dominant – clubs, sports, unions, professions, colleges, churches, companies, and armies. In contrast, the principles of Ubuntu, as noted earlier, are the foundation of a caring and communal society that believes in the contribution of all social actors for the benefit of the wider society. These principles include kindness, empathy, and respect (Luvalo, 2019).

In the 1920s, the term "Ubuntu" was coined by Inkatha, a "cultural movement," to restore respect for Zulu traditions (Bennett, 2011). Ubuntu is a non-racial, communal experience that values treating everyone with respect, an expression of African spirituality and moral philosophy. The Xhosa phrase *"umuntu ngumuntu ngabantu,"* which means, "I am because we are," perfectly captures it. According to Mbigi (1997), Ubuntu represents a deep sense of community and kinship that shapes our existence. As Broodryk (2006) explained, Ubuntu represents an ancient African worldview that is deeply rooted in values such as profound humanness, caring, sharing, and compassion. It is an ideology, a worldview, a philosophy, an ethic, a mentality, a culture, and a theology.

A core element of Ubuntu is equal respect for all stakeholders in the society, premised on the orientation of interdependence. Therefore, women and other minorities are perceived as valuable social actors with the potential to make significant contributions at multiple levels in the society.

Although the Ubuntu ideology is generally acclaimed, several criticisms exist. Bennett (2011), for example, contends that the notion lacks a precise definition, is hardly used in modern society, is not encapsulated in the constitution, its advantages may lead to hierarchies that undermine traditional leadership, and that Ubuntu denies an individual his or her independence. According to Keevy (2009), it is repressive of African women and personifies patriarchy. Similar studies further posit that Ubuntu has a gendered nature and that it is an ideology that may promote patriarchy in African cultures (Hall et al., 2013; Letseka, 2012; Manyonganise, 2015). Others state that Ubuntu is a two-edged sword that supports women on the one hand by arguing for ideals of dignity and equality while oppressing them on the other hand by sustaining male dominance and patriarchal norms (Chisale, 2018; Ngubane-Mokiwa, 2016). Conversely, Ntokozo (2016) asserts that the principle of "human dignity" is at the heart of Ubuntu and that any treatment that is "degrading" cannot be claimed to be Ubuntu. Furthermore, when the definition is considered, Ubuntu emphasises helpfulness, respect, justice, and fairness, among other things, which are contrary to the components of patriarchy (Makoba, 2016).

In reality, the prevalence of patriarchy can be attributed to the emphasis on respect within the concept of Ubuntu, which has historically been more focused on men (Amoah-Boampong & Agyeiwaa, 2019). This has resulted in the overshadowing of the other ideals of equity, which advocate for equal opportunities for both men and women. It is important to note that Ubuntu itself does not promote gender suppression or subjugation, but the disproportionate emphasis on respect has contributed to the prominence of patriarchy (Amoah-Boampong & Agyeiwaa, 2019).

To address the existing gender inequalities, foster inclusion, and challenge the status quo, tourism organisations need to adopt an Ubuntu management philosophy that incorporates all its essential elements. By embracing an Ubuntu management philosophy, these organisations can integrate practices that align with the values of the African culture. This can be achieved by adopting and implementing policies that ensure equal opportunities for both men and women. Additionally, it is important to foster a culture of respect that acknowledges and values the contributions of all individuals. This holistic approach to management draws from the rich cultural heritage of Africa and would address the historical imbalances caused by the overemphasis on respect within Ubuntu. Embracing an Ubuntu management philosophy in tourism organisations would make it possible to challenge patriarchal structures. This will promote the creation of a more balanced and empowering environment for both men and women, young and old. According to Letseka (2013), Ubuntu enables communities to maintain healthy connections with God, their ancestors, themselves, and the rest of the cosmos. The author contends that "young people who are initiated into ubuntu morality have the potential to become citizens that are inclined to treating others with justice and fairness at all times" (Letseka, 2013, p. 351). The virtue of caring (which is also central to the tourism sector) is influenced by the dignity etched in humans, which emphasises Ubuntu as a philosophy of care and support. As pointed out by Mbiti (1969), "I am because we are; and since we are, therefore I am." Moreover, since Ubuntu is a social function rather than an individual one, African epistemologies' moral values, which promote generosity, harmony, and the treatment of others with respect, are entwined with the practice of caregiving, which can support the weakening of patriarchy in African society (Letseka, 2013).

Tourism can either reinforce or challenge patriarchy. In many tourist destinations, women are relegated to low-wage jobs, while men occupy positions of power and authority (Afenyo-Agbe & Adeola, 2020 ; World Tourism Organization, 2019). This gendered division of labour reinforces patriarchy and the idea that men are better suited to positions of leadership. However, tourism can also provide opportunities for women to challenge patriarchy and gain economic independence. For example, women who operate homestays or small businesses in tourism can gain financial autonomy and challenge gendered power dynamics in their communities (Kimbu & Ngoasong, 2016; Kimbu et al., 2021).

Zhang et al. (2020) and Kimbu et al. (2021) suggested that there is a need to deconstruct tourism in Africa from an Anglo-Western orientation and adopt philosophies that fit the African context. Furthermore, the authors found that women can be involved in leadership through an entrepreneurial journey that empowers them. Incorporating gendered policies and practices like Ubuntu in the tourism industry can promote women's leadership and enable them to contribute to improving tourism performance (Ribeiro et al., 2021). The tourism industry holds many benefits for women through tourism entrepreneurship. Incorporating the ideologies of Ubuntu can foster social and collaborative capital required to promote women's entrepreneurial journeys and empowerment in Africa (Kimbu & Ngoasong, 2016; Kimbu et al., 2019).

Despite some criticisms, the Ubuntu ideology can challenge patriarchal power structures by promoting gender equity and recognising women's contributions to tourism development. By implementing an Ubuntu management philosophy, opportunities will abound for women to aspire and seek equal rights in leadership and other areas. The tourism industry holds significant potential to drive women's empowerment in Africa. Incorporating the fundamental principles of Ubuntu as an integral component of the organisational culture would enhance women's participation in decision-making and promotion to leadership positions within organisations. This gradual and systematic approach would contribute to the dismantling of patriarchy not only within the tourism industry but also in society at large.

Ubuntu Philosophy and Its Potential for Promoting Gender Equity in Tourism

In pre-colonial times, African societies were structured by socially driven relations, egalitarian values, balanced gendered roles, and shared authority (Mikidady, 2022; Ntshangase & Tlhakodisho, 2022). Women actively participated in economic, social, and political spheres. Their roles were dynamic and inclusive across various social and economic engagements (Amoah-Boampong & Agyeiwaa, 2019). However, the colonial era and the introduction of religion diminished women's active participation, limiting their involvement primarily to domestic activities. To achieve its objective of producing a new African civilisation, the colonial system utilised all three forms of education: informal, non-formal, and formal education.

Gender disparity in basic school enrolment and lopsided post-secondary distributions are some of the persisting challenges of education in Africa. The goal of implementing a gender-equity strategy underpinned by Ubuntu is to foster capacity building at all levels of society (Assié-Lumumba, 2016). To mitigate gender inequality, a new educational philosophy that is fundamentally inclusive is required. Promoting the principles of Ubuntu in Africa's social institutions, particularly in education, would serve to reinstate egalitarianism. Egalitarian systems provide equal opportunities for all stakeholders, without elevating any gender as superior. The SDGs and the Africa Union's Agenda 2063 both express the importance of interconnectedness and mutuality to the social and larger natural ecosystem (Assié-lumumba, 2018). According to Muxe Nkondo (2007), Africa's educational system needs to be more successful in helping to reorganise the political, social, and economic structures. The author argues that what is required is that education defines a process for creating an Ubuntu social disposition and educational concepts that embrace all students, regardless of their background or gender. Therefore, the educational system must produce lifelong learners who will act in society's best interests rooted in democratic values, equality, human dignity, life, and justice (Letseka, 2012). Ubuntu is a way of being that reinforces identity and the search for meaning as a group effort rather than an individual one. It captures the basic ideals of interdependence and humanity (Letseka, 2000) and argues for all members of society (i.e., individuals and groups) to have equal access to the empowering and emancipating capacity of education for their personal rights and the overall interests of all.

Tourism has the potential to promote gender equity and education by providing opportunities for women and girls to participate in the industry and gain new skills and knowledge (Afenyo-Agbe & Adeola, 2020; World Tourism Organization, 2019). Ubuntu philosophy can provide an alternative framework for tourism that challenges the barriers and promotes gender equity and education through developing inclusive community-led tourism development initiatives that prioritise the empowerment of women and girls (World Tourism Organization, 2019, 2020). Having equal access to education and training (e.g., in areas such as hospitality, marketing, and entrepreneurship) will provide women and girls with the skills and knowledge they need to compete, participate in decision-making processes, operate and succeed in their own businesses, and succeed in the industry, thereby contributing to the development of their communities.

Ubuntu and Inclusivity: Reviving African Traditions for Gender Equality

Traditional cultural practices often reflect the values and ideas that have been passed down through generations within a community. Every society has its own unique traditions and beliefs, some of which are advantageous, and others perceived as detrimental to some groups or minorities (Olatunji, 2013). Since the colonial era, African cultures have evolved significantly (Singha & Kanna, 2022), particularly with the impact of colonialism on values such as collectivism and egalitarianism. African cultures were traditionally collectivist and preoccupied with social well-being, in contrast to Western cultures, which tend to be more individualistic. Post-colonisation, African cultures have become a hybrid of both traditional and foreign characteristics.

Gender discrimination is often occasioned by patriarchal socio-economic structures (Singha & Kanna, 2022), and the denial of equity in the legal, educational systems and other public institutions. Whilst some scholars have advocated for the socio-economic empowerment of women, others have stated that affirmative action and a move away from a culture that supports masculinity are necessary (Olatunji, 2013). The revival of traditions, and cultures, is an essential aspect of tourism product/service development and identity construction at national and local levels. However, this revival must be grounded in a way of life that values civil rights, freedom, and "equality of voice" for all, as espoused in Ubuntu (Shanyanana & Waghid, 2016). Thus, it is necessary to strike a balance between the revival of cultures and traditions within the mutually agreed upon social and legal structures. The exclusion of women is opposed to the spirit of Ubuntu, and as such, it is vital to instil an attitude of acceptance and respect for gender (Nicolaides, 2015). This is even more relevant as evidence suggests the existence of barriers that mitigate efforts towards the inclusion of underprivileged groups, such as women in Africa (Ilesanmi, 2018; Mutume, 2005; Saluja et al., 2023). The concept of "equality of voice" may offer a deconstructed approach to advancing substantive inclusion (White, 2022).

Gender equity is a core value of the Ubuntu philosophy, as it emphasises the importance of recognising and valuing the contributions of all members of the community, irrespective of their gender. In the context of tourism, this means that women should have equal opportunities to participate in the industry and to benefit from its economic and social impacts. Unfortunately, many women in this sector do not fully enjoy this privilege, as previously mentioned. This disparity reinforces patriarchal power structures and undermines efforts to promote gender equity in the tourism industry. To dismantle patriarchal ideologies, practices, and systems in Africa, it is imperative to promote the Ubuntu culture across all organisations and social institutions. By embracing Ubuntu, values such as fairness, equality, and collectivism can be fostered. African nations should prioritise the adoption of Ubuntu as the foundation of society, as doing so will facilitate increased participation of women and other marginalised groups. A cultural shift that incorporates Ubuntu principles within religious settings, education, economic environments, community, and family structures is crucial. This cultural reset will challenge the perception that women's capabilities should be limited to domestic responsibilities alone.

Gender equity can be promoted through community-led tourism development initiatives that prioritise the empowerment of women. This can include initiatives such as training and education programmes that provide women with the skills and knowledge they need to thrive in the industry, as well as initiatives that support women-owned businesses and promote participation in decision-making processes. According to Ramose (2002), the concept of Ubuntu denotes a higher judgement of the community than that of the individual. A person belongs to the community and expresses their uniqueness via other members of the community. This exhibits a culture of reciprocal relationships, sharing and caring for each other. Women should be given the opportunity to fully engage in the conceptualisation, design, production, and marketing of tourism products and services. By emphasising the importance of interconnectedness and community, Ubuntu can also be a base for tourism that focuses on the well-being of communities and the preservation of natural and cultural resources. This can be achieved through initiatives which engage local communities in the planning, decision-making, and execution processes.

However, institutions may resist the adoption of Ubuntu as a cultural and managerial system. Hence, indigenous scholars must continue to raise awareness about the benefits of Ubuntu for social and economic development in Africa. This should encompass both traditional and digital media to enlighten and encourage members of society to embrace Ubuntu as a way of life in both private and public settings. Emphasis should be placed on highlighting the potential positive impact of Ubuntu in social settings and the progress it can drive not only in tourism organisations but also in the wider community. For example, implementing an Ubuntu-led approach to public administration, adopting Ubuntu principles in the tourism industry, and incorporating Ubuntu educational methods, rather than sole reliance on Western curriculum, can steer Africa towards prosperity. Researchers should further explore areas where Ubuntu can contribute to social and economic development, specifically focusing on eradicating inequality and promoting equal opportunities for all. In collaboration with researchers, practitioners must integrate Ubuntu ideology into their operational systems and organisational cultures. The ideology should be embraced by all not only because it is indigenous to Africa but also because it promotes fairness, equity, cooperation, and collaboration, thereby shifting the focus from competition. With these strategies, development will be catalysed from within, ensuring that no stakeholders are left behind in the progress of Africa.

Recommendations

As an African philosophy, Ubuntu has gained traction in the literature but not so much in practice, particularly in terms of its adoption in resolving social challenges. This should be a cause for concern because the values of Ubuntu would be left unrealised if they are not leveraged in different socio-cultural contexts. One of the important areas that this ideology could be effective is its operationalisation in addressing gender disparities and discrimination in African countries. Women continue to experience social and economic limitations that hinder them from maximising their potential to be agents of innovation and development. In sectors such as education, health, and tourism that are believed to be female-driven, the low involvement of women in decision-making and their poor representation in leadership positions necessitate that stakeholders re-examine the status quo. This chapter provides the following recommendations for researchers and practitioners:

a Public and private sector organisations are encouraged to adopt an Ubuntu management system that is equality, integrity, and service-driven. Adopting this management approach will foster the establishment of equitable leadership opportunities and frameworks that are suitable to the African tourism sector and the continent in general.

b Furthermore, Ubuntu can be applied to promote gender equity in Africa through political participation and community building, as women in Africa often face barriers to political participation, such as lack of representation and cultural bias. Strong and inclusive communities are essential for promoting gender equity, and Ubuntu has the potential to encourage the building of networks and the sharing of resources to promote community empowerment. This can be done through initiatives that promote women's representation in governance, broad-based decision-making, as well as programmes that raise awareness about the importance of women's participation in leadership and the importance of gender equity.

c Ubuntu is a philosophy that stresses the interconnectedness and interdependence of all people (Amugongo et al., 2023). This philosophy can be applied to promote equal opportunities and rights for men and women across sectors who often face barriers to economic participation, such as lack of access to credit and limited opportunities for entrepreneurship. Ubuntu encourages the sharing of resources and the building of networks to promote economic empowerment for all. This can be done, for instance, through initiatives that provide women with access to credit and business training, as well as programmes that promote women's entrepreneurship.

d African governments have a significant role to play in the adoption of Ubuntu as a social practice that could reduce gender inequalities. By adopting the Ubuntu philosophy in the higher echelons of government, it would be easier for other institutions to adopt the practice. One of the ways in which Ubuntu can be applied to promote gender equity in Africa is through education. Education is a powerful tool for empowering women and girls, and Ubuntu encourages the sharing of knowledge to promote the education and empowerment of all people. This can be done through programmes that provide girls with access to education and other resources, such as scholarships and mentorship programmes, as well as through initiatives that promote gender equality in the classroom. Furthermore, the academic programmes at both high schools and colleges need to include core elements of Ubuntu. This will help young people to inculcate the ideals of Ubuntu early in life.

e Ubuntu philosophy offers an alternative framework for tourism development that prioritises community-led initiatives, promotes gender equity and education, and challenges patriarchal power structures in Africa. Community-led tourism development initiatives that empower women and girls can play a critical role in preventing violence against them in African tourism destinations. By emphasising the importance of gender equity as a core value, Ubuntu provides a basis for socially and environmentally sustainable tourism development that prioritises the well-being of local communities and the preservation of natural and cultural resources in the continent.

f Tourism remains an important avenue for women to rise to leadership positions in Africa. The ministries of tourism and other government tourism-related agencies need to implement policies that are Ubuntu-driven. Specifically, the Ubuntu management approach should be first tested in the tourism industry to see the outcomes it would provide for women and the sector at large. Government and private sector operators in the tourism sector need to realise that empowering women and placing them in strategic leadership positions, where they would have significant influence, would increase their confidence and efficiency, yielding greater success for the tourism sector and the development of Africa.

Conclusion

This chapter explored the integration of the Ubuntu philosophy for promoting gender equality and inclusion, specifically in Africa's tourism industry. The ideals of the Ubuntu philosophy, which include equity, compassion, care, collaboration, team spirit, sympathy, dignity, and respect for nature, are critical in driving a sustainable tourism sector and beneficial in addressing gender disparities in Africa. Ubuntu enables the creation of equitable societies where girls and women are accepted and provided with the requisite opportunities and resources. Families, local communities, as well as the government and civil society organisations, bear responsibility for transmitting the ideals of Ubuntu.

Most of the literature and research on Ubuntu have been conducted in African settings. Future research on women's leadership in tourism and related sectors could seek to unpack how this philosophy can be implemented in non-African contexts. This is because Ubuntu places strong emphasis on participation, democracy, and communalism. More research is also needed to understand how practitioners can adapt the Ubuntu management perspective discussed in this chapter to promote a system that engenders and offers equal opportunities. Furthermore, research on Ubuntu can adopt an empirical approach in investigating its implications on the empowerment, effectiveness, performance, and progression of women in their work-related roles. The discourse on Ubuntu has to evolve from theory to praxis for it to gain the attention it deserves. Finally, it is imperative for researchers to continue to investigate the socio-economic and political benefits of Ubuntu for the common good. It is crucial to enlighten practitioners and other stakeholders on the benefits that Ubuntu holds in diverse settings and the need to integrate its core values into business, management, and administrative processes.

References

Afenyo-Agbe, E. A., & Adeola, O. (2020). Promoting gender equality and women's empowerment through tourism in Africa: towards agenda 2030. Empowering African Women for Sustainable Development: Toward Achieving the United Nations' 2030 Goals, 121–132.

African women and girls: Leading a continent. https://www.brookings.edu/essay/african-women-and-girls-leading-a-continent/. Accessed on 15th Sept, 2022.

African Union. (n.d). *Our Aspirations for the Africa We Want.* Available online at https://au.int/en/agenda2063/aspirations.

Amoah-Boampong, C., & Agyeiwaa, C. (2019). Women in pre-colonial Africa: West Africa. In O. Yacob-Haliso & T. Falola (Eds.), *The Palgrave handbook of African women's studies.* Palgrave Macmillan. https://doi.org/10.1007/978-3-319-77030-7_126-1

Amugongo, L. M., Bidwell, N. J., & Corrigan, C. C. (2023). Invigorating Ubuntu ethics in AI for healthcare: Enabling equitable care. In *583-592 FAccT '23: Proceedings of the 2023 ACM Conference on Fairness, Accountability, and Transparency.* https://doi.org/10.1145/3593013.3594024

Assié-Lumumba, N. (2016). Evolving African attitudes to European education: Resistance, pervert effects of the single system paradox, and the ubuntu framework for renewal. *International Review of Education, 62*(1), 11–27.

Bekker, C. J. (2008). Ubuntu, kenosis and mutuality: Locating the "other" in Southern African business leadership. *Global Business Review, 1*(2), 18–21.

Bennett, T. W. (2011). Ubuntu: An African equity. *Potchefstroom Electronic Law Journal*, 14(4), 30–61.

Boele van Hensbroek, P. (2001). *African renaissance and Ubuntu philosophy.* https://scholarlypublications.universiteitleiden.nl/access/item%3A2716810/view

Bosha, S. L. (2018). The importance of gender equality and women's inclusion for resolving conflict and sustaining peace. In William D. Joris L, Richard P (Eds), *Just Security in an Undergoverned World*, pp. 0–117, Oxford University Press, 1st Edition.

Broodryk, J. (2006). The philosophy of Ubuntu: Some management guidelines. *Management Today*, 22(7), 52–55.

Brookings. (2022). *African women and girls: Leading a continent.* https://www.brookings.edu/articles/african-women-and-girls-leading-a-continent/. Accessed on 12th Jul, 2023.

Chisale, S. S. (2018). Ubuntu as care: Deconstructing the gendered Ubuntu. *Verbum et Ecclesia*, 39(1), 1–8.

DeKeseredy, W. S. (2020). *Woman abuse in rural places.* Routledge.

Deutsche Welle, 2022: Media repost 2022 (DW). https://www.dw.com/en/global-media-forum-2022/s-60549163https://www.dw.com/en/global-media-forum-2022/s-60549163. Accessed on 15th Sept, 2022.

Dobash, R. E., & Dobash, R. (1979). *Violence against wives: A case against the patriarchy* (pp. 179–206). Free Press.

Eze, M. O. (2008). What is African communitarianism? Against consensus as a regulative ideal. *South African Journal of Philosophy= Suid-Afrikaanse Tydskrif vir Wysbegeerte*, 27(4), 386–399.

Hailey, J. (2008). *Ubuntu: A literature review.* Document. London: Tutu Foundation.

Hall, D., Du Toit, L., & Louw, D. (2013, February). Feminist ethics of care and Ubuntu. In *Obstetrics and gynaecology forum* (Vol. 23, No. 1, pp. 29–33). In House Publications.

Jaiyeola, E. O. (2020). Patriarchy and colonization: The "brooder house" for gender inequality in Nigeria. *Journal of Research on Women and Gender*, 10, 3–22.

Karsten, L., & Illa, H. (2005). Ubuntu as a key African management concept: Contextual background and practical insights for knowledge application. *Journal of Managerial Psychology*, 20(7), 607–620.

Keevy, I. (2009). Ubuntu versus the core values of the South African constitution. *Journal for Juridical Science*, 34(2), 19–58.

Khoza, R. (2012). *Attuned leadership: African humanism as compass.* Penguin Random House South Africa.

Kim, Y. Y. (2007). Ideology, identity, and intercultural communication: An analysis of differing academic conceptions of cultural identity. Journal of intercultural communication research, 36(3), 237–253.

Kimbu, A. N., de Jong, A., Adam, I., Ribeiro, A. M., Adeola, O., Afenyo-Agbe, E., & Figueroa-Domecq, C. (2021). Recontextualising gender in entrepreneurial leadership. *Annals of Tourism Research*, 88C.

Kimbu, A. N., & Ngoasong, M. Z. (2016). Women as vectors of social entrepreneurship. *Annals of Tourism Research*, 60, 63–79.

Kimbu, A. N., Ngoasong, M. Z., Adeola, O., & Afenyo-Agbe, E. (2019). Collaborative networks for sustainable human capital management in women's tourism entrepreneurship: The role of tourism policy. *Tourism Planning & Development*, 16(2), 161–178.

Krishnan, G. G., English, I. M., Campus, A., & Arjun, I. A. (2020). Men in a patriarchal society and issues. *Technology*, 11(11), 511–515.

Krog, A. (2008). 'This thing called reconciliation…'-forgiveness as part of an interconnectedness-towards-wholeness. *South African Journal of Philosophy= Suid-Afrikaanse Tydskrif vir Wysbegeerte*, 27(4), 353–366. Tutu Foundation.

Letseka, M. (2000). African philosophy and educational discourse. *African Voices in Education*, 23(2), 179–193.

Letseka, M. (2012). In defence of Ubuntu. *Studies in Philosophy and Education*, 31(1), 47–60.

Luvalo, L. M. (2019). Patriarchy and Ubuntu philosophy: The views of community elders in the eastern cape province. *e-BANGI*, 16, 1–10.

Magadla, S., & Chitando, E. (2014). The self become God: Ubuntu and the 'scandal of manhood'. *Ubuntu: Curating the archive*, 176192.

Malunga, C. (2009). *Understanding organizational leadership through*. Ubuntu. Adonis & Abbey Publishers Ltd.

Mangaliso, M. P. (2001). Building competitive advantage from Ubuntu: Management lessons from South Africa. *Academy of Management Perspectives, 15*(3), 23–33.

Manyonganise, M. (2015). Oppressive and liberative: A Zimbabwean woman's reflections on ubuntu. *VERBUM et Ecclesia, 36*(2), 1–7.

Manyonganise, M. (2017). Invisibilising the victimised: Churches in Manicaland and women's experiences of political violence in national healing and reconciliation in Zimbabwe. *Journal for the Study of Religion, 30*(1), 110–136.

Mbigi, L. (1997). *Ubuntu: The African dream in management*. Knowledge Resources, Randburg.

Mbigi, L. (1996). Ubuntu: The spirit of African solidarity. *Enterprise, 62–63*.

Mbiti, J. S. (1969). *African religions and philosophy*. New York: Frederick A. Praeger.

Metz, T. (2014). Just the beginning for ubuntu: Reply to Matolino and Kwindingwi. *South African Journal of Philosophy, 33*(1), 65–72.

Mikidady, M. (2022). Gender inequality: An alien practice to African cultural settlement. *Universal Journal of History and Culture, 4*(1), 1–15.

Mutume, G. (2005). *African women battle for equality*. Available online at https://www.un.org/africarenewal/magazine/july-2005/african-women-battle-equality. Accessed on 12ᵗʰ Jul, 2023.

Muxe Nkondo, G. (2007). Ubuntu as public policy in South Africa: A conceptual framework. *International Journal of African Renaissance Studies, 2*(1), 88–100.

Nash, K. (2009). Between citizenship and human rights. *Sociology, 43*(6), 1067–1083.

Ncube, L. B. (2010). Ubuntu: A transformative leadership philosophy. *Journal of Leadership Studies, 4*(3), 77–82.

Ndlovu, P. M. (2016). *Discovering the spirit of ubuntu leadership*. Palgrave Macmillan. https://doi.org/10.1057/9781137526854

Ngubane-Mokiwa, S. A. (2016). Delivering open distance e-learning through ubuntu values. *Open Distance Learning (ODL)*, Nova Science Publishers, Inc (pp. 147–162) ISBN: 9781634854030.

Nicolaides, A. (2015). Gender equity, ethics and feminism: Assumptions of an African Ubuntu oriented society. *Journal of Social Sciences, 42*(3), 191–210.

Ntokozo, M. (2016). Ubuntu: A phantasmagoria in rural Kwazulu-Natal? *African Journal of History and Culture, 8*(5), 41–51.

Ntshangase, M. X., & Tlhakodisho, J. M. (2022). The history of gender inequality: Analysis of gender inequality as a colonial legacy in Africa. *African Journal of Gender, Society & Development, 11*(3), 185.

Olatunji, C. M. P. (2013). An argument for gender equality in Africa. *CLCWeb: Comparative Literature and Culture, 15*(1), 9.

Prescott, J. A., & Madsen, A. M. (2011). *Sexual violence in Africa's conflict zones*. https://www.africabib.org/rec.php?RID=334155959

Ribeiro, M. A., Adam, I., Kimbu, A. N., Afenyo-Agbe, E., Adeola, O., Figueroa-Domecq, C., & de Jong, A. (2021). Women entrepreneurship orientation, networks and firm performance in the tourism industry in resource-scarce contexts. *Tourism Management, 86*, 104343.

Saluja, O. B., Singh, P., & Kumar, H. (2023). Barriers and interventions on the way to empower women through financial inclusion: A 2 decades systematic review (2000–2020). *Humanities and Social Sciences Communications, 10*(1), 1–14.

Sanni, J. S., & Ofana, D. E. (2021). Recasting the ontological foundation of ubuntu: Addressing the problem of gender-based violence in South Africa. *South African Journal of Philosophy, 40*(4), 384–394.

Shanyanana, R. N., & Waghid, Y. (2016). *Reconceptualizing ubuntu as inclusion in African higher education: Towards equalization of voice*.

Singha, R., & Kanna, Y. S. (2022). Physical abuse in the absence of ubuntu. *Journal of International Women's Studies, 24*(4), 9.

Venter, E. (2004). The notion of ubuntu and communalism in African educational discourse. *Studies in Philosophy and Education, 23*, 149–160

Walby, S., & Towers, J. (2017). Measuring violence to end violence: Mainstreaming gender. *Journal of Gender-Based Violence, 1*(1), 11–31.

White, S. K. (2022). Agonism, democracy, and the moral equality of voice. *Political Theory, 50*(1), 59–85. https://doi.org/10.1177/0090591721993862

WHO. (2022). *Violence against women in Africa*. Available online at https://www.who.int/health-topics/violence-against-women. Accessed on 15th Sept, 2022.

World Tourism Organization. (2019). *Global report on women in tourism – Second edition*. UNWTO, Madrid. https://doi.org/10.18111/9789284420384

World Tourism Organization. (2020). *AlUla framework for inclusive community development through tourism*. UNWTO, Madrid. https://doi.org/10.18111/9789284422159

World Tourism Organization. (2023). *Snapshot of gender equality and women's empowerment in national tourism strategies*. UNWTO, Madrid. https://doi.org/10.18111/9789284424207

Zengenene, M., & Susanti, E. (2019). Violence against women and girls in Harare, Zimbabwe. *Journal of International Women's Studies*, 20(9), 83–93.

Zhang, C. X., Kimbu, N. A., Lin, P., & Ngoasong, M. Z. (2020). Guanxi influences on women intrapreneurship. *Tourism Management*, 81. https://doi.org/10.1016/j.tourman.2020.104137

PART IV

Practising Gender in Tourism II
Gendered Tourism Workforce

9

INDIAN FEMALE UNDERGRADUATE STUDENTS' PERCEPTION TOWARDS WORKING IN THE TOURISM AND HOSPITALITY INDUSTRY

Md. Tariqul Islam and Uma Pandey

Abstract

The tourism market in India is expanding rapidly. The tremendous growth in the Indian economy is the main reason for tourism development. The travel and tourism sector in India is one of the most lucrative sectors, generating significant foreign revenue. The travel and tourism industry around the globe directly supports a significant number of employment opportunities. Students enrol or join tourism and hospitality programmes to prepare for careers in the tourism and hospitality sector. The current research aims to identify the perceptions and attitudes of Indian female students towards working in the tourism and hospitality industry. The present study employed the qualitative method by conducting one-on-one in-depth interviews with Indian female students. The findings revealed that the perception of Indian female undergraduate students towards working in the hospitality and tourism industry is mainly related to the hospitality industry's social status, career prospects, work environment, and benefits in the hospitality industry. This study provides several contributions to the policymakers of the hospitality industry. The findings will assist stakeholders in the tourism industry in formulating the appropriate policies. The hospitality industry authorities can modify their policies and work environments based on the perceptions of Indian female students.

Keywords

Hospitality Industry, Hospitality and Tourism Employment, Indian Female Students, Undergraduate Female Students' Perception.

DOI: 10.4324/9781003286721-14

Introduction

Since the occurrence of the COVID-19 pandemic, the tourism and hotel industry has been struggling to pick up the pace (Kaushal & Srivastava, 2021). This economic sector has historically contributed significantly to expanding the global labour force before the advent of COVID-19 (Chang & Tse, 2015). More specifically, India's tourism industry created 87.5 million, accounting for 12.75% of total employment and contributed INR 194 billion (USD 2.36 billion) to the country's GDP before taxes in 2015 (WTTC, 2018). In addition, the industry grew by 3.2% in 2018, with USD 29.9 billion generated in foreign exchange earnings from 10.8 million international tourists visiting India. India's direct contribution to the global travel and tourism industry ranked eighth, with USD 108 billion (FICCI, 2020). However, the Indian tourist and hotel industries, like other nations, suffered greatly due to COVID-19. Many firms were forced to close and incurred losses. Industry stakeholders are working hard to rebuild their operations as the situation is improving. They are seeking solutions to get the industry back on track, and one of the tactics they are using is hiring skilled personnel.

As the industry focuses on recovery, understanding how the tourism and hospitality business can retain highly educated and competent individuals is crucial (Lee et al., 2019). The hospitality and allied services industry was considered one of India's fastest expanding sectors, accounting for 6.8% of the country's total GDP, 39 million jobs, and 8% of overall employment in 2019 (WTTC, 2020). While India's hospitality and tourism sectors were predicted to expand in the post-pandemic period, particularly with the planned extension of the e-visa system by the government, several workplace-related challenges must be addressed. Pittie (2023) reported that the potential contribution of the tourism industry to the country's economy in 2022 was recorded at INR 15.9 trillion (about USD 215 billion), a 1% rise from 2019. In addition, it is estimated that there will be about 39 million jobs in the travel and tourism sector in 2023, a growth of 8.3% over 2022 (Aithal et al., 2023). Tourism development relies heavily on accessible human resources, and there is a global problem of finding qualified workers, which has an impact on businesses across the tourism industry (Griffin et al., 2021). To thrive, businesses in the tourism and hospitality sector need a highly trained, passionate, and dedicated workforce. In India, women should be equally motivated and trained to work in the hospitality industry to address the lack of a skilled workforce (Rinaldi & Salerno, 2020).

In India, several hospitality institutions have been formed to train students for careers in the area, emphasising equal opportunities for students of all genders (Joshi & Gupta, 2021). Understanding these students' attitudes towards tourism and hospitality jobs is crucial for their future success as practitioners (Elhoushy, 2014). However, only a few studies have focused on the perceptions of Indian undergraduate students towards hospitality careers. Datta et al. (2013) studied the effect of internships on hospitality students in India and found that the students were less inclined to continue their careers in the hospitality industry after internships. According to some studies, hospitality graduates frequently choose professions in sectors other than hospitality (Qiu et al., 2017; Wong & Liu, 2010) and have no intention of working in the field (Amissah et al., 2020). Kumar et al. (2014) found that women are less likely than men to pursue a profession in the hospitality industry. Hence, the enrolment of female students is less in hospitality institutes. However, 54% of people who work in tourism are women, and UNWTO (2019) mentioned that women should be trained to promote workplace equality. Given the diminishing interest of female students in the hospitality and tourism industry despite growing opportunities, it is important to understand how they perceive a career in this industry. This study aims to determine the perceptions and attitudes of Indian female undergraduate students towards careers in the hospitality and tourism industries. This study will present the views of undergraduate female students who enrolled in tourism and hospitality education with aspirations for their future careers. This study will provide insight for future researchers to understand female students' perceptions of working in the Indian hospitality industry.

Literature Review

Women in the Hospitality and Tourism Industry

Hospitality work provides considerable employment opportunities, employing 1 in every 11 people globally (Dube et al., 2021). The industry also offers more job prospects, entrepreneurship, and leadership for women than other industries. However, women continue to face pay, employment, and educational disadvantages in the industry. Despite women's remarkable economic advancement, they remain underrepresented in the hospitality industry's upper levels (Catalyst, 2022). There is a widespread misconception that personnel in this industry have a 24-hour job and are required to work on a rotating schedule, and some staff are obliged to work till the early morning hours (Sihite et al., 2019). Most hospitality employees, many of whom are women, work in low-wage, casual occupations, and non-standard work, which are associated with low-quality jobs and insecure employment (Rosemberg et al., 2021; Wolfe et al., 2021). One apparent trend in the hospitality sector is the seeming preference for men in high positions and women in more subordinate roles (Fan et al., 2021). According to previous studies, common stereotypes in the organisation include the belief that women cannot handle the stress of long hours and night shifts nor sustain a healthy work-life balance (Collica-Cox & Schulz, 2021; Das et al., 2021). Developmental disparities and social discouragement further hinder women's leadership growth (Schultheiss, 2021; Silver et al., 2022). Women's traditional "homemaker" position has also evolved, yet societal expectations concerning women's domestic tasks have yet to keep up with the social role needs of women in employment (Kwak, 2022). Women worldwide have lower status and remuneration than males (Williams, 2013), which is considered proportionate to the domestic abilities that they are assumed to possess. According to a report, men held 58% of the 25% of top-paid roles in the travel, hospitality, and leisure industries, whereas women make up 54% of the workforce in the lowest paid 25% of jobs (Lake, 2022). The limitation of employment prospects for women is related to a lack of value put on their actions, whether at home or work (Tejani & Fukuda-Parr, 2021). In many civilisations, the proportion of women in top leadership roles is lower than the number of males (Burke & Major, 2014; Russen et al., 2021). Hospitality and tourism in India are also facing this significant challenge of gender equality. The struggle of Indian women to find their place in the world and advance their careers has been an uphill battle (Wani, 2022). Many Indian women in the middle of their careers are leaving the hospitality industry to start their restaurants, bakeries, confectioneries, and even consulting firms. The Indian Ministry of Tourism launched a study on gender equity in India's hospitality business, which will help in overcoming this challenge (Pradeep, 2022). The study uncovered various biases and obstacles for women aspiring to leadership positions, which were divided into three categories: individual, group, and company. At the individual level, the demand to balance work and life is more significant for women employees than for males on an individual basis. Their networking skills are underrated, and it is considered that they have mobility problems when they travel for business, which is a common misconception. At the group level, stereotypes held by co-workers, preconceived conceptions, and paternalistic supervisor behaviours are significant obstacles. At the company level, gender stereotypes employers hold and the lack of mentorship opportunities are serious problems. The result of their study raises serious concerns about gender equality in the hospitality and tourism industry in India as it may further divert female students to choose a career in the industry.

The gender diversity gap across nations is a fascinating topic that can be explored from a theoretical perspective in several ways. This gender diversity gap is most pronounced in the highest ranks of management and executive positions. There are fewer women in upper-level positions in the hospitality business because there has been a decline in the number of women choosing to make a career out of working in the industry. Women made up just 12% of leaders in the hospitality sector, and the odds of women becoming CEOs were 1 in 20 (The Castell Project, 2020). Hillman et al. (2007) examined the presence of women on boards in the United States through the lens of the resource dependency theory, which was established by Salancik and Pfeffer (1978). They identify several firm-level factors, including industry, size, and interlocking directorates, that may account for this phenomenon. Singh and Vinnicombe (2004) explored the causes of gender diversity through institutional theory. The institutional theory, as defined by Meyer and Rowan (1977), contends that entities operating inside the institutional environment are influenced by the surrounding environment's norms, standards, and perceived beliefs (Weerakkody et al., 2009). The causes of gender discrimination in the workplace have been thoroughly studied and discussed. Prior studies have attempted to pin down the many psychological, organisational, and legal factors that contribute to workplace inequality. Scholars from various fields believe the mechanisms that sustain gender disparities in the workplace are hard to change because they are ingrained in and reinforced by commonplace rules of conduct, knowledge frameworks, and belief systems that are seldom questioned (Tina Dacin et al., 2002). The mechanisms that sustain gender disparities in the workplace are embedded in modern organisations' accepted and legal procedures.

Exploring the Significance of Institutional Theory in the Context of Hospitality and Tourism Industry

The relationship between institutional theory and employment in the hospitality and tourism industries is a subject of intense interest and research. Institutional theory provides an insightful concept for understanding how organisations within the hospitality and tourism industry are influenced by external pressures, norms, and regulations. According to institutional theory, institutions have an impact on society and shape the way businesses operate. As Scott (1995) described, institutions are social institutions that have gained a high level of resilience. Institutional forces, rules, norms, frameworks, or social constructions help to increase corporate leadership and board diversity. Additionally, Scott (1995) suggested the three pillars of institutional theory: cultural-cognitive, normative, and regulative. The researcher stated that the interaction of these parts of an institution offers an understanding of human behaviours and motivations in a society.

The first pillar of institutional theory is cognitive culture. This pillar represents an institutionalised understanding of the country's cultures and traditions (Abadi et al.,

2022). It forms linguistic and cultural role models that influence individual conduct (Scott, 2005). These behaviours are subjectively and gradually developed based on unconscious ideas and ingrained assumptions (DiMaggio & Powell, 1991). Because Indian culture is often perceived as patriarchal and characterised by defined gender roles (Kalra & Bhugra, 2013; Siddiqi, 2021). To understand the acceptance and practice of women in leadership positions in India, studies must focus on this pillar.

The second pillar of institutional theory is normative, consisting of a set of beliefs, norms, and standards of behaviour for persons working inside the institutions (Scott, 2005). It is based on critical aspects related to organisational, social, and professional relationships. Norms, which dictate how things should be done in line with the organisation's values, and values, representing what is considered right, constitute the normative force that enables organisations to establish standards of conformity (Scott, 2014). Organisations may sway individuals by appealing to their sense of social duty (March & Olsen, 1989). In some countries, the rules and traditions facilitate women's leadership, while in others, the process is more challenging, though not illegal. The third pillar of institutional theory is regulative, which manifests as the laws, industrial agreements, and other mechanisms that regulate the institution's operation (Biesenthal et al., 2018; Trevino et al., 2008). This institutional pillar can help ensure legal and ethical compliance in businesses led by women in management.

Female Student's Perceptions and Attitudes Towards a Career in Hospitality

Students' perceptions and attitudes about hospitality occupations may be good, negative, or indifferent (Birtch et al., 2021). Hospitality students hold negative perceptions of jobs in an industry characterised by low earnings (Neequaye & Armoo, 2014; Richardson & Butler, 2012), unpleasant work conditions, dullness, and short work hours (Selçuk et al., 2013). Students' perceptions are influenced by preconceived notions about the industry's, including low compensation, constant effort, and limited opportunities for promotion opportunities (Sihite et al., 2019). As a result, several studies have found that the tourism industry has a negative image among tourism students, which suggests that recruiting and retaining qualified staff may present challenges (Abou-Shouk et al., 2021). Blomme et al. (2013) emphasised that graduates interested in working in the hotel sector should maintain a positive outlook on the industry and appreciate the global aspects of their jobs. Both current students and recent graduates acknowledge the importance of having a genuine sense of belonging and affection for the hospitality industry when pursuing a career in this field.

A pre-pandemic report published by the Institute of Hospitality (2020) demonstrates a shift in public opinions towards the industry, which businesses can leverage to attract and retain talent. Both parents and instructors agreed that schools should give the hospitality industry more attention. According to the report, 86% of parents recommended the inclusion of food-related subjects in the curriculum, while 38% of instructors believe that schools should provide more information on jobs in hospitality industry to meet the demand. In general, young people exhibit a greater willingness to pursue employment opportunities in the hospitality sector. In recent years, one in every five instructors has reported an increase in students requesting information on hospitality careers. According to Alananzeh study (2014) conducted in the Aqaba Economic Zone in Jordan, 82%

of individuals who studied hotel management or food and beverage service expressed intentions to work in the industry completing their education. This inclination to work in business can be due to these students' favourable impressions of industrial professions. However, despite these positive opinions of jobs, there is still a need to address the concerns of more than a third (35%) of parents who perceive hospitality careers as involving anti-social working hours and over a quarter (26%) who believe there is a gap between school and university in terms of career opportunities. According to Barron and Maxwell (1993), first-year students had a more optimistic view of the industry and its job prospects than upper-level students and veterans. It is crucial to highlight that students' conceptions of the sector may need to be more practical (Chen et al., 2022), which may impact their overall opinions of professions in the industry.

According to Sirajuddin et al. (2021), three factors that influence student perspective are physical employment conditions, remuneration and perks, and promotion chances. The study found that first-year students held favourable attitudes towards careers in the hospitality industry. Positive opinions were attributed to factors such as the availability of professional growth, the opportunity to meet new people, and the abundance of job opportunities. Amissah et al. (2020) found that female students held more positive views towards careers in the hotel and tourism industries compared to male students, indicating a substantial difference between gender-based opinions regarding careers in the industry. Post-pandemic, even after witnessing the situation of the downfall of the hospitality industry, many students are still enrolled in hospitality education. Given the potential for these female students to become future leaders in the hospitality industry, it is crucial to investigate their perspectives and aspirations.

Methodology

Research Context

The context of this study is the hospitality industry in India. The hospitality industry in India is a dynamic sector that includes various sub-sectors such as hotels, restaurants, travel and tourism, and event management. The hospitality industry plays a vital role in India's economy by generating employment opportunities and contributing to foreign exchange earnings. Global Hospitality Market (2022) reported a compound annual growth rate (CAGR) of 15.1% between 2021 and 2022, indicating that the global hospitality market grew from $3,952.87 billion in 2021 to $4,548.42 billion in 2022. It is projected that the hospitality market will reach $6,715.27 billion by 2026, representing a CAGR of 10.2%. The consistent economic growth observed in both developed and emerging nations provides a positive outlook for the hospitality industry.

Research Design

This study employed a qualitative approach to identify perceptions of Indian undergraduate female students about working in the hospitality industry. The authors employed a qualitative method to gain insights into the perceptions of female undergraduate students working in the hospitality and tourism industry. Face-to-face one-on-one in-depth interviews were conducted in July 2022 at the School of Hotel Management and Tourism, Lovely Professional University, Punjab, India. A total of

18 respondents participated in the interviews. In-depth interviews were chosen for this study as they provide valuable insights into an individual's thoughts, behaviours, and allow for in-depth exploration of new issues (Boyce & Neale, 2006). All the interviews were conducted one-on-one in an informal environment, each lasting 15–20 minutes. The participants for the in-depth interviews were selected using purposive sampling technique, a non-probability sampling technique that allows researchers to choose individuals who possess specific traits of interest (Sharma, 2017). Only female students pursuing a bachelor's degree in tourism and hospitality and belonging to India were invited for the in-depth interview. Table 9.1 presents the demographic profile of the respondents who participated in the in-depth interview. The demographic profile indicates the respondents were from different demographic regions, belonged to the age group of 17–22, and pursuing a bachelor's degree.

This study intends to identify the perception and attitudes of Indian female students towards working in the tourism and hospitality industry. Therefore, it is crucial to understand the students' perspectives on working in the hospitality and tourism industry as well as their perceptions of the working environment within the industry. The interviews commenced with a mutual introduction between the interviewer and interviewee, followed by one fundamental question, *"What is your perception towards working in the hospitality industry?"* and *"How do you describe the working environment in the hospitality and tourism industry?"* Additional questions were posed to investigate the students' perception of working in the hospitality and tourism industry.

Table 9.1 List of respondents for in-depth interview

Participant	Age	Area of residence	Programme of the study	Year of study
R1	17	Bihar	BSc in Airlines, Tourism, and Hospitality	1st
R2	19	Punjab	BSc in Airlines, Tourism, and Hospitality	2nd
R3	17	West Bengal	BSc in Hotel Management	1st
R4	18	Uttar Pradesh	BSc in Airlines, Tourism, and Hospitality	2nd
R5	19	Uttarakhand	BSc in Airlines, Tourism, and Hospitality	2nd
R6	21	Bihar	BSc in Hotel Management	3rd
R7	19	Jharkhand	BSc in Hotel Management	2nd
R8	18	Maharashtra	BSc in Hotel Management	1st
R9	20	Uttar Pradesh	BSc in Hotel Management and Catering Technology	3rd
R10	21	Jammu and Kashmir	BSc in Airlines, Tourism, and Hospitality	3rd
R11	18	West Bengal	BSc in Airlines, Tourism, and Hospitality	1st
R12	19	Jharkhand	BSc in Hotel Management and Catering Technology	3rd
R13	19	Karnataka	BSc in Airlines, Tourism, and Hospitality	2nd
R14	18	Punjab	BSc in Hotel Management and Catering Technology	2nd
R15	21	Maharashtra	BSc in Airlines, Tourism, and Hospitality	3rd
R16	19	Odisha	BSc in Airlines, Tourism, and Hospitality	2nd
R17	22	Rajasthan	BSc in Hotel Management and Catering Technology	4th
R18	18	Punjab	BSc in Hotel Management	2nd

With the cooperation of the participants, all the interviews were recorded. The researchers adopted a manual method of coding. After the interviews, the interview material, comprising words, sounds, and visuals, was detected and translated into the textual content; then, codes of the transcribed data were continuously generated and attributed to superordinate themes. Following the interviews, codes were shared with each participant to validate the accuracy of their interview records. There was unanimity that the researcher had accurately interpreted their meanings, indicating the study's good accuracy (Volpe & Bloomberg, 2010; Wu & Pearce, 2014). Finally, a comparison was conducted between the detected themes and existing literature, and a synopsis was compiled and translated into English. Content analysis was employed to identify the findings of the in-depth one-on-one interviews.

Results and Findings

The keywords were extracted through content analysis and the results are presented in Table 9.2. The keywords were classified into 16 themes. The identified themes were classified into three categories: "social status," "career prospects," and "work environment and benefits."

Social Status

When discussing the hospitality industry, respondents expressed their views on its growth. According to Respondent 1, the "rising growth of industry" is a vital factor in the hospitality industry. Additionally, Respondent 13 highlighted the critical factor of the industry's emergence. Respondent 3 echoed a similar sentiment, emphasising the industry's progressiveness in their statement, "hospitality and tourism industry is a very progressive industry."

Table 9.2 Content analysis result

Keywords	Themes	Constructs
"emerging," "leading," "rapid," "dynamic," "progressive," "possibilities," and "tactful"	Emerging industry Leading industry Progressive industry A sector with huge possibilities	Social status
"job," "opportunities," "employment," "career," "growth," "future," "networking," "mobility," and "diversity."	Job availability Employment opportunities Diverse working environment Networking opportunities Career growth Job mobility	Career prospects
"Entertainment," "enjoyment," "fun," "tour," "travel," "exposure," "flexibility," "busy," "schedule," "harassment," and "new people"	Work with fun Tours and travel opportunities Flexible location Meeting new people Sexual harassment Busy work schedule	Work environment and benefits

The Indian female students perceive the hospitality industry as emerging and progressive. Furthermore, Respondent 11 emphasised the significance of "creating possibilities" as a crucial determinant. That industry is dynamically progressing, offering vast opportunities for students pursuing hospitality careers. They can be involved in hotels, travel agencies, airlines, cruises, etc. (Benaraba et al., 2022; Wut et al., 2022). The hospitality and tourism industry is expanding its wings every day. Respondent 17 highlighted the significant influx of foreign visitors to India. The hospitality industry contributes significantly to India's annual revenue (Sharma & Kochher, 2018). Students studying the tourism and hospitality perceive the Indian industry as dynamic, budding, and progressive (Kumar et al., 2022).

Career Prospects

In the hotel sector, employment offers flexibility and litheness. Moreover, a creative environment nurtures an optimistic outlook. The workplace provides flexible environment that encourages the exploration of innovative ideas and fosters enthusiasm and dynamism for the future. Furthermore, the hospitality industry offers millions of employment opportunities. In addition, Respondent 5 highlighted "job opportunities" as a critical factor in the hospitality industry, attributing it "investment of hotel brands in the Indian market", as mentioned by Respondent 4. Similarly, Respondent 9 emphasised the expansion of the "aviation industry," while Respondent 16 mentioned "the creation of employment opportunities within the sector". "Career growth" is a significant determinant of working in the hospitality industry, as mentioned by Respondent 15. In addition, Respondent 2 asserted that the "availability of hotels and travel agencies in India" is critical in working in the hospitality industry. Respondent 12 stated the "strong networking opportunity" factor, which will enhance their career growth. Career growth in the hospitality industry is influenced by an individual's skills and level of experience (Morosan & Bowen, 2022; Srivastava & Gupta, 2022). Respondent 6 identified a "diverse environment" as a crucial factor in working in this industry. The Indian hospitality industry is expanding rapidly and provides a diverse working environment. Furthermore, professionals in the hospitality and tourism industry can build networks with their peers, which contributes to their career growth.

Work Environment and Benefits

The word "work environment" refers to the non-financial component that provides suitable surroundings for workers to do their tasks (Jaleta et al., 2019). The work environment in the hospitality industry encompasses various elements such as work motivators, job stress, physical environments, and interactions with co-workers (Chung et al., 2016; Schiffinger & Braun, 2020; Zheng & Montargot, 2022). Respondent 3 indicated that working in the hospitality industry can be demanding and "stressful." Similar factors, "doing overtime shifts," were also indicated by Respondent 11. In addition, the students perceive that the working environment in hospitality is delightful and entertaining, and each will get the opportunity to interact with different people. Respondent 7 stated "entertainment and fun" as a critical factor of working in the hospitality industry. Moreover, "interaction with new people" was determined as a major determinant of working in this industry, as mentioned by Respondent 10, and a similar factor "cultural diversity" was mentioned by Respondent 9. Many students expressed concerns about safety issues in the hospitality industry, particularly regarding women's sexual harassment. They believe that renowned brands are more attentive to addressing these concerns. Respondent 14 stated "zero tolerance for sexual harassment" as a major determinant of working in the hospitality industry, and the factor "sexual assault" was asserted by Respondent 5. Furthermore, individuals working in the airline or the travel sector can travel to different destinations. Respondent 18 highlighted the importance of "flexible work location" as a significant factor in choosing a career in the hospitality and tourism industry. Whereas Respondent 8 stated that "travel opportunities" as a significant determinant of working in this industry. Additionally, renowned hotel brands and travel agencies often have multiple chains and branches, providing employees with the flexibility to choose job locations based on their preferences.

Discussion

The findings revealed that Indian female undergraduate students are positively influenced by the perceived social status of the hospitality industry, considering it as an emerging industry with leadership potential, progressiveness, and significant opportunities, career prospects (job availability, employment opportunities, diverse working environment, networking opportunities, career growth), and work environment and benefits (work with fun, tours, and travel opportunities, flexible location, meeting new people, sexual harassment, job mobility) towards working in the hospitality and tourism industry. Tourism in India is experiencing substantial growth, primarily driven by the country's rapid economic expansion. According to Bodhanwala and Bodhanwala (2022), the travel and tourism sector is considered one of the most lucrative industries in India, making a significant contribution to the generation of foreign currency. In addition, the travel and tourist industry directly supports direct employment opportunities. Within the hospitality industry, cognitive culture plays a vital role in shaping individuals' perceptions of social status. The institutional theory provides insights into how cognitive culture influences individuals' perspectives and perceptions. According to institutional theory, social beliefs, values, and norm systems are ingrained throughout organisations and industries (Amine & Staub, 2009; Moran & Ward-Christie, 2022). These institutionalised ideas affect people's cognitive frameworks and impact how they perceive social status. In the hospitality industry context, the cognitive culture prevalent in society can affect how people perceive working in this industry. For example, if broader cultural beliefs perceive the hospitality sector as an emerging and progressive industry, individuals within that society are likely to hold similar views. The findings of the current study reveal that students perceive the hospitality industry as a rapidly expanding sector with ample employment opportunities, indicating its emergence and progressiveness. This perception is influenced by the social status of the industry in India, as it is a substantial contributor to the country's annual revenue. They may perceive working in the hospitality sector as having a decent social status.

The hospitality industry encompasses various sectors, including hotels, tourist information offices, travel agencies, restaurants, museums, reserved locations such as palaces, national parks, airlines, religious sites, resorts, monuments, and more, offering diverse employment opportunities. Furthermore, the Indian hospitality industry has implemented strict measures to address and prevent incidents of employee sexual harassment. Institutional theory examines how institutions, including regulations, laws, cultural norms, and social conventions, shape the behaviour and organisational structure (Aureli et al., 2020; Deephouse & Suchman, 2008). The theory suggests that organisations are not just influenced by economic and rational factors but also by institutional forces that shape their practices and policies. Within the normative framework of institutional theory, participants' concerns about sexual harassment in the hospitality industry are significantly influenced by their gender beliefs and expectations. These thoughts indicate how gender roles and behaviours should be displayed in society as a whole, and they may impact how individuals perceive and react to incidents of sexual harassment. The regulative component of institutional theory can be employed to address the issue of sexual harassment in the hospitality sector. The term "regulatory mechanisms" encompasses the formal rules, regulations, and policies that govern behaviour within the industry (Karlsson et al., 2020). These systems are intended to encourage conformity and deal with societal difficulties. By aligning normative gender beliefs with effective regulative mechanisms, the hospitality industry can foster an environment where sexual harassment concerns are taken seriously, victims receive support, and all individuals are treated with dignity and respect.

Moreover, each year thousands of students enrol in tourism and hospitality courses to receive training for careers in these fields. Students are drawn to the industry's promising career prospects, which provide a flexible working environment that fosters innovation and enables them to establish valuable connections with the peers. Career advancement is closely tied to an individual's level of skill and experience. This perception is influenced by the institutional context of the industry, which provides diverse working environments and opportunities for career growth. Within the framework of institutional theory, cognitive culture influences individual beliefs and perceptions regarding the desirability, prestige, and potential for advancement within a particular sector This, in turn, shapes their perceived career prospects in that sector (Amine & Staub, 2009; Moran & Ward-Christie, 2022). By understanding and leveraging these cultural influences, organisations and industries can cultivate and promote more positive perceptions of career opportunities, thus attracting and retaining talent in their respective fields.

Additionally, the student's perception of working in the tourism and hospitality industry is influenced by the work environment and the benefits it offers. The students view the working environment in hospitality as delightful and entertaining, providing opportunities to interact with different people, travel to various destinations and work in different locations based on their preferences. This perception is shaped by the institutional context of the industry, which encompasses various factors such as work environment, benefits, and non-financial components that create a conducive setting for workers to do their tasks. Prominent brands prioritise addressing and preventing women's sexual harassment maintaining a zero tolerance towards such incidents. For instance, a leading hospitality group, the Taj Group, stated that they have zero tolerance for sexual harassment, value every employee working there, and wish to guard their dignity (Taj Group, 2023). However, employees face challenges such as managing tight schedule during peak hours and the requirement to work additional hours beyond their regular shifts. Therefore, the institutional context of the tourism and hospitality industry in India shapes the perception and attitudes of Indian female students towards working in the industry. The work environment in the hospitality industry's work environment is influenced by cognitive culture, which encompasses shared thoughts, opinions, and perceptions. Their perception of what makes a good or bad work environment is influenced by this cultural lens, including aspects like job satisfaction, work-life balance, and overall well-being. For instance, individuals are more likely to perceive the work environment positively when the cognitive culture emphasises the hospitality sector as offering enjoyable and fulfilling work experiences. On the other hand, people may have a more unfavourable perception (tight schedule, working overtime) of the workplace if the cultural narrative depicts the hospitality industry as difficult.

Conclusion, Limitations, and Implications

The current study aimed to investigate the perceptions of Indian female undergraduate students' regarding working in the hospitality industry. The findings revealed that these perceptions are influenced by factors such as social status, career prospects, work environment, and benefits. Hospitality management encompasses various establishments and organisations, including management of hotels, restaurants, cruise ships, amusement parks, destination marketing organisations, convention centres, and country clubs. It also pertains to individuals who choose to work in the hotel industry and pursue relevant courses in the field.

The findings of this study have implications for stakeholders in the tourism and hospitality industry. These findings can be used by industry authorities and policymakers to inform policy decisions and improve the work environment in the industry. By understanding the perceptions of Indian female students, industry stakeholders can develop strategies to attract and retain more women in the industry. Recruitment campaigns can be designed to highlight the positive aspects of the industry such as social status, career prospects, and work environment. This can help in attracting more female students to pursue careers in the hospitality industry. Training and development programmes can be tailored to enhance the skills and knowledge of female employees, providing them with opportunities for career growth and advancement within the industry. This can help in improving their career growth prospects and provide them with opportunities to work in different areas of the industry. As the respondents highlighted concerns about safety in the industry, especially for female employees, the hospitality industry should prioritise workplace safety measures. Zero-tolerance policies for sexual harassment and other forms of workplace violence should be implemented and strictly enforced. Flexibility in job locations and working hours is important to many respondents. To accommodate the needs of female employees, the industry can offer flexible work arrangements, such as part-time work, job-sharing, and remote work options.

The hospitality industry can indeed play a crucial role in promoting diversity and inclusion. By creating a welcoming and inclusive workplace culture, the industry can attract and retain a diverse workforce. As for the present study, it serves as a valuable foundation for future research on students' perceptions of working in the tourism and hospitality industry. It is important to acknowledge the limitations of the current study. Future research can adopt mixed methods or quantitative approaches to complement the qualitative findings and provide a more comprehensive understanding of students' perceptions. Additionally, increasing the sample size in qualitative studies can enhance the generalisability of the results and provide a more representative perspective.

References

Abadi, M., Dirani, K. M., & Rezaei, F. D. (2022). Women in leadership: A systematic literature review of Middle Eastern women managers' careers from NHRD and institutional theory perspectives. *Human Resource Development International*, 25(1), 19–39.

Abou-Shouk, M. A., Mannaa, M. T., & Elbaz, A. M. (2021). Women's empowerment and tourism development: A cross-country study. *Tourism Management Perspectives*, 37, 100782. https://doi.org/10.1016/J.TMP.2020.100782

Aithal, R., Anil, R. K., & Angmo, D. (2023). Rural tourism in India: Case studies of resilience during crisis. *Worldwide Hospitality and Tourism Themes*, 15(1), 63–73.

Alananzeh, O. A. (2014). Exploring the factors influencing students in enrolling tourism and hospitality management colleges in Jordan: A case study in Aqaba Economic Zone. *Journal of Management Research, 6*(2), 61–73.

Amine, L. S., & Staub, K. M. (2009). Women entrepreneurs in sub-Saharan Africa: An institutional theory analysis from a social marketing point of view. *Entrepreneurship and Regional Development, 21*(2), 183–211.

Amissah, E. F., Mensah, A. O., Mensah, I., & Gamor, E. (2020). Students' perceptions of careers in Ghana's hospitality and tourism industry. *Journal of Hospitality & Tourism Education, 32*(1), 1–13. https://doi.org/10.1080/10963758.2019.1654884

Aureli, S., Del Baldo, M., Lombardi, R., & Nappo, F. (2020). Nonfinancial reporting regulation and challenges in sustainability disclosure and corporate governance practices. *Business Strategy and the Environment, 29*(6), 2392–2403.

Barron, P., & Maxwell, G. (1993). Hospitality management student's views of the hospitality industry. *International Journal of Contemporary Hospitality Management, 5*(5). https://doi.org/10.1108/09596119310046961

Benaraba, C. M. D., Bulaon, N. J. B., Escosio, S. M. D., Narvaez, A. H. G., Suinan, A. N. A., & Roma, M. N. (2022). A comparative analysis on the career perceptions of tourism management students before and during the COVID-19 pandemic. *Journal of Hospitality, Leisure, Sport & Tourism Education, 30*, 100361. https://doi.org/10.1016/j.jhlste.2021.100361

Biesenthal, C., Clegg, S., Mahalingam, A., & Sankaran, S. (2018). Applying institutional theories to managing megaprojects. *International Journal of Project Management, 36*(1), 43–54.

Birtch, T. A., Chiang, F. F. T., Cai, Z., & Wang, J. (2021). Am I choosing the right career? The implications of COVID-19 on the occupational attitudes of hospitality management students. *International Journal of Hospitality Management, 95*, 102931. https://doi.org/10.1016/j.ijhm.2021.102931

Blomme, R., van Rheede, A., & Tromp, D. (2013). The hospitality industry: An attractive employer? An exploration of students' and Industry workers' perceptions of hospitality as a career field. *Journal of Hospitality & Tourism Education, 21*(2), 6–14. https://doi.org/10.1080/10963758.2009.10696939

Bodhanwala, S., & Bodhanwala, R. (2022). Exploring relationship between sustainability and firm performance in travel and tourism industry: A global evidence. *Social Responsibility Journal, 18*(7), 1251–1269.

Boyce, C., & Neale, P. (2006). *Conducting in-depth interviews: A guide for designing and conducting in-depth interviews for evaluation input* (Vol. 2). Pathfinder International.

Burke, R., & Major, D. (2014). *Gender in organizations*. Edward Elgar Publishing. https://doi.org/10.4337/9781781955703

Castell Project. (2020). *Women in hospitality industry leadership castell project*. Retrieved from https://www.ahlafoundation.org/sites/default/files/202002/Castell_Report_lk_v5_01.20%20%28002%29.pdf

Catalyst. (2022). *Women in the workforce: India*. https://www.catalyst.org/research/women-in-the-workforce-india/

Chang, S., & Tse, E. C.-Y. (2015). Understanding the initial career decisions of hospitality graduates in Hong Kong. *Journal of Hospitality & Tourism Research, 39*(1), 57–74. https://doi.org/10.1177/1096348012461544

Chen, C.-C., Zou, S. (Sharon), & Chen, M.-H. (2022). The fear of being infected and fired: Examining the dual job stressors of hospitality employees during COVID-19. *International Journal of Hospitality Management. 102*, 103131, https://doi.org/10.1016/j.ijhm.2021.103131

Chung, J. Y., Buhalis, D., Chatterjee, P., Ye, Q., Law, R., Gu, B., Chen, W., Toder-Alon, A., Brunel, F. F., Fournier, S., Knoll, J., Proksch, R., Canhoto, A. I., Clark, M., Michaelidou, N., Siamagka, N. T., Christodoulides, G., Trusov, M., Bodapati, A. V., & Dennick, R. (2016). Advantages and disadvantages of internet research surveys: Evidence from the literature. *Journal of Internet Commerce, 15*(1), 815–827. https://doi.org/10.1080/17543266.2019.1572230

Collica-Cox, K., & Schulz, D. M. (2021). Having it all? Strategies of women corrections executives to maintain a work-life balance. *Corrections*, 1–25. https://doi.org/10.1080/23774657.2020.1868360

Das, A. K., Abdul Kader Jilani, M. M., Uddin, M. S., Uddin, M. A., & Ghosh, A. K. (2021). Fighting ahead: Adoption of social distancing in COVID-19 outbreak through the lens of theory of planned behavior. *Journal of Human Behavior in the Social Environment*, *31*(1–4), 373–393.

Datta, A., Biswakarma, S. K., & Nayak, B. (2013). Effect of internship on career perception of hotel management students. *Zenith International Journal of Multidisciplinary Research*, *3*(10), 50–63.

Deephouse, D. L., & Suchman, M. (2008). Legitimacy in organizational institutionalism. In *The Sage handbook of organizational institutionalism* (49–77). SAGE Publications Ltd, https://doi.org/10.4135/9781849200387

DiMaggio, P. J., & Powell, W. W. (1991). *The new institutionalism in organizational analysis*. University of Chicago Press.

Dube, K., Nhamo, G., & Chikodzi, D. (2021). COVID-19 cripples global restaurant and hospitality industry. *Current Issues in Tourism*, *24*(11), 1487–1490. https://doi.org/10.1080/13683500.2020.1773416

Elhoushy, S. (2014). *Hospitality students' perceptions towards working in hotels: A case study of the faculty of tourism and hotels in Alexandria University Hotel Outsourcing View project Food Waste View project*. https://www.researchgate.net/publication/307593920

Fan, X., Im, J., Miao, L., Tomas, S., & Liu, H. (2021). Silk and steel: A gendered approach to career and life by upper echelon women executives in the hospitality and tourism industry in China. *International Journal of Hospitality Management*, *97*, 103011. https://doi.org/10.1016/j.ijhm.2021.103011

FICCI. (2020). *Travel and tourism – Survive, revive and thrive in times of COVID-19*. https://ficci.in/api/press_release_details/3736

Global Hospitality Market. (2022). *Hospitality global market report 2022 – by type (non-residential accommodation services, food and beverage services), by ownership (chained, standalone) – market size, trends, and global forecast 2022 – 2026*. The Business Research Company. https://www.thebusinessresearchcompany.com/report/hospitality-global-market-report

Griffin, M., Hodgson, R., & Sivam, S. (2021). Career incentives, career deterrents, and cultural blocks: An investigation of factors impacting female Emirati students' perceptions of tourism. *Journal of Human Resources in Hospitality and Tourism*, *20*(3), 472–496. https://doi.org/10.1080/15332845.2021.1923945

Hillman, A. J., Shropshire, C., & Cannella, A. A., Jr. (2007). Organizational predictors of women on corporate boards. *Academy of Management Journal*, *50*(4), 941–952.

Institute of Hospitality. (2020). *Research reveals a positive change in perceptions of hospitality careers*. https://www.instituteofhospitality.org/research-reveals-a-positive-change-in-perceptions-of-hospitality-careers/

Jaleta, K. M., Kero, C. A., & Kumera, L. (2019). Effect of non-financial compensation on the employees' job performance: A case of Jimma Geneti Woreda Health Centers in Horro Guduru, Ethiopia. *International Journal of Commerce and Finance*, *5*(2), 31–44.

Joshi, V. A., & Gupta, I. (2021). Assessing the impact of the COVID-19 pandemic on hospitality and tourism education in India and preparing for the new normal. *Worldwide Hospitality and Tourism Themes*, *13*(5), 622–635. https://doi.org/10.1108/WHATT-05-2021-0068

Kalra, G., & Bhugra, D. (2013). Sexual violence against women: Understanding cross-cultural intersections. *Indian Journal of Psychiatry*, *55*(3), 244–249.

Karlsson, I. C. M., Mukhtar-Landgren, D., Smith, G., Koglin, T., Kronsell, A., Lund, E., Sarasini, S., Sochor, J. (2020). Development and implementation of mobility-as-a-service–A qualitative study of barriers and enabling factors. *Transportation Research Part A: Policy and Practice*, *131*, 283–295.

Kaushal, V., & Srivastava, S. (2021). Hospitality and tourism industry amid COVID-19 pandemic: Perspectives on challenges and learnings from India. *International Journal of Hospitality Management*, *92*, 102707. https://doi.org/10.1016/j.ijhm.2020.102707

Kumar, A., Kumar Singh, P., Kumar, A., & Dahiya, S. (2014). An investigation of the perception of hospitality graduates towards the industry: A gender perspective. *African Journal of Hospitality, Tourism and Leisure*, *3*(2), 7–12.

Kumar, M. S., Ramaprasad, B. S., Rao, N., & Jamwal, M. (2022). Hospitality and tourism students' perceptions of effectiveness of entrepreneurship education and its effect on entrepreneurial intentions: A cross-lagged two-wave mediation study involving entrepreneurial self-efficacy. *Journal of Teaching in Travel & Tourism*, 1–28. https://doi.org/10.1080/15313220.2022.2123078

Kwak, A. (2022). The Polish family in transition: A shift towards greater gender equality? *Contemporary Social Science*, 17(4), 340–352. https://doi.org/10.1080/21582041.2022.2077419

Lake, E. (2022). *Hospitality's gender pay gap increases for first time in three years. The Caterer*. Retrieved from https://www.thecaterer.com/news/hospitality-gender-pay-gap-2020-2021

Lee, P. C., Lee, M. J. (MJ), & Dopson, L. R. (2019). Who influences college students' career choices? An empirical study of hospitality management students. *Journal of Hospitality & Tourism Education*, 31(2), 74–86. https://doi.org/10.1080/10963758.2018.1485497

March, J. G., & Olsen, J. P. (1989). *Rediscovering institutions: The organizational basis of politics.* Free Press.

Meyer, J. W., & Rowan, B. (1977). Institutionalized organizations: Formal structure as myth and ceremony. *American Journal of Sociology*, 83(2), 340–363.

Moran, M., & Ward-Christie, L. (2022). Blended social impact investment transactions: Why are they so complex? *Journal of Business Ethics*, 179(4), 1011–1031.

Morosan, C., & Bowen, J. T. (2022). Labor shortage solution: Redefining hospitality through digitization. *International Journal of Contemporary Hospitality Management*, 34(12), 4674–4685. https://doi.org/10.1108/IJCHM-03-2022-0304

Neequaye, K., & Armoo, A. K. (2014). Factors used by Ghanaian students in determining career options in the tourism and hospitality industry. *Worldwide Hospitality and Tourism Themes*, 6(2), 166–178. https://doi.org/10.1108/WHATT-12-2013-0053

Pittie, R. (2023, January 14). The budget push that can make India's tourism sector one of the world's best. *The Economic Times*. Retrieved February 13, 2023, from https://economictimes.indiatimes.com/industry/services/travel/the-budget-push-that-can-make-indias-tourism-sector-one-of-the-best-in-the-world/articleshow/96986632.cms

Pradeep, C. (2022). *Evaluation of the state of gender equity in India's hospitality industry.* https://www.financialexpress.com/lifestyle/travel-tourism/evaluation-of-the-state-of-gender-equity-in-indias-hospitality-industry/2461566/

Qiu, S., Dooley, L., & Palkar, T. (2017). What factors influence the career choice of hotel management major students in Guangzhou? *Independent Journal of Management & Production*, 8(3), 1092. https://doi.org/10.14807/ijmp.v8i3.618

Richardson, S., & Butler, G. (2012). Attitudes of Malaysian tourism and hospitality students' towards a career in the industry. *Asia Pacific Journal of Tourism Research*, 17(3), 262–276. https://doi.org/10.1080/10941665.2011.625430

Rinaldi, A., & Salerno, I. (2020). The tourism gender gap and its potential impact on the development of the emerging countries. *Quality and Quantity*, 54(5–6), 1465–1477. https://doi.org/10.1007/s11135-019-00881-x

Rosemberg, M.-A. S., Adams, M., Polick, C., Li, W. V., Dang, J., & Tsai, J. H.-C. (2021). COVID-19 and mental health of food retail, food service, and hospitality workers. *Journal of Occupational and Environmental Hygiene*, 18(4–5), 169–179. https://doi.org/10.1080/15459624.2021.1901905

Russen, M., Dawson, M., & Madera, J. M. (2021). Gender diversity in hospitality and tourism top management teams: A systematic review of the last 10 years. *International Journal of Hospitality Management*, 95, 102942. https://doi.org/10.1016/j.ijhm.2021.102942

Salancik, G. R., & Pfeffer, J. (1978). A social information processing approach to job attitudes and task design. *Administrative Science Quarterly*, 224–253.

Schiffinger, M., & Braun, S. M. (2020). The impact of social and temporal job demands and resources on emotional exhaustion and turnover intention among flight attendants. *Journal of Human Resources in Hospitality & Tourism*, 19(2), 196–219. https://doi.org/10.1080/15332845.2020.1702867

Schultheiss, D. E. (2021). Shining the light on women's work, this time brighter: Let's start at the top. *Journal of Vocational Behavior*, 126, 103558. https://doi.org/10.1016/j.jvb.2021.103558

Scott, R. (1995). *Institutions and organizations.* Sage Publications.

Scott, W. R. (2005). Institutional theory: Contributing to a theoretical research program. In K. G. Smith & M. A. Hitt (Eds.), *Great minds in management: The process of theory development* (pp. 460–484). Oxford University Press.

Scott, W. R. (2014). *Institutions and organizations: Ideas and interests, and identities.* Sage Publications.

Selçuk, E., Erdogan, M., & Güllüce, A. C. (2013). A research into career considerations of students enrolled in tourism degree programmes. *International Journal of Business and Social Science*, *4*(6), 218–226.

Sharma, G. (2017). Pros and cons of different sampling techniques. *International Journal of Applied Research*, *3*(7), 749–752.

Sharma, A., & Kochher, P. (2018). Taj hotels, palaces and resorts: The road ahead. *Emerald Emerging Markets Case Studies*, *8*(3), 1–21. https://doi.org/10.1108/EEMCS-01-2018-0001

Siddiqi, N. (2021). Gender inequality as a social construction in India: A phenomenological enquiry. *Womens Studies International Forum*, *86*, 102472. https://doi.org/10.1016/j.wsif.2021.102472

Sihite, J., Dewi, T. R., & Widyastuti, N. (2019). The tourism student social status: The role nature of work and career prospect. *European Journal of Business and Management*, *11*(23). https://doi.org/10.7176/EJBM

Silver, E. R., King, D. D., & Hebl, M. (2022). Social inequalities in leadership: Shifting the focus from deficient followers to destructive leaders. *Management Decision*. https://doi.org/10.1108/MD-06-2021-0809

Singh, V., & Vinnicombe, S. (2004). Why so few women directors in top U.K. boardrooms? Evidence and theoretical explanations. *Corporate Governance: An International Review*, *12*(4), 479–488.

Srivastava, S., & Gupta, P. (2022). Workplace spirituality as panacea for waning well-being during the pandemic crisis: A SDT perspective. *Journal of Hospitality and Tourism Management*, *50*, 375–388. https://doi.org/10.1016/j.jhtm.2021.11.014

Taj Group. (2023). *The Taj Group of hotels policy on prevention, prohibition & redressal of sexual harassment at the workplace (POSH)*. Retrieved February 13, 2023, from https://www.tajhotels.com/content/dam/thrp/investors/Taj-POSH-Policy.pdf

Tejani, S., & Fukuda-Parr, S. (2021). Gender and COVID-19: Workers in global value chains. *International Labour Review*, *160*(4), 649–667. https://doi.org/10.1111/ilr.12225

Tina Dacin, M., Goodstein, J., & Richard Scott, W. (2002). Institutional theory and institutional change: Introduction to the special research forum. *Academy of Management Journal*, *45*(1), 45–56.

Trevino, L. J., Thomas, D. E., & Cullen, J. (2008). The three pillars of institutional theory and FDI in Latin America: An institutionalization process. *International Business Review*, *17*(1), 118–133.

UNWTO. (2019). *Global report on women in tourism – second edition*. World Tourism Organization, Madrid. https://doi.org/10.18111/9789284420384

Volpe, M., & Bloomberg, L. D. (2010). *Completing your qualitative dissertation: A road map from beginning to end*. SAGE Publications, Inc.

Wani, A. K. (2022). Spousal support and working woman's career progression: A qualitative study of woman academicians in the University of Kashmir. *Journal of Global Responsibility*. https://doi.org/10.1108/JGR-05-2021-0050

Weerakkody, V., Dwivedi, Y. K., & Irani, Z. (2009). The diffusion and use of institutional theory: A cross-disciplinary longitudinal literature survey. *Journal of Information Technology*, *24*(4), 354–368.

Williams, C. L. (2013). The glass escalator, revisited. *Gender & Society*, *27*(5), 609–629. https://doi.org/10.1177/0891243213490232

Wolfe, R., Harknett, K., & Schneider, D. (2021). Inequalities at work and the toll of COVID-19. *Health Affairs Health Policy Brief* (Issue June). https://www.healthaffairs.org/do/10.1377/hpb20210428.863621/full/health-affairs-brief-covid19-workplace-wolfe.pdf

Wong, S. C., & Liu, G. J. (2010). Will parental influences affect career choice? *International Journal of Contemporary Hospitality Management*, *22*(1), 82–102. https://doi.org/10.1108/09596111011013499

WTTC. (2018). *Travel & tourism economic impact 2018 world*. https://wttc.org/research/economic-impact

WTTC. (2020). *Latest research from WTTC shows a 50% increase in jobs at risk in travel & tourism*. https://wttc.org/news-article/latest-research-from-wttc-shows-a-50-percentage-increase-in-jobs-at-risk-in-travel-and-tourism

Wu, M.-Y., & Pearce, P. L. (2014). Appraising netnography: Towards insights about new markets in the digital tourist era. *Current Issues in Tourism, 17*(5), 463–474. https://doi.org/10.1080/13683500. 2013.833179

Wut, T. M., Xu, B., & Wong, H. S.-M. (2022). A 15-year review of "corporate social responsibility practices" research in the hospitality and tourism industry. *Journal of Quality Assurance in Hospitality & Tourism, 23*(1), 240–274. https://doi.org/10.1080/1528008X.2020.1864566

Yunus, I. Y., Ridzwan, F. S., Azhar, S. N., Rajis, A. S. M., Harun, N. A., & Omar, M. S. (2021). Tourism and hospitality students' perceptions of careers in the industry: A case study of Politeknik Tuanku Syed Sirajuddin. *In International Journal of Advanced Research in Education and Society, 3*(2), 82–89. http://myjms.mohe.gov.my/index.php/ijares

Zheng, L., & Montargot, N. (2022). Anger and fear: Effects of negative emotions on hotel employees' information technology adoption. *International Journal of Productivity and Performance Management, 71*(5), 1708–1727. https://doi.org/10.1108/IJPPM-01-2020-0013

10

QUILOMBO AÉREO

Changing Colour and Gender Relations in Brazil's Sky

Cassiana Panissa Gabrielli, Natália Araújo de Oliveira, Gabriela Nicolau Santos, and Laiara Amorim Borges

Abstract

Considering the absence of reflections on racism and sexism in the Brazilian airline industry, as well as the lack of actions aimed at mitigating such issues, this text proposes to analyse the empowerment of black women workers in this context, especially based on the actions of *Quilombo Aéreo*, a collective of black aeronauts created in 2018 to mitigate the effects of racism in aviation. Considering the individual and collective perspective of the empowerment process, exploratory qualitative research was conducted, which interviewed six black women from this collective. The analysis of the results was structured according to four dimensions of female empowerment (cognitive, psychological, political, and economic), also highlighting the intersectional frame to point out the specific conditions of people who are intersected by distinct social markers, such as race and gender. The results indicate that black Brazilian women crewmembers, when entering a sector hostile to them, have developed strategies based on their experiences, and on the collective experience, which contribute in a practical way to their empowerment and that of other black women.

Keywords

Empowerment; Black women; Intersectionality; *Quilombo Aéreo*; Crewmembers; Brazil.

Introduction

The present work emerges from an interest in the intersectional dialogue between gender and race in tourism, and the awareness of the lack of discussion on the topic in Brazilian literature. It aims at exploring the actions of the collective *Quilombo Aéreo*, which was created in 2018 by a group of black women crewmembers to "mitigate the effects of racism in Aviation" (Quilombo Aéreo, 2022), perceived as a strategy for resistance and empowerment of black women. For this, we analyse the path of these women in a sector that is elitist, sexist, and racist such as the Brazilian aviation industry (Oliveira et al., 2022), bringing up the necessary resilience and combativeness to occupy the sector and remain active in it.

DOI: 10.4324/9781003286721-15

Berth (2018) ponders that empowerment is perceived as a process that aims towards the achievement of autonomy and auto-determination of oppressed groups. To talk about empowerment is to discuss something that is essentially political and involves both personal and collective levels. It is a complex concept, misunderstood since it is often followed by debates that are not critical enough and lack reliable theoretical background, as the author affirms. To step away from such a shallow view, this research brings a theoretical discussion which categorises female empowerment according to the four dimensions which guide this work: the cognitive, psychological, political, and economic dimensions (Stromquist, 2003).

Brazilian official data show that despite being the majority (55.8% of the population), black people have higher rates of illiteracy (9.1% against 3.9% of white people) (IBGE, 2019), also have worse average income (R$1.570.00 against R$2.814.00 for white people) (Equipe Lupa, 2018), and are the majority among murder victims, according to the *Atlas da Violência* (IPEA & Fórum Brasileiro de Segurança Pública, 2021).

Starting from feminist postmodernism together with the feminist standpoint as a methodological resource for the elaboration of this qualitative research, also highlighting the intersectional domains of power (Collins & Bilge, 2021), interviews were conducted with three founders and three members (*aquilombadas*) of the *Quilombo Aéreo*. Upon analysing the narratives of personal and professional experiences, self-perceptions and identity negotiations described through their points of view, we have adopted the research-based *subjects*, as understood by Grada Kilomba (2019), by conducting narrative biographical interviews.

The chapter is organised from the theoretical framework, which is divided into two topics: The first summarises data from the Brazilian airline industry; while the second discusses intersectionality, black women, and the categories of empowerment. The following section details the methodology used, presenting the route for research design and data analysis. In the results and discussions, the speeches of the interviewees are presented, revealing professional stories of black women who found – in themselves and the *Quilombo Aéreo* collective – inspiration and access to personal and collective empowerment. The conclusion reflects on the convergence between the theory and the practice undertaken in this research.

Theoretical Reference

Brazilian Aviation Industry

The airline industry in Brazil is characterised as a public service whose operations are conceded to the private sector by the state. According to the National Bureau of Civil Aviation (ANAC, 2021), air transport carried, in the domestic context, an average of 63 million paying passengers, with three companies dominating approximately 98% of this operation. Regarding the international market, 5 million passengers were transported. In 2021 air transport was responsible for approximately 74% of commercial interstate transport, while road transport corresponded to 26% (ANAC, 2021). It should be noted that these numbers refer to a period in which the Covid-19 pandemic was still ongoing.

Such data synthesise the magnitude of the air sector in Brazil – a country known for its continental dimensions and precarious land transport infrastructure – and they justify the importance of analysing the services and the people who execute them. In this context, regarding the impact of Covid-19 on jobs in the national airlines, Pirro et al. (2022) observed that among the three main national airlines, there were measures related to mass dismissals, incentives for voluntary resignation, reduced wages and working hours, and the application of unpaid leave. In 2022, with the resumption of demand, the authors noted the rehiring of crewmembers laid off in the period, but not yet entirely.

Analysing the bio-psychosocial aspects of Brazilian aeronauts, it was identified that the main reasons for long-term work leave for health treatment among male flight attendants were linked to orthopaedics and psychiatry (Matias, 2015). For female flight attendants, the main reasons for work leave are pregnancy, psychiatry, and orthopaedics. The same research points out that, among the three companies that dominate the national market, only one offers differentiated scales for mothers (id.).

This reveals some hardships that the women in the sector face, besides the different physical and aesthetic demands based on gender, and, also, allows us to relate the absence of institutional and public policies to the work overload of the women and mother crewmembers, and to the frequency of physical and mental illness among the female crew. However, in the Brazilian case, it is not only gender that places women in such a situation. Studies show that mental disorders in Brazil are more prevalent or have greater chances of developing among non-white people (Smolen & Araújo, 2017). These authors consider exposure to stress (especially related to unequal social structure and racism) as a possible causal mechanism.

Following that, it is possible to identify that the maintenance of segregation in the sector is perceivable. Upon analysing the little data made available by the regulating bureau of the sector in Brazil, *Quilombo Aéreo* (2022) identified that among the categories of pilots (commercial, helicopter, and airline) the women (regardless of race or ethnicity) represented only around 2.5% of the acting professionals. Among the flight attendants, women make up a total of 66% of the active workforce. There is no data regarding race, which reinforces the manifestation of structural and institutional racism, facing the absence of internal policies and the interest in data that may be stratified and analysed by race. Through research conducted by the entity, the participation of black people in national civil aviation is about 5% for flight attendants and 2% for pilots. The *Quilombo Aéreo* collective has not identified any black women pilots acting in the national airlines.

Empowerment, Black Women and Intersectionality

According to Stromquist (2003), the origins of the concept of empowerment date back to the popular African-American movements for civil rights that happened in the United States of America in the 1960s, after the efforts of civil disobedience and voter registration to make democratic rights effective. The context in which "Black Power" would come up, as the acknowledgement of one's heritage and the construction of a sense of community among black people, is the ideal which the *Quilombo* wishes to achieve.

In the middle of the 1970s, according to the author, the term "empowerment" started being used by the women's movements. Similar to the oppression experienced by black people, the need to confront the absence of public policies supporting women's harsh life and work conditions needs to be highlighted. Empowerment was defined by Batliwala (1994, p. 130), as "… a process directed towards transformation of nature and direction of systemic forces that marginalized women and other excluded sectors." That is, a process of caring for women and other categories systematically deprived of power, aiming towards the transformation of matrixes and directions of oppressive powers.

Djamila Ribeiro (2018) points out that empowerment implies an action that produces collective and democratic results, and is a process that must be considered both in the personal and collective spheres, depending on the appropriation of each person involved, since it demands their active involvement, in parallel to the opportunity/stimuli that are socially presented. Berth (2018) complements by pointing out that the concept of empowerment is an instrument of political and social emancipation that does not aim to remove power from one and hand it to another – inverting the poles of oppression – but it is a struggle for the elimination of social injustice, since the oppressed themselves must develop power for the change to occur, as power will not be given to them upon request.

Stromquist (2003, p. 23) points to four dimensions of female empowerment, each of them fundamental, but insufficient if not followed by the others. "These are the cognitive dimension (critical understanding of one's reality), the psychological dimension (feeling of self-esteem), the political dimension (awareness of power inequalities and the ability to organize and mobilize) and the economic dimension (capacity to generate independent income)" (Stromquist, 1995 *aped* Stromquist, 2003).

Nelly Stromquist's proposal comes from a feminist perspective, which is dear to us, and enables a pedagogical dialogue, as proposed in this handbook. Besides, this author has a long trajectory of work in Latin America, having theorised on the education of adult women and other awareness-raising projects for this public, especially in Brazil, when the own local gender researchers did not pay attention to such potentialities (Sardenberg, 2006).

The cognitive component refers to the understanding of the causes and conditions of the subordination of women in individual and global social contexts. It involves the need to make choices that can go against cultural expectancies and comprehend behavioural patterns that create dependence, interdependence, and autonomy within family and society in general (Hall, 1992 *aped* Stromquist, 1995).

The psychological component includes the development of a "power-to-act feeling" both on a personal and social level, to improve people's condition, as well as the belief that they can be successful in their efforts to change. The rejection and questioning of stereotypes related to sexual roles are associated with this dimension, for example.

The political component implies the capacity to analyse the surrounding environment in political and social terms; it also means the ability to organise and mobilise for social change, considering that the process of empowerment must involve individual consciousness and that collective action is also fundamental to achieve social transformation (Stromquist, 2003).

Lastly, the economic component presupposes that women are economic agents emancipating themselves financially. Although having a job often means a double workload for women, the author reinforces that access to the work market gives women more autonomy, increasing their economic independence.

Upon verifying that in Brazil black women in general earn on average 54% less than white men (DIEESE, 2021), the importance of disaggregated data production as well as its intersectional analysis should be perceived. Black female empowerment is linked to the debate of intersectionality, a concept officially coined by Crenshaw, in 1991, but already presupposed in practices and discussions of black feminism. According to the author:

The intersectionality (...) treats specifically the way by which racism, patriarchy, class oppression and other discriminatory systems create basic inequalities that structure the relative positions of women, races, and ethnicities, among others.

Besides that, intersectionality deals with the form [of] how actions and specific policies generate oppressions that flow along this axis, constituting dynamic aspects of disempowerment.

(Crenshaw, 2002, p. 177)

There is consensus that intersectionality seeks to exhibit and understand the intertwining between different social markers, especially the ones of gender, race, and class, without hierarchising them. Akotirene (2019) points out that hierarchising suffering is not feasible given the structural load and the heterogeneity of connected oppressions, considering that intersectionality may be a tool for the perception of the modern colonial matrix that focuses on oppressed groups.

Dialoguing with such reflections, Collins and Bilge (2021) propose four power domains, from which it is possible to perceive matrixes of intersectional oppression. These are disciplinary, structural, cultural, and interpersonal. The first one refers to the form in which norms and rules are applied distinctly among people who carry different intersectional social markers. The structural domain regards how society's fundamental structures are crossed by intersectional power relations, as well as constituting the subjects with which they interact. The cultural domain reflects how culture and discourses linked to it are permeated by (and reproduce) the social inequalities, generally legitimating those. Lastly, the interpersonal domain reflects how individuals experience the convergence of the three aforementioned domains, shaping identities and social interactions, highlighting individual experiences given the complexity of the thought identities from intersectionality.

These domains of power, which come from an intersectional perspective, dialogue with those dimensions of female empowerment. While the former allows for the appreciation of the matrices of oppression, the latter starts from the opposite direction, aiming to point out possible spheres of empowerment of marginalised groups based on their experiences. Thus, the psychological, economic, political, and cognitive dimensions of female empowerment blend with the structural, disciplinary, interpersonal, and cultural domains, improving the analysis of oppressions suffered by groups and people in an intersectional perspective, allowing combative actions in favour of these.

Methodology

For this research, the feminist standpoint principles are considered which, freeing us from the supposed neutrality proclaimed in the dominant scientific construction, allows objectivity to be achieved from the explicitness of the researchers' identities and their relationship with the object of investigation. From this perspective comes the validation of the experiences of women and other socially marked groups in contemporary scientific production. We also consider the epistemological current of feminist postmodernism, highlighting the determining character of the discursive constructions and the fluidity of categories related to gender and sexuality in this context. Thus, dialoguing with both enables the interaction with other domains of knowledge in this field, such as black feminism. Through the recognition of the transience and multiple identities of the subjects in contemporaneity, one can perceive the cultural (and discursive) constructions that generically shape singular actors, opening multiple possibilities of understanding and validation of the experiences of the black women interviewed here.

Table 10.1 Identification of the interviewees

Characterisation	Fictitious name	Age	Position they work/worked in
Interviewee 1	Brenda Robinson	44	Former flight attendant
Interviewee 2	Bessie Coleman	33	Flight attendant/aviation student
Interviewee 3	Mae Jemison	35	Flight attendant
Interviewee 4	Madeline Swegle	33	Flight attendant
Interviewee 5	Chipo Matimba	33	Aeroplane pilot (off the job)
Interviewee 6	Elizabeth Petros	34	Aeroplane pilot (off the job)

This qualitative and exploratory research made use of interviews as a technique of data collection, with an intentional sample, mediated by *Quilombo Aéreo* and conducted by virtual means, during the months of October and November 2021. The semi-structured script focused on both the interviewees' personal and professional lives investigating the hardships they faced for being black women acting professionally in a sector mostly composed of men and focused on public attendance, in its almost totality for and by white people. Furthermore, we have also asked about the collective hereby studied.

Among the six interviewees (Table 10.1), three are active flight attendants, while one of them is studying to become a pilot. Two of the interviewees are graduated pilots but are not employed in such occupations. The last interviewee was a flight attendant who quit the job due to medical recommendations. To preserve the identities of the women who participated as research subjects, a fictional name – whose inspiration was taken from black women who were aviation pioneers – was used instead of their real names. All of them have no children, except for Brenda Robinson, who has one daughter.

The data analysis was conducted from a thematic analysis that, according to Gaskell (2008), is conducted when the recurrence of data collected is verified, perceiving the most mentioned topics by the interviewees, seeking patterns and connections that lead to a wider reference framework. Since this was qualitative research with in-depth interviews, the repetition of patterns and information was considered sufficient to limit the sample.

Results and Discussions

Here we intend to show, based on the speech of the interviewees, how actions individually developed by them, as well as by the collective, can be understood as strategies for the empowerment of black women in the aviation industry. Thus, we investigated empowerment at personal and collective levels, which are "two inextricable sides of the same process (...) as an empowered collectively cannot be formed by individualities and subjectivities that are not consciously acting within processes of empowerment" (Berth, 2018, p. 42).

Bringing the dimensions of empowerment pointed out by Stromquist (2003) as support, we begin with the cognitive aspect, which regards the criticism about reality. When inquired: "Who are you?" the interviewees highlighted:

A black *favelada* woman (...) a black child filled with dreams that lived among the militia and knows the pains of growing up in a place like this. A woman who felt prejudice on her skin, especially race prejudice, in a gentrified and glamourized environment, which was aviation at the time when I joined it (....).

(Brenda Robinson)

[A] flight attendant for thirteen years, black *favelada* woman from *Axé* (...).

(Mae Jamison)

(Explanation of a *favelada*: in Brazil, the so-called periferia and the favela are historical and social constructions that relate to economy and culture, materialised in different forms of occupation of the urban space, reflecting the processes of (re)existence of the black population in the living body of cities. We use the term favela to refer to urban territories that were occupied in a way that does not comply with the control of the state. In many cases, they are close to the commercial centres and economic hubs of the cities and came into existence before or together with them. Each favela has its own history and identity. The "periferias" are territories on the outskirts of the cities and towns, distant from the commercial and economic centres. They can spring from spontaneous occupations or population removals and relocations promoted by the state or by private interests Some of these "periferias" have some basic infrastructure. Some favelas are in "periferias" and some "periferias" are favelas.)

In the interviewees' speech, we identified critical consciousness of their social position and violence suffered specifically for being black women. The process of recognising oneself as a black woman is relevant while complemented by observations about Chipo's choice for the aviation career:

(...) I heard a lot: "Wow, but you took a course that has nothing to do with you, it's not for you. Do you know any woman who does that? Are there black women in it?" People want to place their limitations upon us... Until you find that out, you cry a lot, right?

From these brief excerpts, we identified that the dimension of empowerment about the cognitive aspect of critical recognition of reality is contemplated, since in recognising themselves as black women they are invoking the social marks that, intersected, impose specific barriers for them to access and remain in Brazilian aviation.

Bessie Coleman states: "In my imagination, I would never be able to be a pilot, so we thought that it was only for men." When questioned about the profiles of their teachers and instructors throughout their training, there is unanimity among the interviewees regarding the absence of black women, as well as the small presence of white women or black men in those ranks. Upon persisting on admittance into a sector hostile to them, aiming at personal objectives, they encouraged representativeness and the creation of the collective studied here, leading to social impact linked to collective empowerment.

When pondering the psychological dimension of empowerment on the feeling of self-esteem, we have observed its intrinsic interaction with the cognitive dimension, already analysed, bringing the report on the process of recognising oneself as a black woman as an example.

Yeah, look, it is crazy, it was mainly from readings, right? Reading books, even reading Djamila Ribeiro's book, "Who's afraid of Black Feminism" she talks about black women, she mentions several situations and while reading the book I said "Goodness! So, I'm black?" (laughs), it's quite funny. It's because well... my family is religious, they don't have this thing of the... black love, of let's go... let's recognize our characteristics and, well, find out that they are incredible and beautiful, they don't have that.

My father, for example... wow, my father is super ignorant, he says, until now, that my hair looks like a homeless girl's hair who doesn't take a bath. But then there's another thing, reading bell hooks, with my psychologist, indicated by Quilombo Aéreo. She sent me a very interesting text... it's the text "Living to Love". She says that black people have an entire slavery legacy, black people had to hide emotions, [and] didn't learn to love themselves... Well, these people began having descendants and forming, I don't know, they were forming families how? My goodness, you ... need to survive, to give love to someone, no, there's no time, he has to survive, I have to provide you with food, health, a home...So there's no time to give love, that whole thing. So, my family has a lot of that... they didn't have this thing of... politicization, of having all this thing, you know? This thing of love, you are beautiful, your hair is wonderful, you know? You are black, my daughter. There was none of that, so, people, I didn't know I was black, my mother never called us that way. It's crazy. On the contrary, my father used to say my hair was ugly, well, how is a person like that going to think she is black? Nope, it's brunette. And then, reading Djamila Ribeiro and all that, I discovered that, and then my world fell apart. Well, when Marielle Franco was murdered, that was another *boom,* in which I felt like that: guys, something is happening in this country, there is something wrong. Along with my self-discovery as a black woman, there was that, her murder, and then I said – guys, what now, like, everyone that bad-mouthed her got me saying – wait, but you didn't know her, I did not either, but let's research. Then I went, also, recognizing myself in these women, well, mainly in her, and then I began reading a bit more, politicizing myself, knowing a bit more about feminism, reading about Angela Davis as well and... then I didn't stop anymore, I didn't. Well, lots of transformation, huh?! (laughs) (...) well, recognizing myself as a black woman was a process... well... a transforming one, and a very, very deep one for me, deep indeed.

(Madeline Swegle)

This report highlights the importance of the psychological dimension in the process of empowerment, showing that the access to means that allow the process to happen is fundamental. In this case, the professional experiences of the interviewee gave her conditions to access the psychological dimension, as well as the critical view (cognitive). It should be noted that both were mediated by the political (through the *Quilombo Aéreo*) dimension as well as the economic, since the airline crews are considered relatively well-paid in Brazil.

Still, from the perspective of the psychological dimension, we have noticed that the interviewees recognised themselves as role models and have shown pride for having reached their professional positions, reflecting the psychological dimension of empowerment inherent to such recognition. However, they have also pointed out that they feel their mental health is constantly put to test due to the frequent racist and sexist acts they suffer, especially by differentiated demands and treatment. They feel that, for being few black women in such positions, they are held accountable "for the whole," feeling greater pressure not to make mistakes, consequently increasing pressure towards black women. Kilomba (2019) reminds us that racism implies a process of naturalisation in which all members of the group have stereotypes imposed upon them, as well as stigmas that point them as problematic, hard to deal with, dangerous, lazy, or exotic, among others. Thus, collective vigilance among black women is conducted so that they do not make mistakes.

Regarding mental health, Mae Jemison, one of the collective's founders, points out that the organisation has as a future goal to provide group therapies to the *aquilombadas* with the intention of "allowing our pains to be exposed and treated in a safe place, where everyone can be welcomed by black mental health professionals, who are already part of the collective."

Another aspect identified as a possible indicator of the empowerment process these women have been going through, still in the psychological dimension, regards the defence of their natural hair. Brenda Robinson relates how, at the beginning of her career, she succumbed to the impositions of airlines:

(...) I began cutting, cutting, cutting [the hair]. I turned into someone that became such a person just to be accepted (...) I had curly hair down to my waist, full, beautiful, wonderful, powerful, pure power, my crown, which I had to tie in a braid clip, within a bag which fit a tiny bit of hair (...) which they gave me that and said: that's your uniform, do what you can to make it fit or you won't fit in.

Just like Brenda, other interviewees reported having suffered prejudice because of their hair, both for being admitted into the aviation industry and for remaining in it. Mae Jamison, for example, when questioned about advice she would give to black women who intend to be crewmates, shot: "(...) tie your hair to get in because aviation is not ready to accept us with our crowns." However, we perceive indicators of the empowerment process when they begin demanding respect and proper conditions to wear their hair freely, through their work journey, as well as when they refer to their hair as "crowns."

While they understand the need to adapt to the system in order to join it, black women articulate themselves to use their natural hair in the aircraft, promoting collective actions of pressure – which meets Berth's (2018) reflections when the author points out the praise to naturally curly hair as an important factor in the combat against racism, as when black women love their hair, they will throw back offences, rejections, and exclusions that were aimed towards them throughout their lives.

Moving to the economic dimension of empowerment, we have identified that working as a crewmember tends to provide decent payment in the Brazilian context. The starting salary for flight attendants, between minimum wage and hours flown, corresponds to about five national minimum monthly wages and approximately three times the average income of black women in Brazil (IBGE, 2019). Upon achieving the basic income as flight attendants, such women have "freedom" to make choices, without financial dependence. Sometimes, the financial perspective given by the professional placement made it possible for these women to be her family's support, allowing her own ascension and empowerment, as well as that of other black women in her family (Bessie).

Furthermore, relatively stable income influences access to other goods and services, which can favour the process of personal and collective empowerment. Upon disposing of enough income to invest in education, therapy and psychological support, goods and cultural services, legal orientation, physical exercise, and aesthetics, among others, it increases quality of life and facilitates individual and collective empowerment.

In this field, the project *Pretos que Voam* (Blacks who fly), idealised by the *Quilombo Aéreo*, stands out, aiming to favour the graduation of black people to act as flight attendants. Considering the absence of black people in this occupation, assessing the high costs for graduation, and recognising the burden of structural racism that excludes black people from access to certain spaces and positions, the collective created a project based on *match-funding* which funds the training, and in some cases, the necessary documentation for acting in aviation for suburban black people.

The first edition of Pretos que Voam graduated 10 flight attendants (7 women and 3 men) in November 2021, in Porto Alegre (RS). The conditions of eligibility for candidates were as follows: being self-declared black; having graduated from a state high school; proving social vulnerability, among others. The scholarships included the complete flight attendant course in an accredited school, uniforms, an English language mini-course and mentorship with experienced flight attendants.

Two of the female flight attendants were hired, while others are going through selection processes with the airlines. Aiming for the project's sustainability, it is agreed in the contract that people who benefitted from such actions, after the third month acting as crewmates, must return the values invested in their formation, to allow other people to benefit from it, making the formative process of black crewmates into an organic one, considering that "empowerment is a process and not an end in itself" (Berth, 2018, p. 56).

In 2022, a new *match-funding* campaign was developed to create the first anti-racist and Afrocentric Brazilian civil aviation school to provide distance-learning instruction, so that the entire country can be contemplated. This project clearly involves not only the economic dimension – by facilitating access often denied by lack of resources, thus making it possible for people to reach for better-paid positions – but also the other dimensions.

The very own articulation of the collective *Quilombo Aéreo* can be considered both a result and a precursor of the empowerment process, especially through the political dimension. Starting from the ideas of two black women sensitive to racial causes and the need for representation in the air sector, in 2018 they presented three initial guidelines: representativeness – showing that there are black women in the sector, telling their stories and paths, and fighting the Eurocentrist stereotypes that exist in the area; access to formation – bringing information and training to the Brazilian suburbs, since there is lack of knowledge regarding the profession and possibilities of access to such market; and employability – to talk with the airline companies to integrate those professionals into the labour market, bringing about racial literacy for the sector.

Over more than three years of acting, the contribution of the political articulation linked to the actions of the *Quilombo Aéreo* is undeniable, facilitating access to other empowerment dimensions: cognitive, economic, and psychological. We also agree with Berth, when she states:

> When talking about empowerment, we are essentially talking about political work, even though it runs through all areas of an individual's education and all the nuances that involve collectivity. In the same way, when we question the model of power that involves these processes, we understand that it is not possible to empower someone. We empower ourselves and aid other individuals in the process, being conscious that the conclusion will only happen through a symbiosis between the individual and the collective process.
>
> *(2018, p. 130)*

Through political mobilisation around the initial guidelines, the *Quilombo Aéreo*, according to its founders, managed to obtain results such as the Project *Pretos Que Voam*; the creation and dissemination of a resume bank of black people for the airlines; the offer of psychological care for female black crewmates, with conversation and experience exchange groups; contracts for racial literacy for the ANAC workers; visits to state schools in less privileged areas; quantitative and qualitative research about black male and female crewmembers; acting along with the Aeronauts National Union on issues of diversity and inclusion.

Specifically, regarding airlines, the interviewees explained that the *Quilombo Aéreo* has mobilised the insertion of collective members in the committees of the three main national airlines, promoted discussions about the employability of black people in the sector; conducted lectures about "The dark side of aviation" along with the human resources departments; achieved the authorisation for tresses, dreads, and natural curly hairs for black crewmembers; negotiated the implementation of an ethic channel for receiving reports of racism (being developed by one company at the moment); took an active part in the making of the personal presentation manual of one of the airlines; free air tickets for *Pretos que Voam* scholarship students to participate in selection processes in other cities; partnerships with aviation schools, GOL, AZUL, ANAC, and EMBRAER.

When glimpsing the main dimensions of empowerment related to the individual and collective actions analysed throughout the interviews (Table 10.2), we cannot fail to emphasise, once again, the inextricable connection between all the dimensions for effective empowerment. It is also noticeable that some actions reach more than one dimension, only the main one being registered here. Finally, some actions, in a certain dimension, facilitate the implementation of other actions that favour the other dimensions.

From this synthesis it is possible to perceive the relationships between the actions identified in this research with the dimensions of empowerment and, also, the inextricable articulation among them, which is summarised in the dynamics of mutual and necessary influence between such dimensions, as illustrated in Figure 10.1.

Table 10.2 Individual and collective actions linked to the dimensions of empowerment

Cognitive	Psychological	Political	Economic
Critical acknowledgement of the condition of being a black woman	Access to black feminism through readings	Founding and/or participation in the *Quilombo Aéreo*	Access to wages above the national average for black women
Understanding of their representativeness and absence in the sector	Therapy groups for/with black women	CV bank for black people in the aviation sector	Access to stable income
Being a pilot/flight attendant	Recognition as role models	Actions with airlines to include and respect black people and women	Project *Pretos que Voam*
Visits to schools in poor areas	Defence of their natural hair	Race literacy contract with ANAC	Quality of life
		Quantitative and qualitative surveys	
		Acting along with the Aeronauts National Union on issues of diversity and inclusion	
		Insertion of collective members in the committees of national airlines	

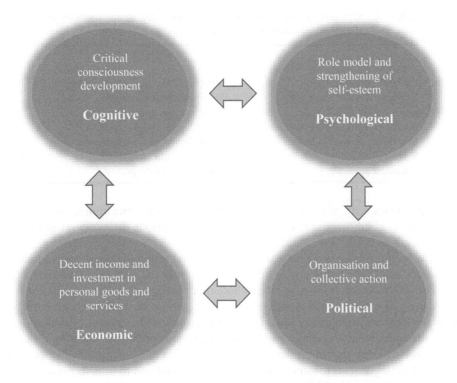

Figure 10.1 Interweaving between the four dimensions of women's empowerment.

In this way it is easy to see that each dimension of empowerment is a "step" to another one, being all of them, in synergy, an essential condition in the empowerment process.

It is worth pondering that such actions made possible the visibility of race and gender questions within the aviation industry and extrapolated the institutional frontiers. According to estimates from the masterminds, there are over 13,000 people following and supporting the collective in its social networks and, most importantly, feeling represented.

Over 300 donations from individuals for the creation of scholarships for affirmative action in the area have happened. There are social partnerships such as the *Fundação Tide Setúbal*, *Fundo Baobá* for racial equity; *Rede de Mulheres Negras*, *Odabá Associação de Afroempreendedorismo*, *Movimento Negro Unificado*, among others still in the process of consolidation.

Through one of its creators, the collective was contemplated to participate in the Baobá Fund's *Programa de Aceleração do Desenvolvimento de Lideranças Femininas Negras Marielle Franco* (Program for the Acceleration of Development of Black Female Leaderships Marielle Franco), supported by Kellogg Foundation, Ford Foundation, *Instituto Ibirapitinga* and Open Society Foundations.

The program promotes personal and organisational financial investment, seeking the consolidation of black women in leadership positions. Such resources have been personally used for the funding of flight hours for the formation of the contemplated student as a female pilot and a post-graduation course. Organisationally the resources were invested in the construction and maintenance of the collective's website, and a few

Table 10.3 Obstacles for black women to remain in the airline industry and domains of power from Oliveira et al. (2022, p. 160)

Obstacles for remaining	Domains of power
Lack of representativity	Structural, cultural, and interpersonal
Different levels of demands	Disciplinary and cultural
Denial of the black body	Interpersonal and cultural
Non-acceptance of the afro hair	Disciplinary and interpersonal
Mental/occupational health	Disciplinary and interpersonal
Sexism	Structural and cultural
Harassment	Cultural and interpersonal

leadership courses with black organisations, thus showing, once more, the articulation between different empowerment dimensions, as well as among the personal and collective dimensions.

The *Quilombo* founders estimate a reach and social mobilisation of approximately 12 communities among the South and Southeast regions of Brazil, and approximately 32 black families directly affected. It is interesting to note that the support network available is made up exclusively of women professionals, mostly black, who voluntarily dedicate their formations in the legal and psychology areas to help the project with contracts, orientations, and care for the *aquilombadas*, besides psychological preparation for the selection processes.

It is possible to note that many actions identified within the dimensions of empowerment are in direct dialogue with the mitigation of the barriers related to intersectional power domains, presented in Table 10.3, that impact the permanence of black female crewmembers in the Brazilian airline industry.

Thus, from an analysis of the domains of intersectional power, we identified how black women live their professional experiences as air crewmembers and, from the dimensions of empowerment, we noticed how their articulation around the collective *Quilombo Aéreo* has enabled, albeit slowly, a change in this scenario.

Final Considerations

Throughout this work, we sought to address how a collective formed by aeronauts has been promoting the empowerment of black women in the Brazilian airline industry. In a context in which black women comprise the base of the social pyramid, national aviation reproduces and reinforces limitations and prejudices imputed to these women, naturalising data showing that in Brazil, only about 5% of the flight attendants are black, that is 2.5% of the pilots are women and approximately 2% are black men, exposing the absence of black women in such position in the Brazilian airlines.

From the actions of the *Quilombo Aéreo*, it has been possible to explore the routes and strategies used for strengthening the presence and qualification of black people, especially women, in the airlines. Founded in 2018, the collective follows an agenda promoting visibility, support, self-care, and equity, as well as articulations seeking individual and collective empowerment in different dimensions.

As explored throughout the text, we have made use of a theoretical framework of women empowerment and intersectionality, both intimately related to black feminist movements, providing a practical analysis of the strategies and individual and collective actions of black women crewmates. The theoretical assumption of intersectionality facilitates the under-standing of the power domains and their corresponding interconnections, leading to the un-derstanding of the places occupied by certain groups, like the black women crewmembers, in a society led by patriarchy-racism-classism (Akotirene, 2019).

The perspective of the empowerment dimensions reveals how, despite the concept having suffered shallow appropriations, it is possible to perceive life experiences and strategies that lead to the empowerment of "excluded" groups. Upon recognising and subverting (cogni-tive) structures and societal troubles which limit them, accessing and remaining in hostile environments, such as the aviation industry, it is possible to legitimate these black women crewmates as role models, strengthening the self-esteem (psychological), stimulating the undertaking of other agenda, such as the positive appreciation of curly hair, or even the creation of the collective itself.

The *Quilombo Aéreo* arises from the sharing of pains of two black women in the avia-tion industry who, aware and empowered, organised an agenda (political) for the collec-tive empowerment of Brazilian black aeronauts. There are many fronts and actions of the *Quilombo* which aim to increase representativeness, promote racial literacy, and have a voice in leadership and decision spaces, among others, which have as their focus the change of power vectors currently perceived. Such political articulation can be seen in practice while observing Bessie's position about necessary flight hours for her formation as a pilot: "Then I explained: I'm here financed by a fund which is for valuing women, I won't come here and give my money to men. It is also a form to encourage the flying club to hire more women. Then I couldn't fly on several occasions, because there were no women."

It is possible to identify, here, both the personal empowerment perspective (Bessie) and the collective one (pressure/incentive for the presence of more women), as well as the im-portance of resources (economic) for envisioning the other dimensions. Such connection be-comes evident while observing the *Pretos Que Voam*, which facilitates the access of *favela* black people, especially women, to a reasonably well-paid job, impacting their income, but also the increase of representativeness and recognition, since the formation is Afrocentric. More than that, it brings intersectional references promoting organicity, and foreseeing the capillary empowerment of even more black women, who have usually occupied, mostly, subservient positions in the tourist sector.

"Gender equality strategies for the tourism sector are vital for women's empowerment, and must be backed by institutional and budgetary support" (UNWTO, 2020, p. 14). However, little has been done in this sense in the context analysed. Companies and agencies responsible for aviation in Brazil should pay attention to actions related to the implemen-tation of efficient channels for the registration and solution of complaints of racism and sexism, both by employees and by customers and partners; promotion of racial literacy programs for all hierarchical levels of the institutions; organisation and stratification of employee data based on gender and race; incorporation of actions aimed at hiring women and black people for functions in which they are still underrepresented, among others that aim at the expansion of diversity in this scenario and the recognition of women and black people in this sector.

In addition, it is worth remembering that the United Nations (UN) launched the 2015 declaration: "Transforming Our World: The 2030 Agenda for Sustainable Development" which, among 17 goals for sustainable development (SDG), brings Gender Equality (SDG 05), the Reduction of Disparities (SDG 10) and Responsible Consumption and Production (SDG 12). The participation of public instances on behalf of the state, as well as the private initiative, here represented by the airlines, is of great relevance for the achievement of such goals. Both spheres are interesting arenas for the incorporation of demands, such as those presented by *Coletivo Quilombo Aéreo*, to promote sustainability and responsibility in the Brazilian airline sector.

For future research on gender in tourism, we recommend attention to other social markers, and the intersectional bias may be useful for more fruitful analyses in specific contexts. In the field of broadening the impacts of academic research, we suggest that these bring critical projections for the alteration of matrixes aiming towards social justice in the sector. Paying attention to the SDGs may be a good strategy to justify the look on gender and race, as well as the expansion of discussions on ESG (environmental, social, and corporate governance), may serve as support to take such proposals to practical ways. Thus, we reinforce the relevance of expanding reflections regarding intersectionality along the tourism studies, as well as the potentiality of empowerment of oppressed groups in this scenario.

References

Akotirene, C. (2019). *Interseccionalidade*. Pólen.

ANAC. (2021, January 11). *Painel de Indicadores do Transporte Aéreo 2019.* https://www.gov.br/anac/pt-br/assuntos/dados-e-estatisticas/mercado-do-transporte-aereo/painel-de-indicadores-do-transporte-aereo/painel-de-indicadores-do-transporte-aereo-2019

Batliwala, S. (1994). The meaning of women's empowerment: New concepts from the action. In G. Sen, A. Germain, & L. C. Chen (Eds.), *Population policies reconsidered: Health, empowerment and rights* (pp. 127–138). Harvard University Press.

Berth, J. (2018). *O que é empoderamento?* Letramento.

Collins, P. H., & Bilge, S. (2021). *Interseccionalidade*. Boitempo.

Crenshaw, K. (2002). Documento para o encontro de especialistas em aspectos da discriminação racial relativos ao gênero. *Estudos Feministas, 171*(1), 171–188. https://www.scielo.br/j/ref/a/mbTpP4SFXPnJZ397j8fSBQQ/?format=pdf&lang=pt

DIEESE. (2021). *Brasil: a inserção da população negra e o mercado de trabalho.* https://www.dieese.org.br/outraspublicacoes/2021/graficosPopulacaoNegra2021.html

Equipe Lupa. (2018, April 18). *Pnad contínua: cinco verdades sobre a renda dos brasileiros em 2017.* Lupa. https://piaui.folha.uol.com.br/lupa/2018/04/13/cinco-verdades-renda-brasil/

Gaskell, G. (2008). Entrevistas individuais e grupais. In M. Bauer & G. Gaskell (Eds.), *Pesquisa qualitativa com texto: imagem e som: um manual prático* (7th ed., pp. 64–89). Vozes.

Hall, Margaret (1992) *Women and Empowerment. Strategies for Increasing Autonomy.* Washington, D.C.: Publishing Corporation.

IBGE. (2019). *Desigualdades sociais por cor ou raça no Brasil. Estudos e pesquisas: informação demográfica e socioeconômica.* 41. https://biblioteca.ibge.gov.br/visualizacao/livros/liv101681_informativo.pdf

IPEA & Fórum Brasileiro de Segurança Pública. (2021). *Atlas da Violência 2021.* https://forumseguranca.org.br/atlas-da-violencia/

Kilomba, G. (2019). *Memórias da plantação: episódios de racismo cotidiano.* Cobogó.

Matias, M. S. (2015). *SNA – Mapeamento biopsicossocial do aeronauta brasileiro – conhecendo os aspectos biológicos, psicológicos e sociais durante o exercício da profissão.* https://www.aeronautas.org.br/images/_sna/noticias/Mapeamento_saude_aeronauta_br.pdf

Oliveira, N., Gabrielli, C., Santos, G., & Amorim, L. (2022). Intersectionality between racism and sexism in the Brazilian airline industry: Perceptions and strategies of Black women crewmembers. In P. Cembranel, J. R. R. Soares, & A. R. Perinotto (Eds.), *Promoting social and cultural equity in the tourism sector.* IGI Global.

Pirro, B. G., Gonçalves, J. P., Silva, M., Delphino, R. B., Oliveira, R., & Lima, R. (2022). O Impacto da Pandemia do Covid-19 para as companhias aéreas e tripulantes. *Brazilian Journal of Production Engineering, 8*(5), 23–27.

Quilombo aéreo. (2022). *Quilombo Aéreo*. https://quilomboaereo.com.br/#

Ribeiro, D. (2018). *Quem tem medo do feminismo negro?* Companhia das Letras.

Sardenberg, M. C. (2006). *Conceituando "Empoderamento" na Perspectiva Feminista*. 12 f. Artigo – NEIM/UFBA.

Smolen, J. R., & Araújo, E. M. (2017). Raça/cor da pele e transtornos mentais no Brasil: Uma revisão sistemática. *Ciência & Saúde Coletiva*, 22(12), 4021–4030. https://doi.org/10.1590/1413-812320172212.19782016

Stromquist, N.(1995) 'The theoretical and practical bases for empowerment', in Carolyn Medel-Anonuevo (ed.) *Women, Education, and Empowerment: Paths towards Autonomy*, Hamburg: UNESCO Institute for Education.

Stromquist, N. P. (2003). Education as a means for empowering women. In J. L. Parpart, S. M. Rai, & K. Staudt (Eds.), *Rethinking empowerment: Gender and development in a global/local world* (pp. 22–38). Routledge.

UNWTO. (2020). *Global report on women in tourism: Second edition*. https://www.unwto.org/publication/global-report-women-tourism-2-edition

PART V

Practicing Gender in Tourism III

Gendered Mobilities

11

SOLO FEMALE TRAVELLERS' EMOTIONS

An Analysis of Specialist Bloggers' Narratives

Marina Abad Galzacorta and Maria Cendoya Garmendia

Abstract

Travelling alone is an essential part of empowering women and enhancing personal development. Although women have been travelling independently for centuries, solo female tourism is currently a booming trend around the world. They tend to experience feelings of freedom, spontaneity, and empowerment but also experience feelings of anger and fear when confronted with perceived barriers. In this context, women's blogs have proliferated over the last few decades, shifting from classic destination-based travel blogs to others that focus on the emotional and safety aspects of female solo travel experiences. The chapter tackles the profiles, emotions, and motivations of solo female travellers from a theoretical point of view and it analyses the discourses of real women's experiences in over 30 specialised Spanish-language travel blogs. The results confirm that women's travel narratives have changed, and that the expression of emotions is one of the dominant themes in the weblogs of women who travel alone.

Keywords

Emotions; Solo traveller; Women; Travel blog; Empowerment; Tourism motivation.

Introduction

Over the last few decades, interest in gender-related issues has evolved significantly, particularly in the field of tourism where women play a vital role in supporting and achieving their empowerment (United Nation World Tourism Organization [UNWTO], 2017). The United Nation (UN) Millennium Development Goals were set at the World Conference on Women in Beijing in 1995 and were signed in 2000. One of these goals is to "promote gender equality and empower women." The UN has made this goal a core part of its agenda, making it one of the 17 SDGs in the 2030 Agenda for Sustainable Development (2015). Since then, the UN World Tourism Organization (UNWTO), in partnership with UN Women, has been working through its 'Ethics, Culture and Social Responsibility Programme' where is the section "Women's empowerment and tourism."

To mention another milestone, in 2007, the UNWTO dedicated World Tourism Day to the theme "Tourism opens doors for women" and organised a Forum on Women in Tourism at the ITB Berlin (Internationale Tourismus-Börse Berlin) tourism trade fair the following year. In 2010, UN Women and the UN Global Compact launched the Women's Empowerment Principles (WEPs) and also the UNWTO published the first edition of the Global Report on Women in Tourism (2010) and later, a second edition of the report (2019). The report, focused on the role of women in the labour market (discrimination, poor payment, or insecurity), concludes that the tourism sector can pave the way for women's professional success (UNWTO & UN Women, 2011, UNWTO, 2019). In short, gender equality is a priority on the 2030 journey where the fifth goal underlines "the capacity of the tourism sector to empower woman (…). So, tourism can be a tool for women to become fully engaged and lead in every aspect of society" (UNWTO, 2017, p. 16).

Another area of interest in gender-related tourism studies is sex tourism and the perception of women as a tourist attraction or product. Related to the concept of holidays being associated with freedom and opportunities for sex, several studies on the subject of sexual harassment have emerged. Research that focuses on demand often deals with reported sexual harassment and therefore focuses on emotional issues such as risk perception or fear and anxiety (Calafat et al., 2013; Vlahakis, 2018).

This chapter, however, focuses on women solo traveller as a niche market where 'solo' is referred to as those women who arrive at the destination alone (McNamara & Prideaux, 2010; Wilson & Harris, 2006). Women control more money today than ever before and nearly two-thirds of today's travellers are women; many of whom have travelled and travelled solo. In fact, 74% of women claim to have travelled solo or are planning to do so (trekksoft. com). Current data confirm that women travellers' profiles have changed and emphasise the growing popularity of this segment market through Internet-based data about solo women travel (Goggle searches, Pinterest, etc.) (Hamid et al., 2021). For instance, in Facebook there are a number of public pages, Solo Travel Society, and a Solo Female Traveller Network, a private group which connects more than 500,000 members. Also, tour companies report that most of their woman travellers are going solo (solotravelerworld.com).

For this reason, many different theoretical approaches have been adopted in the literature to study the motivations of solo female travellers and understand their travel decision-making process. Most research to date has approached the subject by examining perceptions of freedom, empowerment, and barriers and, as such, has examined emotions in female solo travellers (McNamara & Prideaux, 2010; Wilson & Harris, 2006; Wilson & Little, 2005, 2008). Other studies have taken a social identity approach towards understanding the role of gender in travel motivations and solo travellers' behaviour (Bowen, 2005; Ejupi & Medarić, 2022; Yang, 2021). Falconer (2011) pointed to the special issue on Female Travellers published by Tourism Review International in 2005 as a reference for academics interested in gendered approaches to women's experiences and his own research drew on the narratives of female backpackers to examine the emotional conflicts and barriers they experience on their travels. So, as a growing market niche, it has been widely studied mainly in recent years (Ejupi & Medarić, 2022; Pereira & Silva, 2018; Prideaux & McNamara, 2010) even studying in more detail the difference between "solo by circumstances" and "solo by choice" (Yang, 2021).

However, the outcomes are quite negligible and the different results are not conclusive; therefore, it is still necessary to fill the research gap in existing literature by adding important discussions on women solo travel (Hamid et al., 2021). Therefore, the main aim of

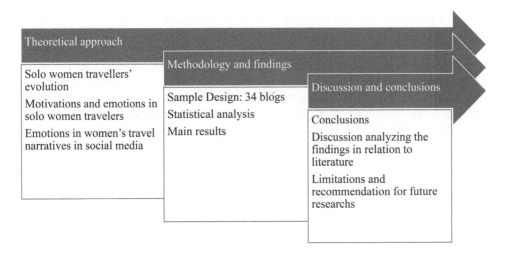

Figure 11.1   Structure of the study. (Source: Own elaboration)

this study is to analyse the profile of solo women travellers today. For the purpose to add to further research on women solo travelling trends, this chapter examines the role emotions play in their travelling experience through their narratives by reviewing a selection of female travel blogs (see Figure 11.1).

Solo Women Travellers' Evolution

Historically, women travellers were often conceived as unconventional, adventurous, intrepid women who chronicled their experiences in a range of narrative forms. The 5th-century pilgrim Egeria is one of the first female travellers on record. Also known as Etheria or Aetheria, she documented her travels in the book "*Itinerarium ad Loca Sancta.*" From the 19th century, other female travellers have also left a legacy of travel memoirs: the Swiss traveller Isabelle Eberhardt (1877–1904), the Belgian-French explorer-spiritualist Alexandra David-Néel (1868–1969), the British explorer, writer, and photographer Isabella Bird (1831–1904), or the North American Elizabeth Jane Cochran Seaman (1864–1922) (all of them cited in Morató, 2003, 2005).

These writings invite many questions such as: How could women travel back then? Wasn't it dangerous to travel unaccompanied? Weren't they afraid? How and why did they travel? Within the realm of cultural studies, Ferrús (2011) synthesised these ideas in her examination of the role of women as travellers and professional writers in relation to the concept of otherness. There is no doubt that they had to overcome significant sociocultural barriers in a male-dominated society where women's roles were clearly defined and strict norms existed regarding what was deemed appropriate behaviour for them (Wilson & Harris, 2007). So, the literature evidences that women have been travelling alone for much of history, but, often, they needed to adopt a masculine role during their voyage. In other words, they refer to women who travelled alone and enjoyed adventure as tomboys, because they had to abandon their femininity in favour of masculine adventure (Falconer, 2011).

McNamara and Prideaux (2010; citing Foo 1999 and Chai 1996) define solo female travellers "as women who arrive in a destination alone and do not travel to a destination as part of a packaged trip, group or tour" (p. 254). In Falconer's study (2011), the women travellers interviewed defined their experiences as opportunities to reinforce their individualised identity and enhance their personal development. Therefore, specialised service industries focused on making access to tourism more equal in terms of gender are proliferated around the world: travel agencies for women as Wild Women Expeditions (Canada), accommodation as the Bella Sky Comwell Hotel with a women-only floor (Copenhagen) or Som Dona hotel (Mallorca, Spain) and women-only taxi companies as Lady Driver (Brazil).

Some studies stress that solo travel is a big opportunity for the empowerment, positive change, or self-realisation of women (Hamid et al., 2021; Levy, 2013) and authors cited by Falconer (2011) have suggested that women are more interested in the quality of the personal and introspective experiences of travelling. However, although the market is growing, as in other areas of leisure, some differences and barriers still need to be overcome (Wilson & Harris, 2007; Wilson & Little, 2005). For instance, Levy (2013) has studied the differences between men and women in the field of leisure and tourism in terms of the quantity and quality of leisure consumption and Yang (2021) has explored the difference between solo and non-solo travellers. In summary, solo women travellers emerge as a growing and influential market segment.

Motivations and Emotions in Solo Women Travellers

Tourism is a hedonic pursuit and as such is a multisensory experience where the senses, cognitive processes, and emotions are engaged (Cohen et al., 2007; Gretzel & Fesenmaier, 2003; Sun & Zhang, 2006). Wilson and Harris (2006) agree on the concept of "meaningful travel" as involving the search for physical, emotional, or spiritual fulfilment through tourism. Regarding the complexity of the tourist experience, Sharpley and Stone (2011) also explored the factors that lead to positive and memorable tourist experiences, including the emergent motivations, behaviours, and responses to such experiences. It is essential, therefore, to understand how women experience destinations – particularly the emotional and affective factors involved – to improve customer satisfaction, tourism management, and planning processes, but also because they play a vital role in the tourist consumption decision-making process (Abad-Galzacorta et al., 2013; Bigné et al., 2008; Gretzel & Fesenmaier, 2003).

This is a complex field of knowledge. While some studies use the terms emotions, moods, and feelings as synonyms of each other, others differentiate them clearly (Cohen et al., 2007; Sun & Zhang, 2006). For instance, Reeve (2005) goes into considerable detail on the subject of emotions, motivations, and the relationship between the two, listing as many as 24 theories in the study of motivation and emotion (p. 18). McNamara and Prideaux (2010, p. 254) carried out a detailed review of the literature from the end of the 20th century to the start of the 21st on the motivations of independent women travellers. Several of these studies focus on the significance of travelling alone in the lives of women, where the perception of control over their own choices is central. They also confirm that women's motivations for travelling solo have changed. While early women travellers were driven by a sense of adventure and daring, women's motivations today are more complex and nuanced. As Wilson and Little (2008) explain, women today want to move beyond their comfort zones and develop feelings of independence and autonomy. They are looking for experiences which can help them come to terms with changes in their lives and develop a new sense of self. In other words, travel is related to the desire to challenge oneself, one's sense of autonomy, empowerment, or self-determination (Wilson & Little, 2005), "although the search for self may not be a stated or recognised motive for these women, and could be something that was discovered along the journey or reflected upon once at home" (Wilson & Harris, 2006, p. 165).

More recently, Pereira and Silva (2018) present a table with the main motivations dimensions of woman solo travellers and theorise about independent women's travel experiences through eight factors: identity and self-development, challenge, autonomy, adventure, learning, new life perspectives, escape and connectedness with others (p. 102). Hamid et al. (2021) present a conceptual model which proposes that self-transformation and positive changes in well-being came out as resultant factors of solo travelling in many cases. Similarly, Ejupi and Medarić (2022) also revise the literature about the motives of solo female tourists confirming that they vary and change over the life course. They include seeking physical, emotional, and spiritual fulfilment, self-confidence, empowerment, and getting out of one's comfort zone.

So, according to today's feminist discourse, the key motivation for women travelling alone is to empower themselves to challenge gendered constraints, resist the gendered "geographies of fear," and enhance their development as liberated women with freedom of mobility (McNamara & Prideaux, 2010; Wilson & Harris, 2006; Wilson & Little, 2008;). This idea is consistent with the cited concept of "meaningful travel" which "involves women searching for an increased sense of self and reconsidering their perspectives of life, society and their relationships with others" (Wilson & Harris, 2006, p. 161). However, some researchers oppose these theories, arguing that women are more deeply constrained, and that their perception of self-empowerment has more to do with fear and risk perception, which determines their leisure activity choices (McNamara & Prideaux, 2010; Wilson & Little, 2008). Elsewhere, Valentine's "geography of women's fear" associates women's fear or danger perception of certain places, such as isolated areas, large cities, and everywhere at night, with determining their use of space (cited in Wilson & Little, 2008).

Emotions in Women's Travel Narratives in Social Media

As mentioned previously, several approaches have been taken in the conceptualisation, classification, and measurement of emotions in academic literature (Abad-Galzacorta et al., 2013; Cohen et al., 2007; Reeve, 2005;). Bigné et al. (2008), who studied the salience of emotions in understanding consumer responses to hedonic services, defend two approaches: the examination of emotional experience and expressions (joy, anger, sadness, etc.) and the study of the dimensions underlying emotions. An interesting debate also exists regarding the words feelings, emotions, and moods, with some authors preferring to use the term "affective dimension" to describe internal feeling states (Cohen et al., 2007; Sun & Zhang, 2006). Following an in-depth review of the literature, Reeve (2005) defined up to eight basic emotions: fear, anger, disgust, sadness, threat and harm, joy, interest and motive involvement, and satisfaction. In short, while researchers are not unanimous on the subject, most authors agree that emotions are complex, and that the interactions between subjective and objective factors are what constitute behaviour, cognitions, physiological changes, and feelings.

Apter's (2005) Reversal Theory is one of the most widely used and well-validated theories in clinical case histories and phenomenological, psychometric, experimental, and psychophysiological studies. Reversal Theory is a psychological theory that deals with subjective experiences and emotions by identifying the individual's current goal states. It proposes four pairs of motivational states that influence felt emotions (telic-paratelic ends-related, conformist-negativistic rules-related, mastery-sympathy transaction-related, and autic-alloic relationship-related) and suggests that people move from one state to another, fluctuating between motivations and emotional experiences according to the moment and personal state. For instance, a person in the conformist state is compliant and agreeable but becomes rebellious, unconventional, and defiant in the opposing negativistic state. Reversal Theory divides emotions into two categories: somatic emotions, which "relate primarily to certain bodily feelings," and transactional emotions, which "relate to the outcome of our actions in relation to other people and things" (Apter, 2005, p. 48). In addition, different combinations of motivational states result in different emotional outcomes, which, according to the theory, can be divided until 16 primary emotions.

As discussed earlier, solo women travellers have been recording their emotions as they travelled the world for centuries. From methodological point of view, interviews have been widely used in research for exploring and analysing women's travel narratives (Ejupi & Medarić, 2022; Falconer, 2011, 2017; McNamara & Prideaux, 2010; Yang, 2021). But nowadays, social media are being studied as a valuable platform for solo women travellers to express their emotions, so they are an important secondary data source. In fact, some studies review a range of relevant social media contents to list motivational factors and reasons for solo women travelling (Hamid et al., 2021).

Social media platforms have also been used to overcome barriers and change women's roles (Weatherby, 2018). Although McNamara and Prideaux (2010) maintain that risks perceived by women do not prevent them from taking part in leisure pursuits, others claim that solo women travellers tend, in general, to be conservative about engaging in this kind of activity – mainly due to safety concerns – and tend to display a high-risk perception when travelling (Falconer, 2011; Wilson & Little, 2005, 2008). Consequently, negative emotional reactions, such as fear, anger, or anxiety, have been reported when situations were perceived as threatening (Falconer, 2011). Indeed, several women, influenced by the narratives of other solo female travellers, perceived travel as a risky, fearful, and dangerous activity and modified their holiday and travel activity behaviour as a result (Wilson & Little, 2005). Falconer (2011) claims that women often feel intimidated and fear for their personal safety when faced with threatening situations on their own. However, their narratives present confused feelings regarding their status as empowered women, and a personal conflict can be appreciated "between how they should respond, and how they feel, towards local strangers as lone women travellers" (p. 78). She refers to this contradiction as "negotiations of feminist identities" (Falconer, 2011, 2017).

In this context, social media platforms have provided an environment where women feel that they can express themselves freely. Women-only pages have become spaces where "they feel comfortable asking questions and felt that they would get appropriate responses" (Weatherby, 2018, p. 85). This author studied the role of social media platforms as discursive devices capable of changing identities and empowering women to venture into traditionally male domains such as the outback or natural wildernesses. So, she claims that online communities provide an environment where women feel comfortable not only asking questions but also sharing their experiences and emotions. Indeed, many studies have revealed that belonging to an online community has the power to reinforce one's sense of identity, recognising online disciplines such as netnography and blogs such as "Geeking communities" as appropriate tools for understanding and expressing individual realities (Kozinets, 2010). Social media platforms eliminate barriers between participants, allow specific questions to be asked, and provide a multifaceted tool for negotiating personal constraints (Wilson & Harris, 2007).

Therefore, many women have used their travels to write blogs where personal or group accounts of their experiences are recorded, and advice is shared with fellow solo travellers motivating women for sole exploration (Hamid et al., 2021; Weatherby, 2018). As Weatherby (2018) points out, blogs, unlike other social media platforms such as Facebook, "create a separation between the presenter and the audience" where "authors have full autonomy over their sites, and the context of posts is often focused on the experiences of one person" (p. 34). In the same way Hamid et al. (2021) cited authors who believe that "blogs can be an important resource in studying psychology and behaviour of travellers" (p. 8).

Social media platforms are also a source of inspiration for people interested in travel and an effective means of showcasing the changing trends in women's travel. And blogs, in particular, are a powerful device for getting more women involved in solo travel (Weatherby, 2018; Wilson & Harris, 2006).

In short, social media, and specific blogs, are effective platforms for addressing the negotiation process that occurs between the motivations and deterrents of travelling alone. The approach adopted in this study is consistent with Falconer's hypothesis (2011) that the framework of confusions and continual negotiations of the feminist identity has an emotional dimension, and that the "travelling arena is a key space in which to explore how feelings of fear, shame, anger and empowerment can fit into wider theoretical frameworks" (p. 65). Therefore, narrated emotions and the way they are understood, resisted, and articulated need to be studied.

Methodology

To analyse the experiences of the different women solo travelling, volunteers willing to tell their stories were needed for the quantitative part of the research. Finding solo female travellers who were willing to share their experiences proved so difficult, and following cited authors (Hamid et al., 2021; Weatherby, 2018), the decision was made to analyse the discourse of a series of readily available online solo female travel bloggers (see Appendix I).

As discussed previously, analysing publishing blog postings is completely unobtrusive way of studying social phenomena (Kozinets, 2010). Hookway (2008) explains the benefits of using blogs as sources for academic research: the availability of large amount of data ready for collection; usually data are already in text form, one adequate format to be analysed; researchers reach out to populations which would be geographically and socially inaccessible to contact otherwise; and they allow collecting sensitive and subjective expression about everyday experiences.

Banyai and Glover's review (2012) of current research on travel blogs reveals that the two most common research methods used to analyse blogs' content are "content analysis" and "narrative analysis." With the aim to get an objective perspective of travellers' speeches and see whether they talk about emotions or not, this chapter suggests measuring the observable data characteristics, as a quantitative approach to explore the emotions of travel blogs. By doing so, it is assumed that the analysis of quantitative content to produce counts of words and measurements is often contested by researchers (Banyai & Glover, 2012) but there is no subjective interference when analysing the results. So, following Krippendorff (2004; cited in Banyai & Glover, 2012), the rationale behind the research design was simple: by analysing the words that appear in travel blogs, whether or not travellers write about their emotions could be easily verified, and if they did, it would be possible to ascertain whether or not the emotions in question were positive or negative.

The data collection started with searching relevant (most visited) travel blogs written in the Spanish language by solo female travellers considering that the top-travel blogs in Spain were written by male solo travellers, couples, or groups of three (Nadal, 2018). The sample design and the results were conditioned by the enormity of blogs existing on the Internet (Hookway, 2008). To select them, blogs were regarded as "suitable" if the blogger identified herself as female solo traveller, and if they were written in Spanish, selecting a total of 34 travel blogs (see Table 11.4 in Appendix). Furthermore, the length of the blogs was not considered when doing the word count, which can affect the results in relative terms.

Equally, the list of chosen keywords can always be disputed, but based on the theoretical background presented earlier, which emphasised emotional concepts, the authors have selected 58 keywords for the study (see Table 11.1). R Project (package sf) was used to carry out a statistical analysis. R was used in favour of other programmes such as ATLAS. ti, because it is free and does not limit its functions or capabilities and this software allows add-on packages to be installed for further research. So, R was the better choice for analysing the large number of documents involved in the study effectively. The process involved

Table 11.1 Selection of keywords for analysis (58)

Source	Key words	Translation
Emotions based on the 16 primary emotions (8 positives and 8 negatives)	*Relajación (Tranquilidad/tranquila) Placidez, Excitación, Provocación, Orgullo, Gratitud, Modestia, Virtud. Ansiedad, Enfado (Enfadada),*	Relaxation (Calmness/Calm), Placidity, Excitement, Mischief, Pride, Gratitude, Modesty, Virtue. Anxiety, Anger (Angry),
Apter (2005)	*Aburrimiento (Aburrida), Hosquedad, Humillación, Resentimiento, Vergüenza, Culpa.*	Boredom (Bored), Sullenness, Humiliation, Resentment, Shame, Guilt.
Motivations for solo travel	*Seguridad, segura/as, Empoderamiento, poder, Soledad, sola, Atreverse, atreví, atrevo., Capacidad, capaz,*	Safety, safe, Empowerment, power, Loneliness, alone/solo, To dare, I dared, I dare Ability, capable
McNamara and Prideaux (2010)	*Miedo/miedos Temor/temores, Libertad, liberador, libre/es*	Fear/fears Dread/dreads Freedom, free
Travel and emotion management	*Satisfacción, satisfecha. Confianza, desconfianza Fuerte/es Prejuicio/os*	Satisfaction, satisfied Confidence, mistrust Strong Prejudice
Levy (2013)	*Respeto*	Respect
Falconer (2017)	*Felicidad, feliz,*	Happiness, happy
Sharpley and Stone (2011)	*Frustración, frustrada Tristeza, triste. Descansar, descanso, descansada*	Frustration, frustrated Sadness, Sad To rest, rest, rested

Source: Own elaboration

four main stages. Firstly, R software was used without filters to download all the information from the travel blogs. Python was used to remove anything other than text, i.e. images, links, plug-in data, and HTML tags, and to merge all the entries from the same blog into a single file per blogger (34 files) where R was used to search for the keywords.

Findings

The results of the blog analysis are shown below. Following the blog analysis, a table with the number of times the keywords appeared ("mentions") on the blogs was created (see Table 11.2). The total number of mentions ranged from 0 to 83.104, with *power* appearing the most often and *sullenness* being the only word that did not appear in any of the blogs. The word that occurred most often in a single blog was *calmness*, with 45.255 mentions in the *Tips de Viajero blog*.

In order to gain more insight into how the words were used in the blogs, some basic statistical measurements were carried out to ascertain the maximum number of times a word was mentioned, the blog it was mentioned on most, the average number of mentions, and the standard deviation in the distribution of the mentions of the word. As the mean is not robust to outliers, the median of the mentions of the keywords in the blogs was also calculated. Thus, the data revealed whether a word is commonly used among female solo travel bloggers in a generalised way, or if only a few bloggers use it regularly and, in reality, the mean is not a valid measurement.

The word *sola* was the second most commonly used word by female travellers, appearing a total of 71,984 times. It is also one of the few words that appeared in every single blog. On average, it was mentioned 2,117.2 times, with a median value of 418.5. By comparison, the word *loneliness* was not used by all the bloggers and appeared much less often (107.41 times on average with a median of 9.55). This case shows how the mean is not always a representative measure.

Alternatively, the word *power* was the most commonly used word; it came up as both a noun (power) and a modal verb (to be able) 83,104 times and was picked up on all the websites. Although the term *empowerment* was only used by one-third of the sample, bloggers did use "being capable" and "daring" to describe situations they experienced during their travels.

All the variations of the word "safety" also appeared on the blogs. The words *safe* (singular and plural form of the adjective safe) appeared a maximum of 7,138 and 418 times respectively on one particular blog and *safety* was mentioned a total of 8,584 times.

Table 11.2 Most mentioned keywords (n)

Key word	Mentions
Power	83,104
Solo	71,984
Calmness/calm	55,448/21,805
Fear	48,607
Strong	32,978
Happy	32,555
Safe/safety	27,959/24,922
Free	22,694

Source: Own elaboration

Although safety is a recurring theme, it is not something that appears to concern everyone, as the averages reflect: 822.3; 49.4; and 703, respectively.

The results also show that the keywords (and associated variables) most closely connected to the leisure experience: *rest, happiness, satisfaction, and calmness* were used in different ways. Although only 9 out of 34 bloggers mentioned the word *rested*, all of them used the variables *rest* and *to rest* which were repeated a total of 28,189 times (counting all the blogs). Something similar happened with the words *happy* and *happiness*, the former appeared much more often (32,555 times) than the latter (4,293 times in total). By contrast, *satisfied* and *satisfaction* were hardly used at all. The blogger who mentioned *satisfied* (adjective) most did so 212 times, while the blogger who most used *satisfaction* (noun) did so 505 times. Meanwhile, the use of *calm* and *calmness* was extraordinarily high, with the adjective *calm* appearing 10,460 times and the noun *calmness* 45,255 times.

When analysing the keywords according to Apter's 16 primary emotions (2005), where the positive emotions are gathered in one group and the negative ones in another, *relaxation* and *pride* were the only words that got over 1,000 mentions in a single blog. But when the means and medians of the other terms were compared, it was clear that these words were not used that much, since the means were very low and the medians were equal to zero. Alternatively, when the basic negative emotions were analysed, while *guilt, shame,* and *anxiety* came up as the words used most often in the blogs, the other negative emotions appeared to a lesser extent. As in the case of the positive emotions, the medians of the non-significant keywords usually tended to zero. Furthermore, the basic statistics showed that the items *freedom, confidence,* and *strength* were mentioned quite frequently. *Sadness* and *frustration* were also mentioned, although to a lesser extent. Finally, despite prejudice not being a dominant theme in the posts, all the bloggers highlighted the issue of respect. Subsequently, an analysis was done to see if clear patterns could be established between bloggers and their use of words (if bloggers are grouped according to certain values) but the results confirm that no pattern could be appreciated.

Finally, correlations between the keywords used for the study are analysed. Five clear blocks emerged in which the words were very closely correlated, meaning that the bloggers tended to use these words together in groups (see Table 11.3).

The first block included the words *sola, sadness, gratitude, satisfaction, frustrated, frustration, empowerment, fear,* and *prejudice* – the plural forms of the last two were also included. Many negative keywords coincided in this group, particularly about feelings experienced before the trip and during the first few days. Therefore, when writing about their first impressions, female solo travel bloggers focus on their own experience (*sola, sadness, prejudice,* and *fears*) and emotions (*frustration* versus *satisfaction* and *empowerment*). The next group included the terms *safe, dare, dread, calm, guilt, power,* and *happy.* Although this group is similar to the first one, positive keywords prevailed in this group and there was very little emphasis on negative aspects. The third and largest group of the five included the words *mistrust, free, boredom, dare, strong, respect, confidence, relaxation, rest, calmness, pride, virtue, capacity, to rest, security, freedom,* and *rested.* The main feature of this group is that most of the words in it were associated with leisure and its objectives (relaxation, rest, and tranquillity, among others). The fourth group contained the words *safe, loneliness, satisfied, capable, bored, happiness, dreads, strong, to dare, frees* (the plural form of the adjective free), *and freedom* (the noun). The words in this block were associated with travelling alone and featured feelings of doubt regarding the unknown (loneliness, fears, boredom) followed by feelings of fulfilment and satisfaction (dare, freedom, happiness, satisfied). The last block featured *anger, sad, shame, mischief, humiliation, excitation,*

Table 11.3 Summary of the main findings

Approach	Results
Basic statistical measurements: number of the mentions based	• The words *sola* and *power* were the most commonly used and they appeared in every single blog. • Although *safety* is a recurring theme, it is not something that appears to concern everyone. • According to Apter's emotions, there are interesting findings but no conclusive outputs.
Correlations between the keywords	Five groups emerged containing the following: 1. words with negative connotations 2. words with positive connotations 3. words related to safety and danger 4. words that enhance empowerment 5. words that refer to relaxation, rest, and tranquillity

Source: Own elaboration

placidity, and *modesty*. Except for *placidity* and *excitement*, the words in this group were interpreted as negative emotions experienced when travelling.

Finally, the analysis also revealed very little correlation between *anxiety, resentment, anger*, and the other keywords. This could indicate either that such feelings are experienced less frequently during trips, or that bloggers choose not to discuss them on their web pages. It could also mean they do not want to convey these sensations to their readers, or that, when they do, they focus their attention entirely on those feelings to the exclusion of all others.

Discussion and Conclusions

This chapter is an exploratory study of a selection of solo female travel blogs whose analysis generated a series of data and allowed some interesting conclusions to be drawn. The last section of the chapter sets out the conclusions regarding the results and the methodology used in the study.

The analysis of the travel blogs demonstrated that emotions are a leading topic of discussion on solo women travellers' platform. Although the limitations of the study are discussed later in the section, it can be affirmed that women's feelings and emotions were mentioned in all the blogs analysed. Therefore, the willingness of solo female travellers to share their travel experiences with acquaintances and strangers is also confirmed. It was also observed that the discourse of travel blogs has evolved. The recommendations about places and activities which were commonplace on early travel blogs have been replaced with descriptions of memorable experiences (Sharpley & Stone, 2011). It can be said therefore that, nowadays, emotions are unquestionably a regular feature of this type of blog. Differences were also observed between the blogs and the authors' narratives.

The terms that appeared most often in the sample were *sola, safe* and *safety, strong, free*, and *power*. These results confirm the findings in the literature review and reveal that solo women travellers support the feminist discourse regarding women's empowerment (Falconer, 2011, 2017; Hamid et al., 2021; Weatherby, 2008; Wilson & Little, 2006). The issues discussed on the blogs were associated with solo travel and sharing experiences and points of view. This finding coincides with the theory that, for women, "meaningful travel" (Wilson & Harris, 2006) involves a search for emotional fulfilment through tourism, reinforces their identity, and empowers them. Other words used included *calm, calmness, happy, satisfied,* and *satisfaction,* which shows that women experience a destination from an emotional point of view, improving and enhancing positive and memorable tourist experiences (Sharpley & Stone, 2011). The most frequently used words in a single blog in this study mirrored the differences Falconer observed in discourses (2017). According to Falconer (2011, 2017), the theories that enable women to overcome difficult obstacles (without showing anger or fear and become what is referred to as emotionally successful) may also, at the same time, inhibit their freedom of mobility. He also noted, however, that feminist narratives do not support these theories and argue that the negotiation of bad emotions provides a powerful insight into women travellers' perceptions. The results of this study show that, while some bloggers stress the relaxation and safety dimension of solo female travel, others insist on highlighting the sense of empowerment through the use of words like *power.* A good example of this is the blog *Mis viajes por ahí (My travels around)* where *power* is used 18,069 times, while *happy* is only mentioned 3,830 times.

The two danger and risk-related emotions mentioned most in the reviewed literature were *anger* and *fear.* This coincides with Falconer's critical view (2017). Falconer argued that, for women to be successful travellers, they must feel and display happiness and good humour but also made the point that "positive thinking" has turned happiness in women's travel into a duty rather than a tool for achieving success. For Apter (2005), *anger* as a basic emotion means "wanting to do what one knows what one should not do, for example in reaction to unfairness. (...) so the high arousal is unpleasant and represents a form of tension" (p. 48). Elsewhere, Reeve (2005) defined it as the ubiquitous and most passionate of all emotions. Nevertheless, *anger* also makes people – especially women – stronger, and more energised, and increases their sense of control, so it can be a productive emotion. Similarly, even though *fear* and *anger* are both high-arousal negative emotions, *fear* tends to activate risk-aversion, whereas *anger* tends to cause risk-seeking because *anger* is associated with situations of high control. *Fear,* by contrast, is associated with situations of low control and uncertainty (Cohen et al., 2007).

Fear has been widely studied in the travel literature and it is almost always risk and danger perception-related. In its positive form, it motivates defence mechanisms. (Reeve, 2005). This research shows that the terms *fear* and *anger* can lead to different behaviour. While *fear* is the fourth most frequently mentioned emotion (55,448 times), anger-related words (*anger, angry*) only appeared significantly in two blogs: "Trajinando por el mundo" (145) and "Mis viajes por ahí" (48). These findings support Falconer's hypothesis of emotional adaptation (2017), where women tend to express happiness and reject anger. So, adaptation is a key factor for travel motivations, where women must learn to negotiate difficult situations, and women who travel a lot need to believe that they are responsible for their own safety, emotional well-being, and happiness in travelling spaces.

While these results might point to the conclusion that basic emotions are not discussed in female solo traveller blogs, this interpretation may be conditioned by the limitations of the quantitative analysis involved in the study. The analysis used in the study did not consider the semantic value of the context of the words. It may also be the case that emotions were mentioned but were expressed using other less formal, more colloquial synonyms. Furthermore, while, in some cases, all derived or related forms of the words were considered (e.g. *safe, safety,* and *security*). This is one important limitation and shows that the meaning of a word without a context provides insufficient information, in the authors' view, to decode the emotion properly, and that a semantic approach is therefore necessary. For this reason, despite being a useful analytical tool, quantitative analysis cannot substitute qualitative analysis for reading and interpreting blogs.

Regarding the methodology of the study, this chapter confirms that research on social media is valid means of exploring social behaviour (Banyai & Glover, 2012; Kozinets, 2010; Yang, 2021). But the results also revealed the aforementioned limitations. While, as the methodological section has explained, text analysis has a lot of advantages, its limitations include having to restrict the study to only one language (Spanish in this case) which involves having to translate etc. Furthermore, the data source was so overwhelming in terms of the number, size, and scope of the online blogs that a selection had to be made. Most of the blogs analysed were "typical blogs," i.e. online journals where women share personal and emotional accounts of their travels and offer advice on different situations. Therefore, solo women travellers write blogs to impact readers' lives, create a travelling community identity, inspire confidence in women, and influence their behaviour.

As mentioned earlier, the findings of this study have to be seen in light of some limitations, two in particular. The first limitation is the sample's design conditioned by the language (culture) and the blog's selection process. The second limitation, also methodological-related, concerns the quantitative approach. So, the recommendations for future research are, on one hand, constructing similar researches in a new type of contents (social media, podcast, etc.) or language. On the other, it should be recommendable a qualitative approach to delve into the meaning of a word in its context. In other words, a semantic approach.

This research also reinforces some key notions about tourism management. As mentioned before, specialised service industries (that focus on making access to tourism more equal in terms of gender) have proliferated in response to research findings on women's feelings of fear and anger, and risk perception. Therefore, operators and destination companies need to explore how to create a friendly and safer environment for solo female travellers. Such measures could include: encouraging more local women to write destination blogs from a female perspective; supporting local female guides; reposting reviews by solo female travellers; creating safe and friendly tours; and making recommendations aimed specifically at new arrivals.

References

Abad-Galzacorta, M., Gil, I., Peralta, M., Reino, S., & Alzua, A. (2013). Acercamiento al estudio de las emociones. Una propuesta metodológica para la medir las emociones en la interacción persona – tecnología. *tourGUNE Journal of Tourism and Human Mobility*, 0, 23–30.

Apter, M. J. (2005). *Personality dynamics. Key concepts in reversal theory*. Apter International Ltd.

Banyai, M., & Glover, T. D. (2012). Evaluating research methods on travel blogs. *Journal of Travel Research*, 51(3), 267–277. doi:10.1177/0047287511410323

Bigné, E., Mattila, A., & Andreu, L. (2008). The impact of experiential consumption cognitions and emotions on behavioral intentions. *Journal of Services Marketing*, 22(4), 303–315. doi:10.1108/08876040810881704

Bowen, H. E. (2005). Introduction: Special issue on female travellers—Part 1. *Tourism Review International*, 9(2), 119–121. doi:10.3727/154427205774791690

Calafat, A. (2013). Sexual harassment among young tourists visiting Mediterranean Resorts. *Archives of Sexual Behavior*, 42(4), 603–613. doi:10.1007/s10508-012-9979-6

Cohen, J. B., Pham, M. T., & Andrade, E. B. (2007). The nature and role of affect in consumer behavior. In P. H. Curtis (Ed.), *Handbook of consumer psychology* (pp. 297–347). Lawrence Erlbaum.

Ejupi, R., & Medarić, Z. (2022). Motives of female travellers for solo travel. *Academica Turistica*, 15(2), 177–185. https://doi.org/10.26493/2335-4194.15.177-185

Falconer, E. (2011). Risk, excitement and emotional conflict in women's travel narratives. *Recreation and Society in Africa, Asia & Latin America, 1*(2), 65–89.

Falconer, E. (2017). 'Learning to be zen': Women travellers and the imperative to happy. *Journal of Gender Studies, 26*(1), 56–65.

Ferrús Antón, B. (2011). *Mujer y literatura de viajes en el siglo XIX: entre España y las Américas*. PUV.

Gretzel, U., & Fesenmaier, D. R. (2003). Experience-based internet marketing: An exploratory study of sensory experiences associated with pleasure travel to the Midwest United States. In A. Frew (Ed.), *Proceedings of the tenth international conference on information and communication technology in tourism, ENTER 03* (pp. 49–57). Springer Verlag.

Hamid, S., Ali, R., Azhar, M., & Khan, S. (2021). Solo travel and well-being amongst women: An exploratory study. *Indonesian Journal of Tourism and Leisure, 02*(1), 1–13. doi:10.36256/ijtl.v2i1.125

Hookway, N. (2008). Entering the blogosphere': Some strategies for using blogs in social research. *Qualitative Research, 8*(1), 91–113.

Kozinets, R. V. (2010). *Netnography: Doing etnographic research online*. Sage.

Levy, D. (2013). Women-only tourism: Agency and control in women's leisure. *Sociation Today* (Fall/Winter).

McNamara, K. E., & Prideaux, B. (2010). A typology of solo independent women travellers. *International Journal of Tourism Research, 12*, 253–264. DOI: 10.1002/jtr.751

Morató, C. (2003). *Viajeras Intrépidas y Aventureras* (1a ed.). Plaza & Janés.

Morató, C. (2005). *Las damas de Oriente: Grandes viajeras por los países árabes*. Plaza & Janes.

Nadal, P. (2018, January 12). Los 25 blogs de viajes más leídos de España. El País https://elpais.com/elpais/2018/01/11/paco_nadal/

Pereira, A., & Silva, C. (2018). Women solo travellers: Motivations and experiences. *Millenium, 2*(6), 99–106.

Reeve, J. M. (2005). *Understanding motivation and emotion* (4th ed.). John Wiley & Son.

Sharpley, R., & Stone, P.(Eds.). (2011). *Tourist experience: Contemporary perspectives*. Routledge.

Sun, H., & Zhang, P. (2006). The role of affect in information system research. A critical survey and a research model. In P. Zhang (Ed.), *Human computer interaction and management information system: Foundations* (pp. 295–329). Sharpe.

United Nation World Tourism Organization (UNWTO) (2017). *Tourism and the sustainable development goals – Journey to 2030*. World Tourism Organization.

United Nation World Tourism Organization (UNWTO) & UN Women (2011). *Global report on women in tourism 2010*. World Tourism Organization. http://www2.unwto.org/en/publication/global-report-women-tourism-2010

United Nation World Tourism Organization (UNWTO) (2019). *Global report on women in tourism 2019*. World Tourism Organization.

Vlahakis, L. (2018, August 2). The #MeToo movement has revealed the pervasiveness of sexual harassment and 'Quick Take on Travel'. Mower. https://www.mower.com/insights/fly-metoo-two-out-of-five-women-report-sexual-harassment-when-traveling-solo/

Weatherby, T. G. (2018). *Beyond the screen: How women's use of social media is changing the ideological American wilderness landscape* [PhD dissertation, College of Environmental Science and Forestry Syracuse, New York].

Wilson, E., & Harris, C. (2006). Meaningful travel: Women, independent travel and the search for self and meaning. *Review, 54*(2), 161–172.

Wilson, E., & Harris, C. (2007). Travelling beyond the boundaries of constraint: Women, travel and empowerment. In A. Pritchard (Ed.), *Tourism and gender: Embodiment, sensuality and experience* (pp. 235–250). CAB International.

Wilson, E., & Little, D. E. (2005). A 'relative escape'? The impact of constraints on women who travel solo. *Tourism Review International, 9*, 155–175.

Wilson, E., & Little, D. E. (2008). The solo female travel experience: Exploring the 'Geography of Women's Fear'. *Current Issues in Tourism, 11*(2), 67–186. doi:10.2167/cit342.0

Yang, E. C. L. (2021). What motivates and hinders people from travelling alone? A study of solo and non-solo travellers. *Current Issues in Tourism, 24*(17), 2458–2471. doi.org/10.1080/13683500.2020.1839025

Appendix I

Table 11.4 Selection of female travel blogs in Spanish (sample)

Blog	Bloggers' name	From	URL
Alicia Sornosa	Alicia Sornosa	2014	www.aliciasornosa.com/blog-alicia-sornosa
Bitácora viajera	Maru Mutti (Mariana)	2011	www.bitacora-viajera.com
Bueno bonito barato	Sarah Yañez-Richards	–	www.buenobonitobarat0.blogspot.com
Camino Salvaje	Julia del Olmo	2015	www.caminosalvaje.org
Crónicas de una argonauta	Irene García	2013	www.cronicasargonauta.com
Dejarlo todo e irse	Patricia Jiménez	2013	www.dejarlotodoeirse.com
En el camino con moonflower	Carol Gutiérrez	2008	www.enelcaminoconmoonflower.com
Judith Tiral	Judith Tiral	2013	www.judithtiral.com
La cosmopolilla	Patricia Rojas	2013	www.lacosmopolilla.com
La mochila de mamá	Marta Aguilera	2011	www.lamochilademama.com
Lápiz nómada	Andrea Bergareche	–	www.lapiznomada.com
La vida nómade	Fran Opazo	2015	www.lavidanomade.com
Los viajes de Ali	Alicia Ortego	2011	www.losviajesdeali.com
Los viajes de Mary	Mary Salas	2013	www.losviajesdemary.com
Los viajes de Nena	Laura Lazzarino	2008	www.losviajesdenena.com
Mariel de viaje	Mariel Galán	2013	www.marieldeviaje.com
Meridiano 180	Laura Fernández	2009	www.meridiano180.com
Mindful travel by Sara	Sara Rodríguez	–	www.mindfultravelbysara.com
Mis viajes por ahí	Inés Fernández	2008	www.misviajesporahi.es
Nomadic chica	Gloria Apara	2014	www.nomadicchica.com
Sin mapa	Verónica Boned	2010	www.sinmapa.net
Solo ida	Claudia Rodríguez	2014	www.soloida.com
Sonia Graupera	Sònia Graupera	2009	www.soniagraupera.com
Tierra sin límites	Paula Mayoral	–	www.tierrasinlimites.com
Tips de viajero	Verónica Marmolejo	2008	www.tipsdeviajero.com
Trajinando por el mundo	Carmen Teira	2009	www.trajinandoporelmundo.com
Viaja el mundo	Adriana Herrera	2011	www.viajaelmundo.com
Viaja en mi mochila	Cristina E. Lozano	2008	www.viajaenmimochila.com
Viajando. Imágenes y sensaciones	Sabela Montero	2005	www.viajandoimagenesysensaciones.com
Viajando por ahí	Aniko Villalba	2010	www.viajandoporahi.com
Viajar alimenta el alma	Ana Gómez	–	www.viajaralimentaelalma.com
Viajar para vivir	Analucía Rodríguez	2014	www.viajarparavivir.com
Viajar sienta bien	Patricia Otero	2014	www.viajarsientabien.com
Vida de viajera	Eli Zubiria	2014	www.vidadeviajera.com

Source: Own elaboration

12

DEVELOPING A RISK PROFILE OF FEMALE BUSINESS TRAVELLERS IN PANDEMIC TIMES

Bingjie Liu-Lastres, Alexa Bufkin, and Amanda Cecil

Abstract

Female business travellers are an essential segment in the global tourism industry. Risk perceptions and safety concerns often play more important roles in their decision-making. This chapter aims to develop a risk profile of female business travellers, including risk perception, perceived safety, and willingness to travel before and after the COVID-19 pandemic. This study utilised the risk perception attitude (RPA) framework as the leading conceptual framework and employed a quantitative approach and collected 402 completed responses through a national survey of US female business travellers in the summer of 2020. The results showed that the participants' concerns had significantly shifted to health and safety following the pandemic. Furthermore, the outcomes of mediation analyses showed that safety perception mediated the relationships between the samples' RPA variables and travel willingness amid a global pandemic. The findings also revealed that safety perception, perceived severity, and self-efficacy determined female business travellers' travel intentions. Based on the findings, this study further discusses the importance of utilising a gendered approach to female business travellers' risk perception. This study also offers implications on how to craft effective marketing messages to encourage female travellers to return to business travel during the times of the COVID-19 pandemic.

Keywords

Female Business Travellers, Risk Perceptions, Risk Perception Attitude, Perceived Safety, COVID-19, Business Travel.

Introduction

Female business travellers are an essential segment in the global tourism industry, especially considering that they now represent nearly one-third of the business travellers (Skift, 2020). In contrast with their male counterparts, female business travellers display distinct

DOI: 10.4324/9781003286721-18

preferences and needs regarding hospitality products (Global Business Travel Association [GBTA] 2018; Yang et al., 2017). For instance, it is noted that female travellers tend to be more risk aversive than their male counterparts and pay more attention to travel safety (Mirehie et al., 2020). It is further noticed that female business travellers display their keen awareness of various issues during their business trips, such as medical emergencies, changes in travel plans, and sexual harassment (GBTA, 2018).

In addition, the recent outbreak of the COVID-19 pandemic has further intensified female business travellers' risk perceptions (Liu-Lastres et al., 2021). Female business travellers have to undertake multiple responsibilities during the pandemic and appear to be more careful, considering their frequent interactions with vulnerable populations such as young children and the elderly. The ongoing pandemic has also added more anxiety and uncertainty to the situation, leading to female business travellers' increasing reluctance to return to business travel (Liu-Lastres et al., 2021).

Although researchers (Figueroa-Domecq & Segovia-Perez, 2020; Yang et al., 2017) advocate addressing the gender perspective in tourism studies, there is limited knowledge regarding the risk perception and perceived safety among female business travellers. Similarly, little is known about their changing perceived risks of business travel in times of a global pandemic. Thus, this book chapter aims to develop a risk profile of female business travellers, including their risk perceptions, safety perceptions before and after the COVID-19 pandemic. This chapter also tested their risk perception attitudes'(RPAs) effects on their willingness to travel during pandemic times. Mainly, this study is led by the following research questions:

- How do female business travellers perceive travel risks before the COVID-19 pandemic?
- How do female business travellers perceive travel risks during the COVID-19 pandemic?
- How do female business travellers' RPAs affect their perceived safety and travel willingness amid the COVID-19 pandemic?

Literature Review

Applying a Gendered Approach to Study Female Business Travellers

Business travel is a key market in tourism and hospitality; 464.4 million business trips were taken by US residents in 2019, establishing $334 billion in total economic impact and $139 billion in tourist income (GBTA, 2018). To better understand tourists' attitudes, behaviours, and decision-makings, segmenting the market according to their gender has become increasingly popular (Swart & Roodt, 2015). The concentration on gender is particularly valuable, considering that businesswomen now encompass nearly one-third of the segment (Skift, 2020). This growing trend has led to accelerated attention towards businesswomen in the travel market.

As a strong indicator of consumer behaviours, gender differences have been extensively studied in the tourism literature (Wilborn et al., 2007). Although it is widely believed that tourists' decision-making varies by gender, inconsistent findings have been noted. For example, Collins and Tisdell (2002) found that when females were the dominant holiday decision-makers, fewer sport-related activities were selected for incoming trips, and more information sources were involved in the planning process. On the other hand, although different choices of leisure activities were noticed among young travellers, Carr (1999) argued that such differences could be attributed to particular life stages rather than gender. Similarly, Harvey et al. (1995) did not find any gender differences in the context of community tourism dependence.

When it comes to the business tourism literature, only a couple of studies have explored gender differences. For example, Ho and McKercher (2014) noted that business tourists' shopping behaviours significantly differ by gender. Additionally, Willis et al. (2017) found that the consequences of frequent business travel are gendered, where female business travellers who took on more business trips are more likely to feel guilty and separated from their families. They also called for more attempts to explore the gendered nature of business travel.

Despite its importance, unfortunately, the majority of business traveller studies have focused on mixed-gender samples and used gender comparison approaches to study business travellers' preferences, perceptions, and behaviours. Knowledge on the perception of risk involving gender and safety remains limited. In the same vein, research investigating female business travellers' safety concerns is scarce. Previous research (Swart & Roodt, 2015) suggested that gender is the primary segmenting variable in studying the business traveller market; however, specifics on a gender's perceptions, attitudes, and behaviours are still premature. Differentiating gender is more essential in travel risk studies because issues like sexual assault are becoming more concerning for female travellers (GBTA, 2018).

Understanding Female Business Travellers' Risk Perception

Risk perception is a key component in tourists' decision-making process, as people naturally want to avoid risky situations (Liu-Lastres et al., 2021). Consistently, existing studies have already noted the gender differences in tourists' engagement in risk reduction behaviours. For example, Mattila et al. (2001) found that gender affected college students' enactment of health-related risky behaviours during spring break. Mitchell and Vassos (1998) also found that female travellers are more likely to opt for safer options for their vacation purchases.

Similarly, there are differences between how men and women perceive risks (Liu-Lastres et al., 2021; Yang et al., 2017). Female travellers had higher safety concerns than their male counterparts (Yang et al., 2017). Transportation incidents, medical emergencies, and sexual harassment are all examples of unsafe risks female travellers endure (GBTA, 2018; Wilson & Little, 2008). Consistently, the tourism literature also finds that women are more sensitive towards certain risks such as sexual harassment and, therefore, tend to prefer risk aversion and avoiding risky situations when travelling (Mirehie et al., 2020).

Furthermore, COVID-19 may have changed an individual's risk perceptions (Liu-Lastres, 2022). For example, the perceived risk among female business travellers may have intensified due to their increasing sense of caring and kindness. Notably, the female's assumed family and work responsibilities lead to a significant concern about carrying the virus and exposing others (Maddy, 2020). Thus, a gender-specific approach is needed to examine people's perceptions and behaviours within tourism and travel.

Risk Perception Attitude (RPA) Framework

Many theories, including protection motivation theory, health belief models, and the theory of planned behaviours, have been used to understand the effects of risk perception on tourists' decisions, attitudes, and behaviours. An overlooked yet suitable theoretical framework in this context is the RPA framework (Liu-Lastres et al., 2021). The RPA framework was developed based on the extended parallel process model and the social cognition theory (Rimal & Real, 2003). The RPA framework proposes that individuals can be divided into different RPA groups based on risk perception and efficacy beliefs. Perceived risk reflects the level of severity and vulnerability perceived by an individual regarding specific issues (Liu et al., 2016). Efficacy beliefs generally indicate one's confidence in enacting certain protective actions and protecting themselves against risks (Liu et al., 2016). These attitudinal differences often lead to subsequent behavioural changes.

As shown in Figure 12.1, four distinct groups are outlined in the RPA framework. The first group is the indifferent group, characterised by low risk and low efficacy, and they usually behave indifferently. The second group is the proactive group, which is characterised by low risk and high efficacy, and they often engage in risk reduction strategies. The third group is the avoidance group, which assumes high risk but low efficacy. People in this group tend to avoid facing the issue directly. The last group is the responsive group, which assumes the high risk and high efficacy. This group is more likely to take action to protect themselves.

In social marketing, segmentation is an essential component that can effectively divide public audiences into distinct segments, amplifying the popularity of the RPA framework (Skubisz, 2014). The RPA framework is also a useful tool for dividing the general market into distinctive divisions, all of which are characterised by different attitudes and behaviours (Liu et al., 2016). People's health behaviours have been the main focus of previous studies using the RPA framework (Skubisz, 2014). Besides dividing the individuals into segments, studies built upon the RPA framework (Rimal & Juon, 2010) also directly measured

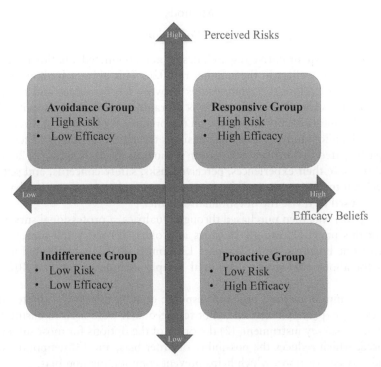

Figure 12.1 RPA conceptual framework, created by the authors.

the relationships between the RPA variables and the outcomes. One shortcoming of the RPA framework deals with the fact that it needs to be contextualised, given the uniqueness of each health issue (Rimal & Real, 2003). The segmenting approach can sometimes also be problematic, considering that people's perceptions and attitudes towards certain emerging issues change over time (Liu et al., 2016).

Tourism scholars have also used the RPA framework to understand people's RPAs and their impacts on preventive behaviours and travel intentions; Liu et al. (2016) investigated the association between US residents' risk perception and the likelihood of them visiting Jordan. Equivalently, Wang et al. (2019) examined adventure tourists' adoption of preventive measures against adventure travel risks through the lenses of the RPA framework. These studies showed that the RPA variables closely relate to tourists' safety concerns and travel intentions. On most occasions, people's perceived safety mediated the relationships between their RPA and travel intentions. Proceeding with these studies leads to the RPA framework's adequacy to explore the interactions among tourists' risk perception, perceived safety, efficacy, and subsequent behaviours. Thus, guided by the RPA framework, this study aimed to explore the relationship involving female business travellers' RPA, perceived safety, and willingness to travel during a global health pandemic. Additionally, this study chose to focus on the specific context involving business travel and a prolonged health pandemic. In addition to a segmenting analysis, this study included insights from a preliminary study and additional statistical analyses. In doing so, the findings of this study not only can test the key assumptions of the RPA framework but also can lead to further theoretical development in tourism risk research.

Methods

Data Collection

This study adopted a quantitative research design and conducted a national survey among US female business travellers in the summer of 2020. This approach allows the researchers to measure and explore the relationships between key variables and test hypotheses developed from relevant theories (Bowling & Ebrahim, 2005). More specifically, the development of the questionnaire was based on the preliminary findings of three focus group discussions and a pilot study of 30 female business travellers. A total of 12 female business travellers participated in the three focus group sessions in the early spring of 2020, and they were asked to share their experiences, perceived risks, safety concerns, and self-protective strategies related to business travel. The focus group findings and the feedback from the pilot test were used to revise the questionnaire further.

An online panel was then purchased through Qualtrics, a professional survey company. Members of this panel (1) are all 18 years and older, (2) work full-time, and (3) have taken at least one business trip in the past 12 months. Data was collected in May 2020 and lasted for about two weeks. The final sample includes a total of 402 completed responses.

To manage common method bias and response quality, this study adopted the following measures: (1) expert opinions and pilot test feedback were used to ensure the quality and clarity of the survey instrument; (2) the order of the options for most survey questions is randomised, which reduces the possibility of order bias; and (3) temporary separations were placed between sections, which helps prevent common method bias.

Measurement

The participants' perceived risk was measured through two parts – one relates to their concerns before the pandemic and the other deals with the COVID-19 pandemic. Regarding their general concerns about business travel before the pandemic, this study examined the seven most common risks reported from the literature and the focus groups, including food safety issues, infectious diseases, severe weather/natural disasters, transport accidents, crimes, terrorism, and sexual harassment. First, the participants were asked to report if they had experienced any of those incidents in their business trips using a binary scale (1 = yes, 2 = no).

Second, their RPA was inspected through the following three dimensions: perceived severity, perceived susceptibility, and self-efficacy (Liu et al., 2016). Perceived severity was measured by asking the respondents to indicate the severity associated with an incident's consequence associated with business trips with a five-point Likert-like scale where "1 = not severe at all" and "5 = very severe." Perceived susceptibility was measured by asking the respondents to indicate the likelihood of an incident happening during their business trips with a five-point Likert-like scale where "1 = not likely at all" and "5 = very likely." Self-efficacy was measured by asking the respondents to indicate the ability to handle various business travel-related incidents with a five-point Likert scale where "1 = not confident at all" and "5 = very confident."

When it comes to their concerns related to the pandemic, the three RPA variables (i.e., perceived severity, perceived susceptibility, and efficacy beliefs) were measured. Similar to the pre-pandemic risks, all these dimensions were measured using a five-point Likert-like scale.

Lastly, safety perception was measured by asking the participants to indicate how safe they felt about business trips before and during the pandemic through a seven-point semantic differential scale where "1 = very unsafe" and "7 = very safe." Their travel intentions were measured by asking the participants to rate their willingness to travel for business in the next three and six months with a five-point Likert scale. Although different scales were used due to the nature of questions, studies (Colman et al., 1997; Dawes, 2008) show that both five- and seven-point Likert scales were reliable, and the results did not change significantly between these two measurements.

Information related to the participants' individual characteristics were also included in the survey. Examples of such information include demographics (e.g., age, education, sexual orientation, and family status), employment information, experiences related to business travel, and business travel styles (e.g., types of destinations, travel frequency).

Data Analysis

The data analysis consists of four steps. To start with, a series of descriptive analyses were conducted to describe the profile of the sample, such as their demographic information and their business travel-related behaviours. Additionally, to answer the first research question, descriptive analysis was conducted to measure business travellers' RPAs regarding different risks before the pandemic. Their perceived safety before and after the pandemic was also measured, respectively. Pearson tests were conducted to measure the correlations between the sample's RPA variables and perceived safety before the pandemic. Lastly, the samples of RPAs related to the pandemic were measured through descriptive analyses. A mediation analysis was conducted to identify the major drivers of the sample's business travel intention amid the pandemic. The mediation analysis treated RPA attitude-related variables as the independent variables, perceived safety as the mediator, and travel intention as the outcome variable. The descriptive analyses and Pearson tests were all conducted using the software Statistical Package for the Social Sciences (SPSS) (V27.0). The mediation analyses were conducted using the PROCESS Macro.

Results and Discussion

Sample Profile

Table 12.1 presents the profile of the sample. A total of 402 completed responses from US female business travellers were included in the sample. The average age of the sample is 36.4 years old, with 37.1% falling into the 25–34 age group. Most (79.3%) of them are heterosexual, and 73.6% are white. The sample is also well educated, where 60.7% of them have bachelor's degrees, and 32.9% have professional or graduate degrees. Nearly half of the sample (52.6%) are married or in domestic relationships, and 51.9% of them have children. Regarding their employers, about half of them (50.5%) work in for-profit organisations, followed by public entities (25.2%) and not-for-profit organisations (20.7%). Most (90.2%) travel for business about one to five times every year, and nearly half (49.3%) have one to five years of business travel experience. When it comes to crisis experiences, nearly a third (30.0%) of the participants have experienced incidents related to severe weather/natural disasters, followed by food safety issues (11.0%), sexual harassment (9.8%), and crime (9.3%). Also, the participants visit domestic destinations (M = 3.97) at a higher frequency for their business trips. On the other hand, the frequency of travelling alone (M = 2.89) was about the same as group travels (M = 2.45).

Female Business Travellers' Perceived Risk Attitudes (RPA) in Pre-Pandemic Times

The first research question concerns female business travellers' perceived risk attitudes before the pandemic. Figure 12.2 shows a breakdown of the samples' RPAs regarding each type of risk. In terms of perceived severity, the sample indicated that the consequences of the following incidents are comparatively severe, including terrorism (M = 3.91, SD = 1.36), infectious diseases (M = 3.48, SD = 1.27), and sexual harassment (M = 3.40, SD = 1.34). When it comes to perceived susceptibility, the sample felt that the following incidents have a relatively higher probability of occurring during their business trips: transport incidents (M = 2.5, SD = 1.06), crime (M = 2.49, SD = 1.05), food safety issues (M = 2.46, SD = 1.08), and infectious diseases (SD = 2.40, SD = 1.13). Regarding self-efficacy, it seems that the participants are more confident in their ability to manage issues related to food safety incidents (M = 3.55, SD = 1.08), sexual harassment (M = 3.05, SD = 1.24), and severe weather (M = 3.04, SD = 1.13). Lastly, regarding travel safety, the participants felt relatively safe when taking business trips before the pandemic (M = 5.37, SD = 1.50).

Table 12.2 presents the results of correlation tests, and it is shown that the sample's perceived safety of business travel before the pandemic was related to almost all the risks mentioned except for natural disasters and severe weather. Furthermore, all three RPA variables were related to perceived safety to an extent. Particularly, it is noticed that the perceived threat of terrorism and sexual harassment significantly affected the sample's perceived safety.

Table 12.1 Sample profile (n = 402)

Variable	Frequency	Valid %
Age (M = 36.38, SD = 11.31)	55	13.1
18–24	156	37.1
25–34	117	27.9
35–44	49	11.7
45–54	43	10.2
Sexual orientation		
Heterosexual	333	79.3
Homosexual	16	3.8
Bisexual	34	8.1
Others/prefer not to say	37	8.8
Ethnicity		
White	309	73.6
African American	57	13.6
Asian	33	7.2
Hispanic	30	6.6
Others	27	6.2
Education level		
High school or less	27	6.4
Bachelors' degree	255	60.7
Professional/graduate degree	138	32.9
Marital status		
Never married	164	39.0
Married/domestic relationships	221	52.6
Others (i.e., divorced, widowed, and separated)	35	8.3
Have children		
Yes	218	51.9
No	202	48.1
Type of employee		
For-profit organisation	212	50.5
Not-for-profit Organisation	87	20.7
Public entities (i.e., government)	106	25.2
Others	15	3.6
Average number of business trips per year		
1–5 times	379	90.2
6–15 times	36	8.6
More than 16 times	5	1.2
Years of taking business trips		
Less than 1 year	95	22.6
1–5 years	207	49.3
6–10 years	60	14.3
More than 10 years	58	13.8
Past experiences with business travel incidents		
Food safety issues	46	11.0
Infectious diseases	31	7.4
Severe weather/natural disasters	126	30.0
Transport accidents	37	8.8
Crime	39	9.3
Terrorism	22	5.3
Sexual harassment	41	9.8
Destination type		
Domestic destinations	Mean = 3.97, SD = 1.28	
International destinations	Mean = 1.50, SD = .79	
Travel preference		
Travel alone	Mean = 2.89, SD = 1.34	
Travel in groups	Mean = 2.45, SD = 1.19	

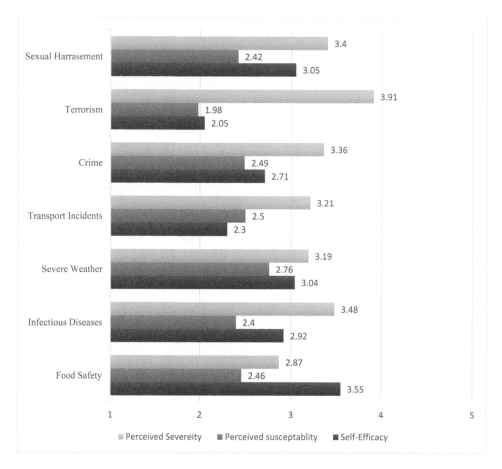

Figure 12.2   Female business travellers' risk perception attitudes in pre-pandemic times.

The results here revealed some interesting insights. First, although the risk associated with extreme weather/natural disasters is one of the most common incidents, it does not necessarily affect the participants' perceived safety of business travel. The participants seemed quite confident in managing these situations. This is consistent with the general tourism literature (Liu et al., 2016), suggesting that past crisis experiences enhance one's awareness and capability to manage a similar issue upon its reoccurrences.

Moreover, it seems that issues related to sexual harassment and terrorism are most concerning for the sample, given their statistically significant correlations with perceived safety. It is noted that most participants reported a relatively low score in the self-efficacy dimension of terrorism. In other words, the participants tend to find terrorism-related issues threatening, and they felt powerless as there was nothing much they could do with the situation. Lastly, the notion of sexual harassment is unique to female travellers and has been repeatedly mentioned in both academic studies and industry reports (GBTA, 2018; Mirehie et al., 2020). Accordingly, Yang et al. (2017) pointed out that this is a combined result of social norms and the gendered tourism space in contemporary society.

Table 12.2 Correlation table between RPA variables and perceived safety before the pandemic

Risk type	RPA variables	Perceived safety of business travel
Food safety issues	Perceived susceptibility	−.15**
	Perceived severity	.02
	Self-efficacy	.13**
Infectious diseases	Perceived susceptibility	−.23**
	Perceived severity	.05
	Self-efficacy	.09
Severe weather and natural disasters	Perceived susceptibility	−.07
	Perceived severity	−.04
	Self-efficacy	.07
Transport accidents	Perceived susceptibility	−.13*
	Perceived severity	.10
	Self-efficacy	.05
Crime	Perceived susceptibility	−.11*
	Perceived severity	.09
	Self-efficacy	.02
Terrorism	Perceived susceptibility	−.23**
	Perceived severity	−.15**
	Self-efficacy	−.01
Sexual harassment	Perceived susceptibility	−.16**
	Perceived severity	−.13**
	Self-efficacy	.16**

$**p < .01, *p < .05$

Female Business Travellers' Risk Perception Attitudes (RPA) during the Pandemic Times

The second research question concerns the participants' RPA regarding the pandemics. The participants reported a lower level of safety in business travel following the COVID-19 pandemic (M = 3.03, SD = 1.64). The results of the paired-sample t-test revealed statistical significance (t = 22.72, p < .01), suggesting that the participants felt more unsafe taking business trips now compared to the pre-pandemic times.

In terms of the RPA variables pertinent to COVID-19, the participants reported a moderate level of perceived severity (M = 3.41, SD = 1.90) and susceptibility (M = 3.25, SD = 1.09), and a low level of self-efficacy (M = 2.85, SD = 1.12). These results here imply that the participants found the COVID-19 situation dangerous because they thought it was highly likely to happen during the business trip, the consequence would be severe, and they did not know how to respond to the issue properly.

The third research question aims to test the effects of the participants' RPA on their business travel intentions via perceived safety. A mediation analysis was conducted using the PROCESS Macro (V4.1, Model 4, Bootstrap n = 5000). As showed in Figure 12.3, the overall model was statistically significant (F(5,377) = 36.22, p < .01) and explained 32.45% of the variance of the outcome variable (travel intentions). Particularly, self-efficacy (B = .25, SE = .05, p < .01, 95%CI[.16, .35]), perceived severity (B = −.10, SE = .04, p = .02, 95%CI[−.19, −.01]), and perceived safety (B = .26, SE = .03, p < .01, 95%CI[.19, .32] are all significant predictors of travel intentions. Furthermore, self-efficacy (B = .37, SE = .07,

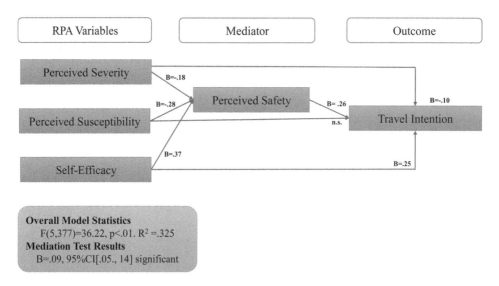

Figure 12.3 Mediation test results.

p < .01, 95%CI[.23, .52]), perceived susceptibility (B = −.28, SE = .08, p < .01, 95%CI[−.44, −.13]), perceived severity (B = −.18, SE = .07, p<.01, 95%CI[−.31, −.04]) are all significant predictors of perceived safety. Finally, the indirect coefficient was significant (B = .09, 95%CI[.05, .14]), suggesting that the mediation hypothesis was supported. The results showed that perceived safety mediated the relationships between participants' RPA variables and travel intentions.

Overall, the findings showed that female business travellers' safety concerns had considerably shifted from issues such as sexual harassment and terrorism to the ongoing pandemic. This is consistent with the literature, suggesting that individuals' perceived risk is typically influenced by current events (Vasvári, 2015). Thus, it is imperative to acknowledge the evolving nature of female business tourists' perceived risks and safety concerns and develop an active response that recognises and addresses emerging issues.

Additionally, the results of the mediation analyses indicate that the sample's perceived safety mediates the relationships between their RPA variables and their willingness to take business trips in the wake of the pandemic. This finding aligns well with previous studies (Liu et al., 2016; Wang et al., 2019). More specifically, the three RPA variables represent the cognitive dimensions of perceived risk, and the perceived safety reflects the affective aspect of risk perception. The mediation relationships show that the rational analysis of the situation is manifested through their affective responses, which subsequently affect their behavioural intentions.

Finally, a closer examination also indicates that female business travellers are more likely to fall into the proactive group, characterised by low-risk perceptions and high-efficacy beliefs (Liu et al., 2016). This group is unique in that they are mostly motivated by the desire to remain risk free and believe that there are steps to take to mitigate the threat (Rimal & Real, 2003). In a tourism context, this group of individuals is more likely to be affected by their sense of safety (Liu et al., 2016), which is affective in nature. Thus, how to create a safe travel climate and how to effectively educate them about self-protective measures should be prioritised in an attempt to relieve their concerns and increase their confidence.

Conclusion

This study aims to understand the RPAs of female business travellers before and during a global pandemic. From a theoretical perspective, this study supports the suitability of the RPA framework to the current study context, which mainly involves female business travellers. This scope echoes the need to employ a gender-specific approach in tourism risk studies (Yang et al., 2017). Unlike previous attempts (Liu-Lastres et al., 2021) featuring RPA segments, this study revealed the relationships between female business travellers' RPA, perceived safety, and travel intentions. This new attempt expands the utilisation of the RPA framework in tourism studies.

Furthermore, this study brings attention to the business travel market, an overlooked research area in tourism risk studies. Different from the previous literature (e.g., Phadungyat, 2008; Wilson & Little, 2008), which focused on female business travellers' safety concerns over a specific area, this study assumes a broader scope and explores their perceptions of a wide array of risks. More specifically, this study extends our knowledge of female business travellers' particular needs and requirements, both in regular and pandemic times. The results showed that before the pandemic, the participants assessed the risks differently, and only specific issues, such as terrorism and sexual harassment, were related to their perceived safety. There were also gaps between their perceptions and actual experiences, where encounters with incidents in their past business travel did not necessarily define their perceived safety. In contrast, the lack of belief in themselves regarding the management of incidents significantly reduced their perceived business travel safety.

Last but not least, the COVID-19 pandemic has changed the perception of safety among female business travellers, and the willingness to return to business travel has been reshaped. As suggested by the study findings, they felt unsafe because they were fully aware of the severe health consequences, the highly contagious nature of the virus, and the lack of confidence in self-protection. Because of this feeling of unsafety, female business travellers have been reluctant to travel in these trying times. There might also be an increase in uncertainty in the situation as a result of the unprecedented nature of the pandemic. Meanwhile, female business travellers might have been discouraged from returning because they lack information and assurance. Thus, to regain female travellers' confidence in business travel, it is essential to provide them with accurate information, resources for self-protection, and, most importantly, a safe climate for tourism and travel.

When it comes to practical implications, the global tourism industry cannot ignore the value of female business travellers. In light of this, the findings of this study offer several practical implications customised to female business travellers' risk perceptions. Female travellers require increased organisational support when travelling, these include communication, assurance, training, additional information, and funding support. To prevent dangerous business trips for female travellers, discussions need to be held on the utilisation of telecommunication or virtual technology. Also, businesswomen's needs and fears in both pre- and post-pandemic worlds need to be continuing to be uncovered. Futhermore, with the strive to narrow and the push to close the gap between perception and actual experiences, additional pre-trip information/training and in-route/emergency support are needed. Regular and persistent research focused on acknowledging and addressing unique risks perceptions of female business travellers need to continue within the industry.

Finally, this study is not without its limitations. First, this study was conducted in the summer of 2020, when many uncertainties were involved. Any new intervention later on, such as the vaccine and treatment, might all have changed people's risk perception and opinions on this issue. Second, this study is limited to a cross-sectional design, while the changing nature of people's feelings and perceptions might not be fully captured. Future studies should consider re-test the relationships and conceptual model with a longitudinal design. Third, this study employed a purposive sampling method. Future studies should consider using other forms of methodology and/or sampling methods, such as mixed methods or random sampling. Lastly, this study measured most of the constructs with single items, which aimed to increase the response rate. However, future studies should consider adopting multiple-item measurements for reliability purposes.

References

Bowling, A., & Ebrahim, S. (2005). Quantitative social science: The survey. In Handbook of *h*ealth *r*esearch *m*ethods: Investigation, *m*easurement and *a*nalysis (pp. 190–214). Open University Press.

Carr, N. (1999). A study of gender differences: Young tourist behaviour in a U.K. coastal resort. *Tourism Management*, *20*(2), 223–228.

Collins, D., & Tisdell, C. (2002). Gender and differences in travel life cycles. *Journal of Travel Research*, *41*(2), 133–143.

Colman, A. M., Norris, C. E., & Preston, C. C. (1997). Comparing rating scales of different lengths: Equivalence of scores from 5-point and 7-point scales. *Psychological Reports*, *80*(2), 355–362.

Dawes, J. (2008). Do data characteristics change according to the number of scale points used? An experiment using 5-point, 7-point and 10-point scales. *International Journal of Market Research*, *50*(1), 61–104.

Figueroa-Domecq, C., & Segovia-Perez, M. (2020). Application of a gender perspective in tourism research: A theoretical and practical approach. *Journal of Tourism Analysis: Revista de Análisis Turístico*, *27*(2), 251–270.

Global Business Travel Association (GBTA). (2018). *83 Percent of female business travelers report safety concern or incident in past year*. Available at https://www.gbta.org/blog/83-percent-of-female-business-travelers-report-safety-concern-or-incident-in-past-year-2/

Harvey, M. J., Hunt, J., & Harris Jr, C. C. (1995). Gender and community tourism dependence level. *Annals of Tourism Research*, *22*(2), 349–366.

Ho, G., & McKercher, B. (2014). A comparison of long-haul and short-haul business tourists of Hong Kong. *Asia Pacific Journal of Tourism Research*, *19*(3), 342–355.

Liu, B., Schroeder, A., Pennington-Gray, L., & Farajat, S. A. (2016). Source market perceptions: How risky is Jordan to travel to? *Journal of Destination Marketing & Management*, *5*(4), 294–304.

Liu-Lastres, B. (2022). Beyond simple messaging: a review of crisis communication research in hospitality and tourism. *International Journal of Contemporary Hospitality Management*, *34*(5), 1959–1983.

Liu-Lastres, B., Mirehie, M., & Cecil, A. (2021). Are female business travelers willing to travel during COVID-19? An exploratory study. *Journal of Vacation Marketing*, *27*(3), 252–266.

Maddy, S. (2020). How Covid-19 is changing women's lives. Available at https://www.bbc.com/worklife/article/20200630-how-covid-19-is-changing-womens-lives

Mattila, A. S., Apostolopoulos, Y., Sonmez, S., Yu, L., & Sasidharan, V. (2001). The impact of gender and religion on college students' spring break behavior. *Journal of Travel Research*, *40*(2), 193–200.

Mirehie, M., Liu-Lastres, B., Cecil, A., & Jain, N. (2020). Business travel, risk, and safety of female university faculty and staff. *Annals of Leisure Research*, *26*(3), 1–19.

Mitchell, V. W., & Vassos, V. (1998). Perceived risk and risk reduction in holiday purchases: A cross-cultural and gender analysis. *Journal of Euromarketing*, *6*(3), 47–79.

Phadungyat, P. (2008). *Factors influencing the selection of serviced apartments by female business travellers*. Srinakharinwirot University.

Rimal, R. N., & Juon, H. S. (2010). Use of the risk perception attitude framework for promoting breast cancer prevention. *Journal of Applied Social Psychology*, *40*(2), 287–310.

Rimal, R. N., & Real, K. (2003). Perceived risk and efficacy beliefs as motivators of change: Use of the risk perception attitude (RPA) framework to understand health behaviors. *Human Communication Research*, *29*(3), 370–399.

Skift. (2020). *The status of business travel 2020*. Available at https://skift.com/insight/new-report-the-state-of-business-travel-2020/

Skubisz, C. (2014). Risk perception attitude framework. In *Encyclopedia of health communication* (pp. 1187–1118). Sage.

Swart, M. P., & Roodt, G. (2015). Market segmentation variables as moderators in the prediction of business tourist retention. *Service Business, 9*(3), 491–513.

Vasvári, T. (2015). Risk, risk perception, risk management – A review of the literature. *Public Finance Quarterly, 60*(1), 29–48.

Wang, J., Liu-Lastres, B., Ritchie, B. W., & Mills, D. J. (2019). Travellers' self-protections against health risks: An application of the full protection motivation theory. *Annals of Tourism Research, 78*, 102743.

Wilborn, L. R., Brymer, R. A., & Schmidgall, R. (2007). Ethical decisions and gender differences of European hospitality students. *Tourism and Hospitality Research, 7*(3–4), 230–241.

Willis, C., Ladkin, A., Jain, J., & Clayton, W. (2017). Present whilst absent: Home and the business tourist gaze. *Annals of Tourism Research, 63*, 48–59.

Wilson, E., & Little, D. E. (2008). The solo female travel experience: Exploring the 'geography of women's fear'. *Current Issues in Tourism, 11*(2), 167–186.

Yang, E. C. L., Khoo-Lattimore, C., & Arcodia, C. (2017). A systematic literature review of risk and gender research in tourism. *Tourism Management, 58*, 89–100.

PART VI

Practising Gender in Tourism IV

Gender and Entrepreneurship

13

GENDER NUANCES IN TOURISM BUSINESS OPERATIONS

A South African Perspective

Nompumelelo Nzama and Ikechukwu O. Ezeuduji

Abstract

Every country benefits significantly from entrepreneurship in terms of socioeconomic development. In contrast to their male colleagues, more women are starting businesses now than ever before, but most of them fail to make their businesses successful. This chapter presents a recent study which explored if the performance of tourism-related ventures varies depending on gender in KwaZulu-Natal's Durban Central Business District (CBD), South Africa. Male and female entrepreneurs working in the tourism industry were purposefully chosen, and data were gathered via a structured questionnaire survey. This exploratory study did not come to a firm conclusion that gender significantly influences business performance or success. The traditional perception regarding gender difference in entrepreneurial success is different from reality. Business failure has no bearing on the manager's gender. Business management and marketing capabilities were found to exert much influence on business success. It is established in this study that female business owners can locate start-up finance more easily. This study suggests that mentorship and training programmes for entrepreneurs should encourage the growth of business skills among tourism entrepreneurs. Additionally, these programmes should help entrepreneurs become more knowledgeable about business financing, and the value inherent in the formation of entrepreneurial networks to especially support women tourism entrepreneurs.

Keywords

Business capabilities; Business performance; Business operations; Tourism business; Societal perception; South Africa

Introduction

Entrepreneurship significantly contributes to every nation's socioeconomic development. The South African government has emphasised tourism as a significant driver of economic growth (Tshabalala & Ezeuduji, 2016); as a result, tourism entrepreneurship is

DOI: 10.4324/9781003286721-20

highly regarded. Although tourism entrepreneurship adds value to the country's socioeconomic development, research indicates that most women in tourism-related firms mainly occupy supporting roles (such as housekeeping, reception, and waitressing) rather than leadership roles (Nzama & Ezeuduji, 2021; Tshabalala & Ezeuduji, 2016). Nonetheless, Kimbu et al. (2019) emphasised that due to its capacity to generate employment opportunities, reduce poverty, and advance women's emancipation, women's contributions to entrepreneurship are both socially and economically advantageous (Mkhize & Cele, 2017), especially in South Africa's tourism industry, which has significant potential for economic growth.

According to research on women's entrepreneurship, there are some barriers that women must overcome while founding and running businesses, including a lack of financial resources, management skills gaps, negative social perceptions, and limited entrepreneurial networks (such as Nxopo & Iwu, 2016; Nzama & Ezeuduji, 2020a, 2020b; Tshabalala & Ezeuduji, 2016). Despite the significant increase of women start-ups in sub-Saharan Africa (Global Entrepreneurship Monitor [GEM] Women's Report, 2012), the failure rate for women-owned businesses is significant (GEM Report, 2014). Much earlier research investigated the obstacles and perceptions of female entrepreneurs about business operations, but they did not compare perspectives of both genders in the same study (using the same variables or constructs) to uncover disparities if they do exist. Some recent studies (such as Figueroa-Domecq et al., 2022a, 2022b) however compared both genders to comprehend the nuanced roles that sustainability and gender play at various phases of tourism entrepreneurship, and to understand how gendered inequalities are reproduced. Therefore, using Durban Central Business District (CBD), KwaZulu-Natal, as a case, the study findings in this chapter sought to explore differences between male and female tourism entrepreneurs' perceived business performance within South Africa and add to the recent literature on gender nuances in tourism-related entrepreneurship.

Overview of Literature

One of the most effective global development techniques, entrepreneurship, has been praised for its major socioeconomic achievements. (González-Sánchez, 2015; Hassan et al., 2014; Okeke-Uzodike et al., 2018; Todorović et al., 2016). The concept of women's entrepreneurship has undoubtedly attracted the attention of many researchers since the first academic papers on it were published in the 1970s because of its significant role in socioeconomic development and the academics' curiosity to explore the differences in business performance between male and female entrepreneurs.

Governments, non-governmental organisations, and academics have noticed the significance of the role that women entrepreneurs play (Moses et al., 2016; Tajeddini et al., 2017). Women's entrepreneurship is widely acknowledged as a catalyst for economic growth (Ceptureanu & Ceptureanu, 2016). Witbooi and Ukpere (2011) found that many women entrepreneurs have the knowledge, abilities, and creativity needed to launch and manage their businesses. Despite this, women entrepreneurs frequently lack the same level of marketing, management, informational access, and educational and financial resources as men do (Kokotović et al., 2016; Todorović et al., 2016). An overview of the research on the issues impacting women-owned enterprises is presented.

Traditional Societal Perceptions

In general, the term business management has been used to refer to skills required for a particular venture to be successful (Ezeuduji & Ntshangase, 2017a, 2017b). Nonetheless, research suggests that the nature of an individual entrepreneur's demographic characteristics might predict failure and/or success in business management (Arora, 2014; Ghiat, 2018; Nsengimana et al., 2017). As evident in most developing nations, male and female roles are well-defined and may impact the business success (Duflo, 2012); however, according to Ghiat (2018), gender roles and the local environment have less of an impact on female entrepreneurs in developed nations. The findings of Vázquez-Carrasco et al.'s (2012) study in Spain revealed that no substantial differences exist between male and female business managers.

African culture views women's tasks to be mostly domestic (Idris & Agbim, 2015; Netshitangani, 2018). According to Ghiat (2018), historically, entrepreneurship has been viewed as a male-dominated profession. Masculine traits like aggression, dominance, independence, little need for help, and a strong propensity for taking risks are considered social expectations of someone who desires to engage in business (Chasserio et al., 2014; Omerzel, 2016; Tlaiss & Kauser, 2019). Whereas women's traditional identities are related to obedience, reliance, and submissiveness (Basargekar, 2007). Then, it is assumed that women are incapable of beginning and running their own enterprises. Similarly, Tshabalala and Ezeuduji (2016) found that in the entrepreneurial world, some tourism firms are deemed inappropriate for women to operate, due to the common cultural and traditional perceptions, especially in the African continent that women are not "natural risk-takers." Examples of such tourism-related enterprises include transportation, tour companies, and wildlife safaris. Women are therefore engaged in support roles in these companies (considered suited for male entrepreneurs), such as bookkeeping, reception, and reservations.

Literature highlights several unfavourable conditions that women entrepreneurs are faced with, such as juggling work and family obligations (a task made challenging by the lack of support from partners and the corporate groups they interact with), negative traditional beliefs, and gender discrimination (for example, general society's and finance providers' perceptions of low efficiency and risk-taking propensity towards female entrepreneurs) and sexual harassment (Ahmed, 2018; Singh et al., 2010; Tlaiss & Kauser, 2019). As a result, the influence of conventional patriarchal norms and restrictions on gender remains powerful, particularly in African countries.

Managerial Capabilities

According to Wessels et al. (2017), management skills are the primary source of managerial competencies. According to the authors of this chapter, managerial capability or competency refers to the capacity to coordinate and combine resources (human, material, and intangible) in order to increase corporate value.

Compared to male entrepreneurs, female entrepreneurs are more likely to be revolutionary leaders. According to Guillet et al. (2019a), male managers usually come across as highly forceful and condescending rather than courteous and eager to listen, whereas female managers frequently place a great value on their personnel and invest in them by offering training and possibilities for growth. Researchers (Schaap et al., 2008) discovered that women business owners support training for their staff members to help them understand how to carry out tasks and meet performance goals set by the organisation; men, on the other hand, are more likely to abuse their position authority and become more domineering. Therefore, it appears that women lead by mentoring, affiliating with, and promoting their subordinates. Men and women entrepreneurs seem to employ various management philosophies in terms of interpersonal abilities. Men and female entrepreneurs communicate differently (Sudarmanti et al., 2013); this suggests that, in contrast to female entrepreneurs, men entrepreneurs utilise power as a tactic to influence their subordinates. Female managers, according to Schaap et al. (2008), believe that although they feel like team members and support the work of the team, their authority derives from themselves and their positions. Manzanera-Román and Brändle (2016) assert that since these traits help women foster entrepreneurship, they provide them an advantage over their male counterparts (Guillet et al., 2019b). Women entrepreneurs are more inclined to choose (or are required to work in) smaller enterprises since they have less experience, talents, and management expertise than male entrepreneurs (Sudarmanti et al., 2013).

The use of different guanxi-based social networks by women managers to build their own identities within those networks and increase their chances of moving up the corporate ladder as female intrapreneurs in China was examined by Zhang et al. (2020). Chinese guanxi, which enables women managers to undertake particular kinds of women intrapreneurship projects in their organisations, is defined by the authors as socially rooted personal relationships built for the aim of trading favours. According to the authors, guanxi can be a socio-culturally constructed concept that supports long-term, reliable, adaptable, risk-oriented, peaceful, and result-oriented women intrapreneurship in a non-Western context rather than just being a problematic and unethical phenomenon as is sometimes misunderstood.

Marketing Capabilities

According to previous studies (Kimosop et al., 2016; Nzama & Ezeuduji, 2021; Welsh et al., 2017), entrepreneurs' marketing skills and business performance have a substantial positive link. Strong marketers may take advantage of possibilities to grow their companies and become more competitive (Welsh et al., 2017). Making a company recognised to potential clients is one of the most crucial aspects of business success. According to research (Yadav, 2018; Zeng & Gerritsen, 2014), expanding the number of guests, particularly tourists visiting a place, depends much on having effective marketing skills. Tshabalala and Ezeuduji (2016), who carried out their research in South Africa, pointed out that women-owned companies have weak marketing strategies, pointing out that female business owners (craft vendors) are scattered without any signs announcing their presence. Their research also showed that many female company owners lack the necessary connections to the tourism sector to publicise their enterprises, making it challenging to attract clients.

Due to the advancement of information and communication technology, the old framework for marketing has changed to smart marketing. The tourism industry's stakeholders were urged by Gidarakou (2015) and Zeng and Gerritsen (2014) to embrace social media as a marketing tool because the market is dependent on consumer perception, destination reputation, word-of-mouth marketing, and advertising. Women have not used technology as much as men have, despite the fact that they are viewed as being especially skilled on social media (Orser & Riding, 2018). Women, on the other hand, are less likely to use their knowledge in social media to help their companies expand (Orser & Riding, 2018). The study came to the additional conclusion that women are less assured in their technological abilities than males are. Access to some technologically advanced markets, including the tourism industry, is generally hampered by this attitude towards technology (Witbooi & Ukpere, 2011).

Formal Education and Business Networks

Many scholars (such as Anwar et al., 2022; Bazkiaei et al., 2020; Iwu, 2022; Nhleko & van der Westhuizen, 2022; Ramadani, 2015) emphasised the value of formal education in problem-solving and corporate management. Due to the perception that those with secondary and higher education backgrounds have more access to knowledge, educational aspects are viewed as significant in entrepreneurial management (Jiyane et al., 2013). According to Iwu and Nxopo (2015), information is seen as a key resource for beginning and running a firm. Tajeddin et al. (2017) contend that since education gives the necessary information for launching a firm, it can aid entrepreneurs, particularly female ones who may have limited access to resources (Zlatkov, 2015). Education does not define entrepreneurial qualities, according to McGowan et al. (2015); nevertheless, Ali (2018) asserts that education is linked to knowledge and abilities, self-assurance, problem-solving, willpower, and discipline, all of which affect entrepreneurial performance. Numerous other African researchers have claimed that formal education aids in the growth and process of entrepreneurship (see, for instance, Ezeuduji & Ntshangase, 2017a; Iwu et al., 2016). The lack of education is one of the main issues limiting the growth of female entrepreneurs in Africa (Chinomona & Maziriri, 2015; Daniyan-Bagudu et al., 2016; Nsengimana et al., 2017). According to several studies (Jiyane et al., 2012; Nxopo & Iwu, 2016; Sudarmanti et al., 2013), women entrepreneurs had less education than their male counterparts. In addition, women are said to have less business training and experience than men, according to Chirwa (2008).

According to Greenberg and Mollick (2017), men and women both have a network of entrepreneurs to learn from, but men have a wider variety of mentors. Women however participate in informal (family and friends) networks more than men do (Jha et al., 2018; Santos et al., 2019). Studies done in developing countries and other emerging nations have dissimilar findings regarding the impact of informal networks on entrepreneurship support. Some of the studies highlighted the critical role played by family and friends and community networks in not only providing seed capital but in supporting other aspects of business growth for many small and micro enterprises (such as Kimbu et al., 2019; Ngoasong & Kimbu, 2016; Ngoasong & Kimbu, 2019). Although they offer emotional support (Welsh et al., 2017), family and friends are viewed in some research as a weak entrepreneurial network since they are not particularly beneficial for business growth or a good start-up (Vossenberg, 2013). Having a support system of family members and friends who are independent contractors is essential, nevertheless (Alam et al., 2012).

The literature overview in this section can be synthesised in Figure 13.1. It shows the links between tourism business performance and some of the factors that impact on it. In developing or emerging countries such as South Africa, female entrepreneurs are likely to be more disadvantaged regarding these factors (traditional societal perceptions, managerial capabilities, marketing capabilities, formal education, and business networks) in comparison to their male counterparts. Male entrepreneurs are therefore more likely than female entrepreneurs to have more prosperous enterprises. These relationships are further explored in the subsequent empirical sections of this chapter.

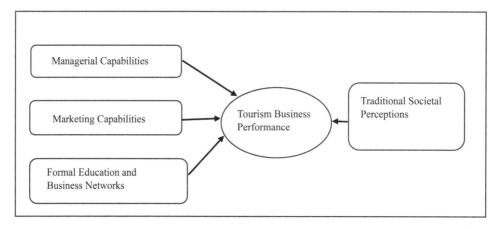

Figure 13.1 Tourism business performance and some influencing factors.

Relevant Theoretical Frameworks

This study looked at how performance in the tourism industry varied depending on gender; it required theoretical frameworks to support the research process. Relevant entrepreneurship process theories found in the literature include "self-efficacy theory" (Baron et al., 2016), "enculturation theory" (Grusec & Hastings, 2015), "enculturation and trait developments" (Buss, 2016; Rigney, 2011; van Oers, 2010), and "impression management" (Ho et al., 2017). These theories connect to the study's main topic: gender differences in business performance in the tourism industry (see Moudrý & Thaichon, 2020).

According to gender identity and enculturation theories, people eventually accept the fundamental beliefs and practices of their society since they are constantly exposed to them (Grusec & Hastings, 2015). This can further clarify how boys and girls begin to exhibit different behaviours at an early age, which eventually have an impact on how they run their businesses as adults. Enculturation is the process of imparting intrinsic and unspoken cultural elements from one person to another by emulating the influencer's social behaviours, beliefs, expectations, values, and a variety of other traits, which are later incorporated or accepted by the influenced person. According to Grusec and Hastings (2015), enculturation and socialisation lead to the formation of behavioural discrepancies between cultures as well as behavioural similarities within them. According to Moudrý and Thaichon (2020), the younger the individual, the quicker and easier it is to enculturate explicit cultural elements. This explains how each feature of society will grow into a person's unique feature.

Enculturation is seen as one component in the production of phenotypes, including gender-specific personality traits, although it is not the only one. The majority of male children are at lower risk of hardship than female children in historically entrenched patriarchal societies, such as Africa (Hofstede Insights, 2019), because they are socialized to believe that getting hurt "is part of being a boy, and boys don't cry," and because their parents were less concerned about them than they were about their daughters (Buss, 2016; Rigney, 2011; van Oers, 2010). Furthermore, according to Buss (2016), "guys" preferred competitive war games more than "girls," which gave them a competitive advantage as adults. Furthermore, the personality dark triad of psychopathy, narcissism, and Machiavellianism is employed to engage in aggressive, dishonest, manipulative, domineering, and exploitative behaviour against others (Buss, 2016; Dant et al., 2013). As mentioned by Moudrý and Thaichon (2020), this may show some slight benefits of entrepreneurial success that men may have over women without such characteristics.

Boys and girls exhibit different adult behaviours as a result of their different upbringings, which is reflected in how each gender manages its own businesses. Because they were taught to be cautious as children, the majority of young ladies avoid risks as a result (Rigney, 2011; van Oers, 2010). According to this study's findings, minimising risks often makes starting a business more difficult (Stark & Zawojska, 2015). This chapter argues that men who exhibit the relevant masculine traits will be more risk-tolerant than people who do not exhibit the relevant masculine traits. Risk-taking is one of the key characteristics of entrepreneurship and is likely to make it more likely for women to recognise fewer opportunities and pursue fewer entrepreneurial paths than men. The good news is that enculturation can be altered by life experience (Moudrý & Thaichon, 2020). For instance, Looi and Khoo-Lattimore (2015) hypothesised that children's learning may be positively impacted by their entrepreneurial parents' role models. According to these researchers, up to 83% of children from entrepreneurial households are actively involved in business and are more likely to choose an entrepreneurial profession. As a result, they are cultivated by their parents and their immediate environment with male traits. Accordingly, it is acceptable to expect that different features can be learnt, developed, and implemented at a later stage, according to Kasen et al. (2006).

According to Baron et al. (2016), the definition of self-efficacy is having confidence in one's ability to carry out particular tasks. According to earlier research, setting smaller goals than necessary hinders an entrepreneur's ability to achieve established objectives or launch new ventures (Baron et al., 2016). This could result in disappointment, a reduction in knowledge acquisition, growth inhibition, demoralisation, or even the failure to launch a firm (Moudrý & Thaichon, 2020). Poor self-efficacy is a result of some known feminine traits, like being gullible, introverted, and not being aggressively competitive (Arévalo Avalos & Flores, 2016; Baron et al., 2016). One skill that female entrepreneurs who exhibit feminine features lack is negotiating self-efficacy, which is crucial for starting and maintaining a new business. In the early phases of new entrepreneurial endeavours, shrewd negotiation skills are necessary for business negotiations, company finance, the hiring of key staff members and managers, and the procurement of other resources. A negotiator will achieve greater results if they are persuasive, macho, and self-assured (Guerrero & Richards, 2015). According to Guerrero and Richards (2015) and Stark and Zawojska (2015), female entrepreneurs with feminine features have a tendency to accept less, which puts them at a competitive disadvantage to advance their new venture idea beyond the conceptual stage and causes it to fail at the crucial start-up phase.

Self-presentation, commonly referred to as impression management, is crucial for affecting how others see you (Ho et al., 2017). According to Nagy et al. (2012), entrepreneurs constantly use impression management techniques to create, maintain, and change their image in order to affect others' attitudes. This ability is essential for businesses because it enables them to get important resources, such as financial and human assets, through improved outcomes in negotiations (Nagy et al., 2012). Therefore, it was claimed that businesses owned by men had higher yearly gross sales than those owned by women (Miller et al., 2003). Shyness, risk aversion, and agreeableness are regarded as detrimental in impression management since pitching and selling venture ideas demand assertive and even confrontational skills while speaking to other parties. Some female business owners worry about possible backlash or other negative effects. Women negotiate more cooperatively because they are more sensitive to others' sentiments or emotions (Moudrý & Thaichon, 2020). Male characteristics are more likely to employ more persuasive techniques to persuade others, which leads to more favourable business outcomes (Guerrero & Richards, 2015), yet this is not a bad trait.

Research Design and Methods

This study attempts to determine whether there are gender inequalities among tourism business owners that can have an impact on the success of their enterprises. Traditional societal perceptions, financing availability, level of formal education, and development of business networks are given special attention. Therefore, the assumptions, convictions, expectations, and ideals of the positivist paradigm serve as the driving force behind this study (Creswell, 2014; Kivunja & Kuyini, 2017). For research that aims to explain observations in terms of facts or quantifiable entities, the positivist paradigm is the chosen worldview (Kivunja & Kuyini, 2017). Because differences can be evaluated and defined more accurately through a questionnaire survey (a quantitative research method), this study's aims had to be addressed in that way (see Veal, 2011).

By selecting men and women who are eager to engage and who have specific understanding of tourism entrepreneurship (those who own and manage tourism businesses), the researchers distributed a structured questionnaire using a non-probability purposive sampling technique (Etikan et al., 2016; Nardi, 2018). A tourism business was defined as an organisation that is involved in the following industries: travel, hospitality and events, business and leisure, sports and recreation, conservation, gaming, government tourism, and tourism support services. Tourism-related entrepreneurs were polled regarding their business operations, in Durban CBD, KwaZulu-Natal Province, South Africa. A total of 150 respondents (comprising 75 male and 75 female entrepreneurs) were surveyed. These respondents have decision-making and line management responsibilities and can respond to the questionnaire based on their business ownership and/or management knowledge (Cave & Kilic, 2010; Mattis, 2004). In order to produce balanced views from both male and female entrepreneurs and the way they manage company success determinants (variables emanate from previous scholars), special attention was taken when distributing surveys.

These business performance factors include "traditional and societal perceptions" (Ahmed, 2018; Daniyan-Bagudu et al., 2016; Ezeuduji & Ntshangase, 2017a; Fairlie & Robb, 2009; Singh et al., 2010; Tlaiss & Kauser, 2019; Tshabalala & Ezeuduji, 2016), "access to financial capital" (Ahmed, 2018; Fairlie & Robb, 2009; Singh et al., 2010; Tlaiss & Kauser, 2019; Tshabalala & Ezeuduji, 2016), "managerial capabilities" (Fairlie & Robb, 2009; Manzanera-Román & Brändle, 2016; Sudarmanti et al., 2013), "marketing capabilities" (Kimosop et al., 2016; Nzama & Ezeuduji, 2021; Tshabalala & Ezeuduji, 2016; Welsh et al., 2017), "level of formal education" (Chinomona & Maziriri, 2015; Fairlie & Robb, 2009; Nsengimana et al., 2017; Ramadani, 2015), and "ability to form business network" (Jha et al., 2018; Santos et al., 2019; Sarfaraz et al., 2014).

Version 25 of IBM's Statistical Package for Social Sciences (SPSS) software was used to analyse the data that had been gathered. Descriptive data on research constructs, including percentage frequencies and mean response scores, and other inferential analyses (namely, Spearman's rank correlation, Mann-Whitney U; Pearson's chi-square; and Cronbach's alpha's reliability tests) were used to explore relationships that will address the study objectives (gender differences regarding business performance factors). All these statistical tests (depending on the type of variables being analysed) explored relationships between study variables and constructs. It is not the intention of this chapter to dwell much on the statistical justifications and presentations of results riddled with statistical relationships and numbers. As a result, the next section describes the main results of this study with respect to earlier research and their implications for practice.

Key Findings and Discussion

Entrepreneurial Success Level versus Gender

Previous research (such as those of Ahmed, 2018; Chasserio et al., 2014; Vrbanac et al., 2016; Wankel, 2008) reveal that women are mostly seen as "not fit" to start and manage an enterprise (reflecting the theories of enculturation and gender identity, and self-efficacy); however, the results of this study suggest that the entrepreneurs think a person's performance is not always influenced by their gender. It found no statistical difference between genders (males and females) regarding responses on the level of entrepreneurial success achieved. The level of formal education is the only sociodemographic factor that significantly affects "entrepreneurial success level." Regardless of gender, respondents who had tertiary education (mean = 2.12) thought they were more successful than respondents without higher education (mean = 2.79) (Mann-Whitney U test, ** $p < 0.01$). The study compared gender identity with the answers to "general entrepreneurial statements"; this study did not find any appreciable variations between male and female respondents. Therefore, neither gender may assert that they are more enterprising.

Entrepreneurial Success Level versus Traditional Societal Perceptions and Start-Up Motivation

This study found that a sizeable percentage of participants agreed that an entrepreneur's home environment (64%) and the "belief system of the society" (60%) determine entrepreneurial performance (embedded in the theories of enculturation, gender identity, and trait formation that may affect self-efficacy). The small size of their businesses may make African women happier than African males because society expects them to manage small businesses, which may have an impact on this outcome. However, there were no statistically significant differences between the genders in this study's findings for business performance or willingness to establish a tourism-related venture. Fisher (2011) and Kokotović et al. (2016) assert that business owners who launched a company to seize an opportunity exhibit trait of entrepreneurs who are growth-oriented. According to these scholars, growth entrepreneurs establish a firm with long-term value and continuously try to make it bigger and more successful because they are driven by the competitive nature of business. Growth entrepreneurs are more likely to launch a company with higher sums of money. This study confirms the findings of a study conducted in Spain (Vázquez-Carrasco et al., 2012), which found no appreciable differences in male and female business managers' start-up motivation as opportunity or survival entrepreneurs or growth entrepreneurs.

Formal Education and Business Training versus Gender

According to this study, having formal education and training is a significant advantage when starting and operating a firm (77.4% of respondents strongly agreed or agreed). Having tertiary education may drive entrepreneur's impression management to achieve better self-efficacy. This supports Ramadani's (2015) and Zlatkov's (2015) position that one of the most important factors in entrepreneurship success is education. Furthermore, the claim by Chirwa (2008) that female entrepreneurs do not have sufficient business training and expertise in comparison to male entrepreneurs is refuted by this study's findings. Many respondents (65.3%), irrespective of gender, strongly disagreed or disagreed with this notion.

Managerial and Marketing Capabilities versus Gender

The results of the study also did not reveal any differences in response mean values between male and female entrepreneurs regarding other important business operation factors, such as managerial (mean value for males = 2.5; mean value for females = 2.7) and marketing capabilities (mean value for males = 1.9; mean value for females = 2.1). Welsh et al. (2017) made the case that having marketing skills helps business owners expand and stay competitive (leading to better impression management and higher chances of self-efficacy). According to a recent study (Mersha & Sriram, 2019), since African entrepreneurs are more likely to launch enterprises due to socioeconomic challenges, they lack business management skills. This study disputes that assertion and reveals that most respondents (92%), regardless of gender, perceive that they have adequate skills regarding running a tourism business. Business owners/managers who were surveyed (72%) revealed that they arrange workshops for employees to receive business training.

Accessing Start-Up Finance versus Gender

This study's discovery that female entrepreneurs had easier access to start-up finance than their male counterparts is among its more intriguing findings. The results of Meunier et al.'s (2017) and Vossenberg's (2013) studies, which contend that female business owners and managers have more difficult access to financing than their male counterparts, are at odds with the results of this study. One of the main causes of women entrepreneurs' poor performance in operating their enterprises is perceived to be their difficulty in obtaining financial resources (Marlow & McAdam, 2013). The study's findings, however, suggest that women are generally more informed of where to find start-up money than males are (mean value for males = 3.35; mean value for females = 2.96; Mann-Whitney U test, *, $p < 0.05$), and they find it simpler to access this capital (mean value for males = 4.23; mean value for females = 3.95; Mann-Whitney U test, *, $p < 0.05$). This finding is consistent with earlier finding (De Vita et al., 2014) which indicates that financial institutions favour supporting female entrepreneurs over male entrepreneurs. De Vita et al. also revealed that women are permitted to get loans under more lenient conditions with regard to collateral security and payment periods. Contrarily, Sattar et al. (2016) observed that female entrepreneurs in poor nations have more difficult access to financial capital compared to their counterparts in developed nations. The study's findings show that female entrepreneurs have an advantage over male entrepreneurs when it comes to obtaining cash because more of them concurred that finding finance is often straightforward. According to the researchers' understanding, women entrepreneurs in South Africa are often more favoured than men when it comes to getting funding to launch a firm.

With grants and incentives, a range of funding choices are provided for small businesses in South Africa via SME South Africa's (2019) Guide to Government Funding for Small and Medium Enterprises, with each funding programme offering some form of financial support for business owners. Even though none of these government funding programmes specifically target male business owners, programmes like the Tourism Transformation Fund aim to change the tourism sector and support the creation of a new generation of Black-owned youth, women, and community-based enterprises. The National Empowerment Fund (NEF), which focuses on empowering women especially, also aims to fund black women-owned enterprises more quickly through its NEF Women Empowerment Fund. In South Africa, women entrepreneurs – particularly young Black women – are given precedence in all government business funding programmes, even though their names may not specify their desired gender, age range, or race. This is so because South Africa's National Development Plan 2030 (National Development Plan, 2012) places a strong emphasis on empowering women and young people.

Formation of Business Networks versus Gender

Having formal and informal business networks can optimise self-efficacy and foster entrepreneurs' impression management. According to Zhang et al. (2020), women managers in China used guanxi-based social networks to create their own identities inside such networks and increase their chances of moving up the corporate ladder. But Daniyan-Bagudu et al. (2016) point out that few female business owners are aware of the networks for other female business owners. Therefore, it is not surprising that more male respondents acknowledged that "they are active members of tourism business networks," compared to the females, based on the study's findings (mean value for males = 2.57; mean value for females = 3.19; Mann-Whitney U test, **, $p < 0.01$), additionally, they connect with other business owners at social events (mean value for males = 1.80; mean value for females = 2.20; Mann-Whitney U test, *, $p < 0.05$). Gidarakou (2015) points out that women have a lesser social status than men do, which may limit their access to important data and resources. According to Klyver and Grant (2010), women entrepreneurs are less likely to be a part of entrepreneurial networks, and these study's findings support this theory. Entrepreneurial networking is crucial for locating resources, promoting businesses, and acquiring knowledge (Hodges et al., 2015; Klyver & Grant, 2010; Lans et al., 2015; Srećković, 2018).

Level of Business Success versus Statements on Business Operation

The association between respondents' opinions on the "perception of entrepreneurial success level" and "business operation" statements was examined using Spearman's rank correlation (two-tailed) test. Based on the study's findings, it is clear that the perceived level of business success and gender-specific company operating characteristics do not correlate (no significant results at 95% confidence interval). "Marketing capability" is where the majority of the individual statements that were discovered to be positively correlated with perceived business success can be located. These statements are (1) "I use Internet to market my business" (**, $p < 0.01$), (2) "I rely on different sources of media to get my business known" (**, $p < 0.01$), (3) "I do research to find out the new market trends" (**, p < 0.01), (4) "I know very well how to offer my products and deliver services to meet customer need (Product)" (*, p < 0.05), and (5) "I use different promotion techniques to sell my products and services (Promotion)" (**, p < 0.01). The fact that all the variables under this dimension have, for the most part, very substantial positive connections with company performance suggests that possessing strong marketing skill is essential to achieve business success. This study findings corroborate previous studies' results, such as Welsh et al. (2017) and Kimosop et al. (2016), who posit a direct correlation between business success and marketing capabilities. According to Welsh et al. (2017), strong marketing abilities equip business owners to recognise and seize chances for business growth and market competitiveness. This is consistent with the other research findings.

Moreso, study statements that have to do with "managers' level and impact of formal education" (*, p < 0.05), "prior business training" (**, p < 0.01), "membership of business networks" (*, p < 0.05), and "socialising with business networks" (*, p < 0.05) correlated positively with business success in this study. This demonstrates that proper formal education, business training, and participation in business networks all have a favourable impact on the degree of entrepreneurial success in the research field. Scholars emphasised the importance of formal education in giving entrepreneurs the skills and knowledge they need to manage the day-to-day operations of their businesses (Iwu & Nxopo, 2015; Nxopo & Iwu, 2016; Rao, 2018). Given the type of the statements on which men agreed more frequently than women, it is noteworthy that male respondents are more engaged in business management dynamics that support corporate operations systems (Ramadani, 2015; Tajeddin et al., 2017; Zlatkov, 2015). Males were more likely than females to think that the performance of the tourism business is greatly influenced by the manager's level of formal education. (mean value for males = 1.80; mean value for females = 2.20; Mann-Whitney U test, *, p < 0.05). The next section concludes this chapter.

Limitations of Study and Future Research

Three limitations apply to this study. (1) The study results cannot be applied to the complete research population due to the use of non-probability sampling in data collection. The results presented in this chapter only apply to the sample studied, and at the time they were studied. (2) The relatively small sample size used for this research is also a limitation, as this sample size and the distribution of responses (non-normal distribution of data) did not allow for more complex statistical analyses, which may give rise to more accurate findings or support predictions. (3) This study therefore used mostly bivariate analyses, instead of the more complex multivariate analyses to address research objectives.

As this study is more exploratory than conclusive, these study limitations are envisaged, but the findings of this foundation research can support more conclusive studies where probability sampling, larger sample size, and more sophisticated statistical analyses can be employed.

It is expected that future conclusive studies generate and present model estimates that can predict gender nuances in tourism-related business performance, and unearth factors affecting or mediating current and future tourism business operations and business performance based on gender identity.

Conclusions and Recommendations

This study did not come to the definitive conclusion that, contrary to popular belief, gender or any other demographic factors are extremely important in determining company success or performance. The level of formal education is the only sociodemographic factor that significantly affects "entrepreneurial success level." Respondents who had tertiary education thought they were more successful than those who did not. The differentiating factor defining a business' performance must then be the manager's skill set in terms of operations, marketing, and general management. In terms of their answers to questions about "entrepreneurial success," "managerial and operational capabilities," "marketing capability," and "general entrepreneurial statements," male and female respondents to this survey did not demonstrate any discernible differences. The management of a firm is not improved by an entrepreneur's gender or gender identity.

As a result, this study suggests that KwaZulu-Natal, South Africa, implements more training and mentorship initiatives in entrepreneurship programmes. These campaigns should emphasise the need for formal education for business success, not gender; and should encourage women business owners to actively participate in relevant business networks that foster their companies' expansion and development. Entrepreneurial networking is crucial for collecting knowledge, resources, and business promotion, as was already said. These networks should not only serve social and information-sharing purposes (for example, information regarding business funding). In order to provide young and upcoming women entrepreneurs with the abilities to efficiently manage business operations, they can also offer business training and mentorship programmes. These will support and maintain the acquisition of business management dynamics and intellectual capital that support innovative business operations systems, in a knowledge-based economy.

At the time this chapter was written, the South African tourism businesses are recovering from the negative economic effects due to the global outbreak of coronavirus disease, and the subsequent crisis of the rising cost-of-living has led to further differences in entrepreneurial endeavours of both women and men in South Africa. According to Rigoni et al. (2021, p. 2), "women across the world are disproportionately affected by the COVID-19 pandemic, reversing gains in gender equality made in recent decades. Women-led businesses have been more negatively impacted economically, especially in sectors hardest hit by the pandemic." Therefore, it is essential that the South African government act right away to ensure gender equity by providing financial support for women-owned enterprises in the post-COVID period. According to Rigoni et al.'s (2021) recommendation, there is a critical need for public-private sector collaboration projects that would encourage and hasten the development of digital skills among women entrepreneurs. Additionally, it is critical to actively encourage and improve financial and economic support for women-owned enterprises.

References

Ahmed, N. (2018). Socio-economic impact of women entrepreneurship in Bangladesh. *Sociology and Anthropology*, 6(6), 526–533. https://doi.org/10.13189/sa.2018.060602

Alam, S. S., Senik, Z. C., & Jani, F. M. (2012). An exploratory study of women entrepreneurs in Malaysia: Motivation and problems. *Journal of Management Research*, 4(4), 282–297. https://doi.org/10.5296/jmr.v4i4.2377

Ali, R. S. (2018). Determinants of female entrepreneurs' growth intentions: A case of female-owned small businesses in Ghana's tourism sector. *Journal of Small Business and Enterprise Development*, 25(3), 387–404. https://doi.org/10.1108/JSBED-02-2017-0057

Anwar, I., Thoudam, P., & Saleem, I. (2022). Role of entrepreneurial education in shaping entrepreneurial intention among university students: Testing the hypotheses using mediation and moderation approach. *Journal of Education for Business*, 97(1), 8–20. https://doi.org/10.1080/08832323.2021.1883502

Arévalo Avalos, M. R., & Flores, L. Y. (2016). Non-traditional career choices of Mexican American men: Influence of acculturation, enculturation, gender role traits, self-efficacy, and interests. *Journal of Latina/o Psychology*, 4(3), 142–157. http://dx.doi.org/10.1037/lat0000048

Arora, N. (2014). A social perception towards women entrepreneurs in India – From perception to reality. *Sai Om Journal of Commerce & Management*, 1(2), 44–49.

Baron, R. A., Mueller, B. A., & Wolfe, M. T. (2016). Self-efficacy and entrepreneurs' adoption of unattainable goals: The restraining effects of self-control. *Journal of Business Venturing*, 31(1), 55–71. https://doi.org/10.1016/j.jbusvent.2015.08.002

Basargekar, P. (2007). Women entrepreneurs: Challenges faced. *The Icfai Journal of Entrepreneurship Development*, 4(4), 6–15. https://www.coursehero.com/file/64257155/Challenges-face-by-women-entrepreneurspdf/

Bazkiaei, H. A., Heng, L. H., Khan, N. U., Saufi, R. B. A., & Kasim, R. S. R. (2020). Do entrepreneurial education and big-five personality traits predict entrepreneurial intention among university students? *Cogent Business and Management*, 7(1), 1–18. https://doi.org/10.1080/23311975.2020.1801217

Buss, D. M. (2016). *Evolutionary psychology: The new science of the mind* (5th ed.). Routledge.

Cave, P., & Kilic, S. (2010). The role of women in tourism employment with special reference to Antalya, Turkey. *Journal of Hospitality Marketing & Management*, 19(3), 280–292. https://doi.org/10.1080/19368621003591400

Ceptureanu, S. I., & Ceptureanu, E.-G. (2016). Women entrepreneurship in Romania: The case of Northeast development region. *Management and Economics Review*, 1(1), 20–32. http://mer.ase.ro/files/2016-1/2.pdf

Chasserio, S., Pailot, P., & Poroli, C. (2014). When entrepreneurial identity meets multiple social identities: Interplays and identity work of women entrepreneurs. *International Journal of Entrepreneurial Behaviour & Research*, 20(2), 128–154. https://doi.org/10.1108/IJEBR-11-2011-0157

Chinomona, E., & Maziriri, E. T. (2015). Women in action: Challenges facing women entrepreneurs in the Gauteng Province of South Africa. *International Business & Economics Research Journal*, 14(6), 835–850. https://doi.org/10.19030/iber.v14i6.9487

Chirwa, E. W. (2008). Effects of gender on the performance of micro and small enterprises in Malawi. *Development Southern Africa*, 25(3), 347–362. https://doi.org/10.1080/03768350802212139

Creswell, J. W. (2014). *Research design: Qualitative, quantitative, and mixed methods approaches* (4th ed.). SAGE Publications, Inc.

Daniyan-Bagudu, H., Khan, S. J. M., & Roslan, A. H. (2016). *The issues and challenges facing the female entrepreneurs in Lagos State, Nigeria*. 3rd Kanita Postgraduate International Conference on Gender Studies (16–17 November) 149–155. Universiti Sains Malaysia, Penang.

Dant, R. P., Weaven, S. K., & Baker, B. L. (2013). Influence of personality traits on perceived relationship quality within a franchisee-franchisor context. *European Journal of Marketing*, 47(1/2), 279–302. https://doi.org/10.1108/03090561311285556

De Vita, L., Mari, M., & Poggesi, S. (2014). Women entrepreneurs in and from developing countries: Evidence from the literature. *European Management Journal*, 32, 451–460. https://doi.org/10.1016/j.emj.2013.07.009

Duflo, E. (2012). Women empowerment and economic development. *Journal of Economic Literature*, 50(4), 1051–1079. https://doi.org/10.1257/jel.50.4.1051

Etikan, I., Musa, S. A., & Alkassim, R. S. (2016). Comparison of convenience sampling and purposive sampling. *American Journal of Theoretical and Applied Statistics*, 5(1), 1–4. https://doi.org/10.11648/j.ajtas.20160501.11

Ezeuduji, I. O., & Ntshangase, S. D. (2017a). Entrepreneurial intention: South African youth's willingness to start tourism businesses. *Acta Universitatis Danubius Œconomica*, 13(5), 48–58. http://journals.univ-danubius.ro/index.php/oeconomica/article/view/4165/4414

Ezeuduji, I. O., & Ntshangase, S. D. (2017b). Entrepreneurial inclination: South African youth's mental attitude towards starting tourism business. *Journal of Economics and Behavioural Studies*, 9(4), 144–152. https://doi.org/10.22610/jebs.v9i4(J).1829

Fairlie, R. W., & Robb, A. M. (2009). Gender differences in business performance: Evidence from the characteristics of business owners survey. *Small Business Economics*, 33, 375–395. https://doi.org/10.1007/s11187-009-9207-5

Figueroa-Domecq, C., de Jong, A., Kimbu, A. S., & Williams, A. M. (2022). Financing tourism entrepreneurship: A gender perspective on the reproduction of inequalities. *Journal of Sustainable Tourism*. https://doi.org/10.1080/09669582.2022.2130338

Figueroa-Domecq, C., Kimbu, A., de Jong, A., & Williams, A. M. (2022). Sustainability through the tourism entrepreneurship journey: A gender perspective. *Journal of Sustainable Tourism*, 30(7), 1562–1585. https://doi.org/10.1080/09669582.2020.1831001

Fisher, G. (2011). Which type of entrepreneur are you? Retrieved from http://www.entrepreneurmag.co.za/advice/growing-abusiness/performance-and-growth/which-type-of-entrepreneur-are-you/ [Accessed 01.07.2019].

Global Entrepreneurship Monitor (GEM) (2012). *Global Entrepreneurship Monitor 2012 Women's Report*. Retrieved from https://www.gemconsortium.org/report/gem-2012-womens-report [Accessed 22.07.2019].

Global Entrepreneurship Monitor (GEM) (2014). *Global Entrepreneurship Monitor. 2014 Global Report*. Retrieved from http://www.baruch.cuny.edu/news/documents/GEM2014.pdf [Accessed 22.07.2019].

Ghiat, B. (2018). Social attitudes towards women entrepreneurs in Algeria. *Global Journal of Women Studies*, 1(1), 1–6. https://doi.org/10.33152/jmphss-1.1.1

Gidarakou, I. (2015). Women's entrepreneurship in rural Greece. *International Journal of Business and Management*, 10(10), 129–142. http://dx.doi.org/10.5539/ijbm.v10n10p129

González-Sánchez, V. M. (2015). Factors promoting entrepreneurship in European countries: Unemployment, taxes, and education. *Journal of Promotion Management*, 21(4), 492–503. https://doi.org/10.1080/10496491.2015.1051405

Greenberg, J., & Mollick, E. (2017). Activist choice homophily and the crowdfunding of female founders. *Administrative Science Quarterly*, 62(2), 341–374. https://doi.org/10.1177/0001839216678847

Grusec, J. E., & Hastings, P. D. (2015). *Handbook of socialization: Theory and research* (2nd ed.). The Guilford Press.

Guerrero, V., & Richards, J. (2015). Female entrepreneurs and negotiation self-efficacy: A study on negotiation skill building among women entrepreneurs. *Journal of Entrepreneurship Education*, 18(2), 17–28. https://www.abacademies.org/articles/jeevol1822015.pdf

Guillet, B. D., Pavesi, A., Hsu, C. H. C., & Weber, K. (2019a). Is there such a thing as feminine leadership? Being a leader and not a man in the hospitality industry. *International Journal of Contemporary Hospitality Management*, 31(7), 2970–2993. https://doi.org/10.1108/IJCHM-06-2018-0486

Guillet, B. D., Pavesi, A., Hsu, C. H. C., & Weber, K. (2019b). What can educators do to better prepare women for leadership positions in the hospitality industry? The perspectives of women executives in Hong Kong. *Journal of Hospitality & Tourism Education*, 31(4), 197–209. https://doi.org/10.1080/10963758.2019.1575751

Hassan, F., Ramli, A., & Desa, N. M. (2014). Rural women entrepreneur in Malaysia: What drives their success? *International Journal of Business and Management*, 9(4), 10–21. https://doi.org/10.5539/ijbm.v9n4p10

Hodges, N., Watchravesringkan, K., Yurchisin, J., Karpova, E., Marcketti, S., Hegland, J., Yan, R., & Childs, M. (2015). Women and apparel entrepreneurship: An exploration of small business challenges and strategies in three countries. *International Journal of Gender and Entrepreneurship*, 7(2), 191–213. https://doi.org/10.1108/IJGE-07-2014-0021

Hofstede Insights (2019). What about New Zealand? https://www.hofstede-insights.com/country/new-zealand/ [Accessed on 25.11.2019].

Ho, T. H., Tojib, D., & Khajehzadeh, S. (2017). Speaking up against service unfairness: The role of negative meta-perceptions, *Journal of Retailing and Consumer Services*, 35, 12–19, https://doi.org/10.1016/j.jretconser.2016.11.002

Idris, A. J., & Agbim, K. C. (2015). Effect of social capital on poverty alleviation: A study of women entrepreneurs in Nasarawa State, Nigeria. *Journal of Research in National Development, 13*(1), 208–222. https://www.ajol.info/index.php/jorind/article/view/120625

Iwu, C. G. (2022). Entrepreneurship education challenges in the African setting. *Academia Letters, 2,* 1–9.

Iwu, C. G., Ezeuduji, I. O., Eresia-Eke, C., & Tengeh, R. (2016). The entrepreneurial intention of university students: The case of a university of technology in South Africa. *Acta Universitatis Danubius Œconomica, 12*(1), 164–181. http://journals.univ-danubius.ro/index.php/oeconomica/article/view/3129/3216

Iwu, C. G., & Nxopo, Z. (2015). Determining the specific support services required by female entrepreneurs in the South African tourism industry. *African Journal of Hospitality, Tourism and Leisure, 4*(2), 1–13. http://www.ajhtl.com/uploads/7/1/6/3/7163688/article22vol4(2)july-nov2015.pdf

Jha, P., Makkad, M., & Mittal, S. (2018). Performance-oriented factors for women entrepreneurs – A scale development perspective. *Journal of Entrepreneurship in Emerging Economies, 10*(2), 329–360. https://doi.org/10.1108/JEEE-08-2017-0053

Jiyane, G. V., Majanja, M. K., Mostert, B. J., & Ocholla, D. (2013). South Africa as information and technology society: The benefit to informal sector women entrepreneurs. *South African Journal Libs and Information Science, 79*(1), 1–12. https://doi.org/10.7553/79-1-115

Jiyane, V. G., Ocholla, D. N., Mostert, B. J., & Majanja, M. K. (2012). Contribution of informal sector women entrepreneurs to the tourism industry in eThekwini Metropolitan Municipality, in KwaZulu-Natal: Barriers and issues. *African Journal for Physical, Health Education, Recreation and Dance, 18*(4), 709–728. https://www.ajol.info/index.php/ajpherd/article/view/83836

Kasen, S., Chen, H., Sneed, J., Crawford, T., & Cohen, P. (2006). Social role and birth cohort influences on gender-linked personality traits in women: A 20-year longitudinal analysis. *Journal of Personality and Social Psychology, 91*(5), 944–958. http://dx.doi.org/10.1037/0022-3514.91.5.944

Kimbu, A. N., Ngoasong, M. Z., Adeola, O., & Afenyo-Agbe, E. (2019). Collaborative networks for sustainable human capital management in women's tourism entrepreneurship: The role of tourism policy. *Tourism Planning & Development, 16*(2), 161–178. https://doi.org/10.1080/21568316.2018.1556329

Kimosop, J., Korir, M., & White, M. (2016). The moderating effect of demographic characteristics on the relationship between strategic capabilities and firm performance in women-owned entrepreneurial ventures in Nairobi, Kenya. *Canadian Journal of Administrative Sciences, 33*(3), 242–256. https://doi.org/10.1002/cjas.1399

Kivunja, C., & Kuyini, A. B. (2017). Understanding and applying research paradigms in educational contexts. *International Journal of Higher Education, 6*(5), 26–41. https://doi.org/10.5430/ijhe.v6n5p26

Klyver, K., & Grant, S. (2010). Gender differences in entrepreneurial networking and participation. *International Journal of Gender and Entrepreneurship, 2*(3), 213–227. https://doi.org/10.1108/17566261011079215

Kokotović, D., Rakić, B., & Kokotović, T. (2016). Female entrepreneurship: Main challenges and the impact of the gender gap. Proceedings of the XV International Symposium of Organisational Sciences: *Reshaping the Future through Sustainable Business Development and Entrepreneurship,* 504–512. Belgrade, Serbia.

Lans, T., Blok, V., & Gulikers, J. (2015). Show me your network and I'll tell you who you are: Social competence and social capital of early-stage entrepreneurs. *Entrepreneurship & Regional Development, 27*(7–8), 458–473. https://doi.org/10.1080/08985626.2015.1070537

Looi, K. H., & Khoo-Lattimore, C. (2015). Undergraduate students' entrepreneurial intention: Born or made? *International Journal of Entrepreneurship and Small Business, 26*(1), 1–20. http://dx.doi.org/10.1504/IJESB.2015.071317

Manzanera-Román, S., & Brändle, G. (2016). Abilities and skills as factors explaining the differences in women entrepreneurship. *Suma De Negocios, 7*(15), 38–46. https://doi.org/10.1016/j.sumneg.2016.02.001

Marlow, S., & McAdam, M. (2013). Gender and entrepreneurship: Advancing debate and challenging myths; exploring the mystery of the under-performing female entrepreneur. *International Journal of Entrepreneurial Behaviour & Research, 19*(1), 114–124. https://doi.org/10.1108/13552551311299288

Mattis, M. (2004). Women entrepreneurs: Out from under the glass ceiling. *Women in Management Review, 19*(3), 154–163. https://doi.org/10.1108/09649420410529861

McGowan, P., Cooper, S., Durkin, M., & O'Kane, C. (2015). The influence of social and human capital in developing young women as entrepreneurial business leaders. *Journal of Small Business Management, 53*(3), 645–661. https://doi.org/10.1111/jsbm.12176

Mersha, T., & Sriram, V. (2019). Gender, entrepreneurial characteristics, and success: Evidence from Ethiopia. *Thunderbird International Business Review, 61*(2), 157–167. https://doi.org/10.1002/tie.21984

Meunier, F., Krylova, Y., & Ramalho, R. (2017). Women's entrepreneurship: How to measure the gap between new female and male entrepreneurs? *Policy Research Working Paper 8242,* 1–28. World Bank Group. Retrieved from https://elibrary.worldbank.org/doi/pdf/10.1596/1813-9450-8242 [Accessed 02.04.2020].

Miller, N. J., Besser, T. L., Gaskill, L. R., & Sapp, S. G. (2003). Community and managerial predictors of performance in small rural US retail and service firms. *Journal of Retailing and Consumer Services, 10*(4), 215–230. https://doi.org/10.1016/S0969-6989(02)00012-7

Mkhize, G., & Cele, N. (2017). The role of women in tourism in KwaZulu-Natal. *Agenda, 31*(1), 128–139. https://doi.org/10.1080/10130950.2017.1371527

Moses, C. L., Olokundun, M., Falola, H., Ibidunni, S., Amaihian, A., & Inelo, F. (2016). A review of the challenges militating against women entrepreneurship in developing nations. *Mediterranean Journal of Social Sciences, 7*(1), 64–69. DOI: 10.5901/mjss.2016.v7n1p64

Moudrý, D. V., & Thaichon, P. (2020). Enrichment for retail businesses: How female entrepreneurs and masculine traits enhance business success. *Journal of Retailing and Consumer Services, 54,* 1–12. https://doi.org/10.1016/j.jretconser.2020.102068

Nagy, B. G., Pollack, J. M., Rutherford, M. W., & Lohrke, F. T. (2012). The influence of entrepreneurs' credentials and impression management behaviors on perceptions of new venture legitimacy. *Entrepreneurship Theory and Practice, 36*(5), 941–965. https://doi.org/10.1111/j.1540-6520.2012.00539.x

Nardi, P. M. (2018). *Doing survey research: A guide to quantitative methods* (4th ed.). Routledge.

National Development Plan. (2012). National Development Plan 2030: Our Future – Make it Work, 15 August 2012. Retrieved from https://www.gov.za/sites/default/files/gcis_document/201409/ndp-2030-our-future-make-it-workr.pdf [Accessed 02.04.2020].

Netshitangani, T. (2018). Constraints and gains of women becoming school principals in South Africa. *Journal of Gender, Information and Development in Africa (JGIDA), 7*(1), 205–222. http://dx.doi.org/10.31920/CGW_7_1_18

Ngoasong, M. Z., & Kimbu, A. N. (2019). Why hurry? The slow process of high growth in women-owned businesses in a resource-scarce context. *Journal of Small Business Management, 57*(1), 40–58. https://doi.org/10.1111/jsbm.12493

Ngoasong, M. Z., & Kimbu, A. N. (2016). Informal microfinance institutions and development-led tourism entrepreneurship. *Tourism Management, 52,* 430–439. https://doi.org/10.1016/j.tourman.2015.07.012

Nhleko, Y., & van der Westhuizen, T. (2022). The role of higher education institutions in introducing entrepreneurship education to meet the demands of industry 4.0. *Academy of Entrepreneurship Journal, 28*(1), 1–23. https://www.abacademies.org/articles/The-role-of-higher-education-institutions-in-introducing-entrepreneurship-education-to-meet-the-1528-2686-28-1-116.pdf

Nsengimana, S., Iwu, C. G., & Tengeh, R. K. (2017). The downside of being a female entrepreneur in Kigali, Rwanda. *The Scientific Journal for Theory and Practice of Socio-Economic Development, 6*(12), 151–164. DOI:10.12803/SJSECO.61203

Nxopo, Z., & Iwu, C. G. (2016). The unique obstacles of female entrepreneurship in the tourism industry in Western Cape, South Africa. *Commonwealth Youth and Development, 13*(2), 55–71. DOI:10.25159/1727-7140/1146

Nzama, N., & Ezeuduji, I. O. (2020a). Gender nuances in tourism-related entrepreneurship in Kwazulu-Natal, South Africa. *African Journal of Gender, Society and Development (AJGSD), 9*(4), 109–137. https://doi.org/10.31920/2634-3622/2020/v9n4a5

Nzama, N., & Ezeuduji, I. O. (2020b). Nuanced gender perceptions on the influences of formal education and business networks on tourism-related business operations: Kwazulu-Natal, South Africa.

EuroEconomica, *39*(2) (Special Issue), 70–84. http://dj.univ-danubius.ro/index.php/EE/article/view/292/477

Nzama, N., & Ezeuduji, I. O. (2021). Nuanced gender perceptions: Tourism business capabilities in Kwazulu-Natal, South Africa. *GeoJournal of Tourism and Geosites*, *35*(2), 372–380. https://doi.org/10.30892/gtg.35215-661

Okeke-Uzodike, O. E., Okeke-Uzodike, U., & Ndinda, C. (2018). Women entrepreneurship in Kwazulu-Natal: A critical review of government intervention politics and programs. *Journal of International Women's Studies*, *19*(5), 147–164. http://hdl.handle.net/20.500.11910/12319

Omerzel, D. G. (2016). The impact of entrepreneurial characteristics and organisational culture on innovativeness in tourism firms. Managing global transitions. *International Research Journal*, *14*(1), 93–110. https://ideas.repec.org/a/mgt/youmgt/v14y2016i1p93-110.html

Orser, B. J., & Riding, A. (2018). The influence of gender on the adoption of technology among SMEs. *International Journal of Entrepreneurship and Small Business*, *33*(4), 514–531. https://doi.org/10.1504/IJESB.2018.090341

Ramadani, V. (2015). The woman entrepreneur in Albania: An exploratory study on motivation, problems and success factors. *Journal of Balkan and Near Eastern Studies*, *17*(2), 204–221. https://doi.org/10.1080/19448953.2014.997488

Rao, D. K. (2018). Growth and development of women entrepreneurs in India: Challenges and empowerment. *International Journal of Advanced Research and Development*, *3*(1), 235–242. http://www.advancedjournal.com/archives/2018/vol3/issue1/3-1-90

Rigney, D. (2011). Boys vs. girls. *Contexts*, *10*(4), 78–79. https://doi.org/10.1177/1536504211427893

Rigoni, G., Herrmann, K., Lyons, A. C., Wilkinson, B., Kass-Hanna, J., Bennett, D., Stracquadaini, J., & Thomas, M. (2021). The economic empowerment of women entrepreneurs in a post-covid world. A policy brief of the Task Force 5, 2030 Agenda and Development Cooperation. Milan, Italy: G20 Insights. Retrieved from https://www.g20-insights.org/policy_briefs/the-economic-empowerment-of-women-entrepreneurs-in-a-post-covid-world/ [Accessed 13.01.2023].

Santos, G., Marques, C. S., & Ratten, V. (2019). Entrepreneurial women's networks: The case of D'Uva – Portugal wine girls. *International Journal of Entrepreneurial Behaviour & Research*, *25*(2), 298–322. https://doi.org/10.1108/ijebr-10-2017-0418

Sarfaraz, L., Faghih, N., & Asadi Majd, A. (2014). The relationship between women entrepreneurship and gender equality. *Journal of Global Entrepreneurship Research*, *2*(6), 1–11. http://www.journal-jger.com/content/2/1/6

Sattar, A., Dewri, L. V., & Ananna, S. A. (2016). Working environment for women entrepreneurs in developing countries: An empirical study of Bangladesh. *International Journal of Business and Management*, *11*(12), 197–206. https://doi.org/10.5539/ijbm.v11n12p197

Schaap, J. I., Stedham, Y., & Yamamura, J. H. (2008). Casino management: Exploring gender-based differences in perceptions of managerial work. *International Journal of Hospitality Management*, *27*(1), 87–97. https://doi.org/10.1016/j.ijhm.2007.07.004

Singh, S., Mordi, C., Okafor, C., & Simpson, R. (2010). Challenges in female entrepreneurial development – A case analysis of Nigerian entrepreneurs. *Journal of Enterprising Culture*, *18*, 435–460. http://dx.doi.org/10.1142/S0218495810000628

SME South Africa's (2019). A guide to government funding for SMEs. Retrieved from https://smesouthafrica.co.za/guide-government-funding-smallbusinesses/ [Accessed 02.04.2020].

Srećković, M. (2018). The performance effect of network and managerial capabilities of entrepreneurial firms. *Small Business Economics*, *50*, 807–824. https://doi.org/10.1007/s11187-017-9896-0

Stark, O., & Zawojska, E. (2015). Gender differentiation in risk-taking behavior: On the relative risk aversion of single men and single women. *Economics Letters*, *137*, 83–87. https://doi.org/10.1016/j.econlet.2015.09.023

Sudarmanti, R., Van Bauwel, S., & Longman, C. (2013). The importance of fieldwork research to reveal women entrepreneurs' competence in communication. *Journal of Women's Entrepreneurship and Education*, *3*(4), 74–87. https://econpapers.repec.org/RePEc:ibg:jwejou:y:2013:i:3-4:p:74-87

Tajeddini, K., Ratten, V., & Denisa, M. (2017). Female tourism entrepreneurs in Bali, Indonesia. *Journal of Hospitality and Tourism Management*, *31*, 52–58. https://doi.org/10.1016/j.jhtm.2016.10.004

Tlaiss, H. A., & Kauser, S. (2019). Entrepreneurial leadership, patriarchy, gender, and identity in the Arab world: Lebanon in focus. *Journal of Small Business Management*, *57*(2), 517–537. https://doi.org/10.1111/jsbm.12397

Todorović, I., Komazec, S., Miloš, J., Obradović, V., & Marič, M. (2016). Strategic management in the development of youth and women entrepreneurship – Case of Serbia. *Organizacija, 49*(4), 197–207. https://doi.org/10.1515/orga-2016-0018

Tshabalala, S. P., & Ezeuduji, I. O. (2016). Women tourism entrepreneurs in KwaZulu-Natal, South Africa: Any way forward? *Acta Universitatis Danubius Œconomica, 12*(5), 19–32. http://journals. univ-danubius.ro/index.php/oeconomica/article/view/3336/3666

van Oers, H. (2010). Children's enculturation through play. In L. Brooker & S. Edwards (Eds.), *Engaging play* (pp. 195–209). McGraw Hill.

Vázquez-Carrasco, R., López-Pérez, M. E., & Centeno, E. (2012). A qualitative approach to the challenges for women in management: Are they really starting in the 21st century? *Quality & Quantity, 46*, 1337–1357. https://doi.org/10.1007/s11135-011-9449-6

Veal, A. J. (2011). *Research methods for leisure and tourism: A practical guide* (4th ed.). Financial Times Prentice Hall.

Vossenberg, S. (2013). Women entrepreneurship promotion in developing countries: What explains the gender gap in entrepreneurship and how to close it? *Working Papers from Maastricht School of Management*, No 2013/08. Retrieved from https://econpapers.repec.org/paper/msmwpaper/2013_2f08. htm [Accessed 2.2.2020].

Vrbanac, M., Milovanović, M., & Perišić, J. (2016). Women's entrepreneurship - A global perspective and current state in Serbia. Proceedings of the XV International Symposium of Organisational Sciences: *Reshaping the Future through Sustainable Business Development and Entrepreneurship*, 522–528. Belgrade, Serbia.

Wankel, C. (2008). *21st century management. A reference handbook*. Sage Publication.

Welsh, D. H. B., Kaciak, E., Memili, E., & Zhou, Q. (2017). Work-family balance and marketing capabilities as determinants of Chinese women entrepreneurs' firm performance. *Journal of Global Marketing, 30*(3), 174–191. https://doi.org/10.1080/08911762.2017.1317894

Wessels, W., Du Plessis, E., & Slabbert, E. (2017). Key competencies and characteristics of accommodation managers. *SA Journal of Human Resource Management, 15*, 1–11. https://doi.org/10.4102/ sajhrm.v15i0.887

Witbooi, M., & Ukpere, W. (2011). Indigenous female entrepreneurship: Analytical study on access to finance for women entrepreneurs in South Africa. *African Journal of Business Management, 5*(14), 5646–5657. DOI:10.5897/AJBM10.1161

Yadav, O. P. (2018). Study on internet marketing practices of the tourism industry by travel agency of Nepal. *Pravaha Journal, 24*(1), 137–146. https://doi.org/10.3126/pravaha.v24i1.20233

Zeng, B., & Gerritsen, R. (2014). What do we know about social media in tourism? A review. *Tourism Management Perspectives, 10*, 27–36. http://dx.doi.org/10.1016/j.tmp.2014.01.001

Zhang, C. X., Kimbu, A. N., Lin, P., & Ngoasong, M. Z. (2020). Guanxi influences on women intrapreneurship. *Tourism Management, 81*. https://doi.org/10.1016/j.tourman.2020.104137

Zlatkov, C. M. (2015) *Women entrepreneurship in Serbia: A qualitative study of the perceived enabling factors for female entrepreneurship*. [Master's thesis, Sveriges Lantbruksuniveritet]. Swedish University of Agricultural Sciences. https://stud.epsilon.slu.se/7637/7/ZlatkovCvetkovi%C4%87_M_150219.pdf

14

STRENGTHENING WOMEN'S TOURISM ENTREPRENEURSHIP IN RURAL UGANDA

Brenda Boonabaana

Abstract

While tourism is a sector that employs more women than any other sector, it has also been critiqued for reinforcing gender inequalities. The persistent gender inequalities in tourism are perpetuated by the long-held patriarchal systems, structures, and norms that define many societies of the developing world. This chapter highlights the gender constraints faced by rural Ugandan women tourism entrepreneurs, and their expressed priority needs necessary for strengthening their entrepreneurial opportunities. It also suggests a gender-transformative approach to addressing the underlying gender inequalities prevalent in the community. Drawing on qualitative methods, findings are based on two women groups (Ride 4 a woman and Buhoma women's group) whose members are actively implementing community-based tourism initiatives. The chapter opens new avenues for addressing gender inequalities in the tourism sector.

Keywords

Women; Tourism; Entrepreneurship; Transformative; Uganda

Introduction

The growing scholarship on gender and tourism points to the critical contribution of women in the tourism and hospitality sector. Globally, women form 54% of tourism workers (World Tourism Organization, 2019). In the accommodation domain, women represent 57% in the Americas, 53% in Europe, 53% in Asia and the pacific, 69% in Africa, and 9% in the Middle East (World Tourism Organization, 2019). Women's visibility in tourism has been galvanised by the United Nation's Sustainable Development Goals (SDGs), one of which (Goal 5) explicitly targets the promotion of gender equality and women empowerment. However, gender gaps remain prevalent across all development sectors, including tourism. Some scholars have posited that without gender equality, genuine sustainable development may never be realised (Gutierrez & Vafadari, 2022; Scheyvens & van der Watt, 2021).

DOI: 10.4324/9781003286721-21

Several scholars have articulated the various gender constraints affecting women in tourism, ranging from the burden of unpaid care roles to the limited skills, mobility, and financial access. Women's work remains concentrated in low-ranking and low-remunerated positions of tourism employment compared to men (Baum, 2013; Ferguson, 2011; World Tourism Organization, 2011, 2019). Jackman (2022, p. 5) has equated this to the "feminization of tourism work," reflective of the already deep-rooted gender inequalities being reproduced and reinforced through tourism (Tucker & Boonabaana, 2012). For example, the World Tourism Organization (2019) estimates that women in tourism earn about 10–15% less than their male counterparts.

Yet, most developing countries continue to take an instrumentalist approach to gender equality (Ferguson, 2010a, 2010b), rather than tackling the fundamental feminist goal of gender equality and equity. Arguably, such a narrow framing of gender equality inherent in the capitalist agenda prevents countries from stepping into the journey of genuine equality and transformation (Ferguson, 2007b). Studies by Ferguson (2007b, 2010a, 2010b, 2011) in Latin America (Honduras, Costa Rica, Belize) indicate that some microenterprise tourism developments and policies that aimed at integrating women into the tourist economy were instead keeping traditional structures of reproduction and burdening women with both productive and reproductive work.

A number of women targeted tourism microenterprises in Central America instead benefitted men and a few better-off women (Ferguson, 2010a). Jackman (2022) has hypothesised that there is a positive linkage between tourism and gender equality in the developed countries, an inverse relationship in sub-Saharan Africa and a "no impact" in the Middle East and North Africa. Scheyvens and van der Watt (2021) have called for the need to confront the persistent unequal power relations by integrating a transformative lens to the tourism and hospitality sector.

Nonetheless, studies conducted Uganda (Boonabaana & Ochieng, 2022; Tucker & Boonabaana, 2012); South Africa and Tanzania (Scheyvens, 2000, 2002); Turkey (Tucker, 2007) and Cameroon (Kimbu & Ngoasong, 2016) have demonstrated the potential of the tourism and hospitality sector to lift women out of poverty and strengthening their agency. However, as Jackman (2022) argues, gender equality and women empowerment go beyond the number of women employed, to shifting the underlying barriers that promote and reinforce gender stereotypes and inequalities in decision-making power, control of resources, voice, and choice (Kabeer, 1999, 2001; Van Eerdewijk et al., 2017).

This chapter unpacks rural women's lived experiences with community tourism initiatives in respect to the existing barriers, needs, and preferences. It also calls for a gender-transformative approach to strengthening rural women tourism entrepreneurship. The chapter has five sections: section one presents current debates on women and tourism entrepreneurship; section two provides the methodology; section three presents key findings; section four discusses findings in relation to extant literature while the last section presents conclusions and recommendations.

Literature Debates

Women and Tourism Entrepreneurship

Tourism entrepreneurship and employment have recently been emphasised as key pathways for enhancing women's agency, equity, and empowerment (Boonabaana & Ochieng, 2022; Ferguson, 2011; Scheyvens & van der Watt, 2021). For many developing contexts, rural, small-scale community-based and cultural tourism enterprises have particularly been attractive for women (Al-Dajani & Marlow, 2013; Boonabaana & Ochieng, 2022; Buzinde et al., 2017; Kimbu & Ngoasong, 2016; Tucker & Boonabaana, 2012; World Tourism Organization, 2019). These include handcraft production, cultural performances, guiding, and accommodation businesses. Africa has about 31% women tourism entrepreneurs mainly running hotel and restaurant businesses, especially in Botswana, Ethiopia, Seychelles, and Mauritius (World Tourism Organization, 2019). As such, there is growing policy emphasis towards accelerating women's entrepreneurship across all development sectors, to address poverty and stimulate economic growth (Guloba et al., 2017). In Latin America, over 50% of tourism businesses are owned by women (International Finance Corporation, 2017).

Nevertheless, gender-based constraints continue affect women's engagement within the tourism space in various ways. These include women's occupation of low-status and low-paying jobs (Baum, 2013; Ferguson, 2010b; Handaragama & Kusakabe, 2021; Jackman, 2022; Kituyi, 2018; Staritz & Reis, 2013; World Tourism Organization, 2019), to limited mobility and the heavy burdens of care roles that leave women with very limited time to meaningfully engage in tourism entrepreneurship (Boonabaana, 2014; Ferguson, 2007a; Handaragama & Kusakabe, 2021; Tucker & Boonabaana, 2012). Staritz and Reis (2013) have also pointed out the vice sexual harassment that women often experience in the tourism workspaces.

Women's entrepreneurship is further constrained by the limited access to resources and assets such as land, finances, skills, and information (Kituyi, 2018; Nomnga, 2017; Staritz & Reis, 2013; Tshabalala & Ezeuduji, 2016). For many women, their inability to easily acquire business loans automatically excludes them from starting and sustaining lucrative tourism business. The prerequisites for accessing formal loans are often tied to property ownership such as land or houses, which the majority of the women lack (Handaragama & Kusakabe, 2021).

Gender inequalities are intricately intertwined with other forms of inequalities such as ethnicity, race, nationality, and age to create different layers of gendered opportunity structures and constraints. In South Africa, black African women face both racial and gender discrimination, with some types of businesses such as transport, tour operation, and wildlife safari perceived to be unfit for women (Parashar, 2014). In Kenya, similar racialised and gendered hierarchies and inequalities persist, with majority of tour companies being male dominated, with ownership and management largely in the hands of white and Asian Kenyans (Kituyi, 2018; Staritz & Reis, 2013; World Tourism Organization, 2019).

Nonetheless, in some contexts, tourism entrepreneurship has offered possibilities for empowerment and social change (Boonabaana & Ochieng, 2022; Ferguson, 2011; Kimbu & Ngoasong, 2016; Scheyvens, 2000, 2002; Tucker, 2007; Tucker & Boonabaana, 2012). Kabeer (1999) defines empowerment as a multi-dimensional process that enables women to gain the ability to make decisions and choices for themselves. Scheyvens and van der Watt (2021) have explained that empowerment is a context-specific process that demands redistribution of power to marginalised groups. It involves agency, autonomy, authority, and challenging of inequitable structures (Cole et al., 2014). Scheyvens and van der Watt (2021) provide a comprehensive framework for enabling tourism-based empowerment across the economic, social-cultural, psychological, political/leadership, and environmental domains.

Ferguson's (2011) work in Costa Rica and Honduras demonstrates the value of tourism in facilitating women's mobility, ownership of assets, and education advancement. Similarly, in Uganda, rural women tourism entrepreneurs have acquired assets like land and rental housing for the first time, due to tourism income (Tucker & Boonabaana, 2012). A recent study by Gutierrez and Vafadari (2022) shows that women's participation in tourism positively influences their psychological and economic empowerment although their social and political empowerment remains largely limited. Interestingly, Kimbu and Ngoasong's (2016) study in Cameroon shows that women tourism entrepreneurs prioritise community and social transformation goals, in addition to personal commercial gains. Previously studies have shown that tourism offers the possibility that gender roles and relationships be re-negotiated (Gibson, 2001; Meethan, 2001; Scheyvens, 2000; Sinclair, 1997a,b; Tucker, 2007). Jackman (2022, p. 3) has termed such opportunities "…a mixed blessing for gender equality in the labour market."

However, Jackman (2022) warns that the impact of tourism on gender equality varies across countries. Europe and Central Asia have experienced more opportunities, especially, in senior roles while East Asia and the Pacific countries are experiencing income equality. South Asia, Latin America, and the Caribbean countries continue to grapple with the negative impacts of tourism, and in sub-Saharan Africa, tourism has reinforced gender inequalities, especially, with career progression (Jackman, 2022). As such, she postulates that the status of a society's gendered power relations and ideologies fundamentally affects the likely impacts of tourism on gender equality.

This chapter builds on these fundamental discussions and calls for a more transformative and sustainable approach to tourism development in patriarchal settings. It also foregrounds women's priorities and needs necessary for strengthening their entrepreneurship in the tourism sector. Attention to gender-transformative approaches (GTAs) is critical for strengthening women's inclusion in the tourism entrepreneurship space.

Gender-Transformative Approaches (GTAs)

GTAs focus on social and gender norms that are deep rooted. Hillenbrand et al. (2015) explain that GTAs move beyond individual self-improvement to transformation of power dynamics and structures that serve to reinforce gendered inequalities. They target to transform unequal power relations by recognising, questioning, and challenging unequal distribution and misuse of power in society towards social justice (Poulsen, 2018).

There is a complex connection between social and gender norms, power and agency. The different forms of power include: "*Instrumental agency*" or "*power to*" which has been defined as one's ability to enact personal goals such as decision-making authority and power to solve problems (Malapit et al., 2019; Rowlands, 1995, 1997), including resistances therein (Scheyvens & van der Watt, 2021). On the other hand, "*power within*" or "*intrinsic agency*" constitutes intangible assets such as self-esteem, self-respect, self-confidence, self-awareness, assertiveness, and an individual's sense of self-worth (Malapit et al., 2019; Rowlands, 1997, 1995). "*Power with*" is linked to "*collective agency*" and comes with collective strength towards a common goal.

Finally, "*Power over*," which is a more negative form of power, is connected to the use of force, abuse, discrimination, and oppression, including exclusion of family or community members from making certain decisions or accessing key resources. GTAs provide a platform for challenging the negative forms of power while enhancing women's different forms of agency. While GTAs are increasingly being applied in agricultural communities and interventions, they have rarely been applied in the tourism and hospitality sector.

Examples of successful GTAs in agricultural communities include (i) *community conversations* which focus on deep-rooted social norms and taboos; (ii) *journeys of transformation*; and (iii) *engaging men as allies in women's economic empowerment*. These address constraints around unequal workload, unpaid care work, including men's caregiving, as well as reduced intimate partner violence; (v) *the gender model family* tackles unequal power relations and decision-making amongst couples/partners in households; the gender household approach that targets unequal decision-making and planning; while (vi) *nurturing connections* addresses women's lack of voice in decision-making, childcare, and health seeking, among others (FAO et al., 2020). Tapping into such novel approaches and methodologies is likely to go a long way to challenge the underlying barriers that affect women's entrepreneurship in tourism. The chapter addresses the questions of: What might work better to strengthen women's tourism entrepreneurship in a rural and tourism-dependent community of Uganda? How would such "strengthening" look like to stimulate and sustain women's engagement and agency? Results have implications for tourism entrepreneurship policy and programming, especially for the developing contexts. It also contributes to contemporary debates on gender, women entrepreneurship, and sustainable tourism.

Methodology

The study was conducted between August and September, 2018, in Buhoma, a rural community close to Bwindi Impenetrable National Park (BINP), Southwestern Uganda, a popular international gorilla ecotourism destination. It utilised a mix of qualitative ethnographic methods, namely, semi-structured in-depth interviews with women group members (22); key informant interviews with national park officials and community leaders (6); focus group discussions with women group members (2); informal interviews and observations. A total of 30 interviews and discussions were held. These methods were relevant for eliciting women's in-depth lived experiences as community tourism entrepreneurs and the meanings they attributed to their experiences (Denzin & Lincoln, 2005; Phillimore & Goodson, 2004). Purposive and snowball sampling techniques were applied to identify relevant participants for interviews and discussions. The sampling process continued until "data saturated" or when no new insights came up (Fossey et al., 2002; Jennings, 2005). Interview notes were taken verbatim, transcribed, reviewed, coded, categorised, and analysed using the inductive thematic analysis method that enabled generation of key themes from the data. Because the interview numbers were manageable, the author used the manual analysis following the previous analytical steps.

Study Context

Tourism Development in Uganda

The Government of Uganda has prioritised tourism as a key sector for promoting social economic development and poverty alleviation (National Planning Authority [NPA], 2020). Uganda's tourism is highly dependent on nature-based resources, especially, the endangered mountain gorillas found in Bwindi Impenetrable and Mgahinga National Parks in Southwestern Uganda, as well as Chimpanzees in Kibale National Park in Western Uganda (Tumusiime & Vedeld, 2015). Gorilla ecotourism contributes over 60% of revenue collected by the Uganda Wildlife Authority (Ministry of Tourism, Wildlife, and Antiquities [MTWA], 2020).

Prior to the COVID-19 pandemic, Uganda received over 1.5 million tourists who contributed about US$1.6 billion in foreign exchange that translated into 7.7% of the gross domestic product (GDP) for the country (MTWA, 2020). The country's conservation and tourism policies encourage sustainable approaches, particularly, community participation and entrepreneurship, with both men and women encouraged to engage (MTWA, 2014). This has triggered programmes and actions geared at community-based tourism initiatives by local men, women, and the youth who live close to the national parks (Ayorekire et al., 2020; Boonabaana, 2012).

The National Tourism Development Masterplan provides a roadmap for local community participation in, and promotion of, innovative enterprises in the local economies (MTWA, 2014). Further, the country's gender policy (2007) emphasises the promotion of gender equality and women's empowerment in all sectors, including tourism (Ministry of Gender, Labour and Social Development [MGLSD], 2007). There are several gender-focused initiatives such as the Uganda Women Entrepreneurship Programme (UWEP) that support women's engagement in different businesses, including financial literacy and access to microfinance. Nevertheless, gender-based constraints remain prevalent and continue to affect women's opportunities across all development sectors.

Ecotourism Development around Bwindi Impenetrable National Park

The popularity of BINP in Southwestern Uganda started when the forest was gazetted a national park for gorilla ecotourism in 1991 and later declared a UNESCO Natural Heritage site in 1994. Since then, it has become Uganda's most sought-after destination (Tumusiime & Vedeld, 2015).

BINP park which covers 321 square kilometres is bordered by Rwanda in the South and the Democratic Republic of Congo in the West (Ampumuza, 2021). The first inhabitants of the forest were the *Batwa*, also derogatively known as *pygmies*, that occupied the forest between 32,000 and 47,000 years ago, surviving as hunter-gatherers (Ampumuza, 2021; Mukasa, 2014). However, with the introduction of tourism, the *Batwa* were evicted from their forest homeland (Kagumba, 2021), eventually, turning into "conservation refugees" (Harper, 2012). The *Batwa* now live as squatters depending on handouts, casual farm labour, illegal hunting, cultural entertainments, and handcraft sale (Ampumuza, 2021; Tumusiime & Vedeld, 2015). While the *Bakiga* (the dominant tribe) have been able to combine their traditional agricultural livelihoods with the new tourism opportunities, this has not been the case for the *Batwa* (Ampumuza, 2021).

While the Uganda Wildlife Authority and some non-governmental organisations (NGOs) have come out to support the *Batwa*, they remain vulnerable with possibilities for extinction. Batwa experience cultural stereotypes, sexual and physical brutality perpetuated by the non-Batwa communities (Kagumba, 2021). They are often described by the dominant *Bakiga* communities and NGOs alike, as primitive, unable to save food and money, and predominantly dependent on handouts (Ampumuza et al., 2020). Conversely, tourism promotional materials portray their "exotic images" as authentic lifestyle and their experience, as a "must do" activity after gorilla tracking (Ampumuza et al., 2020). Gorilla ecotourism has therefore become a "blessing" and "curse" depending on one's ethnicity, gender, age, and role.

Case Studies: Membership and Operations

Case Study One: Ride 4 a Woman

Ride 4 a woman is a women-focused NGO found in Buhoma village, the closest community to BINP. It was founded in 2009 by a local woman and tourism graduate, born, and raised in Bwindi with passion for gender equality and women empowerment. It aims at empowering women with income and skills, as well as enabling a critical mass of rural women who are able to tap into the vibrant tourism economy in their community. The organisation has about 300 women members who are mainly poor, illiterate, widowed, HIV positive, as well as school dropouts. Most members care for several children, grandchildren, and other family members.

In terms of membership fees, each member pays about UGX 2000 (less than half a dollar) per year to be able to benefit from services offered by the organisation. They operate multiple enterprises ranging from accommodation facilities (guest house) craft businesses, hiring out bicycles to tourists, microfinance, bakery, as well as provision of safe water and agricultural inputs such as seed and fertilisers. The guest house doubles as a temporary shelter for women experiencing domestic violence as well as a paid accommodation facility for tourists. While women contribute fees to run the organisation, it benefits from financial funding from former tourists, especially, the accommodation and microfinance projects. The organisation is also supported by government entities such as the Uganda Wildlife Authority that offers marketing, advisory, and capacity development support. Some women (about 50 out of 300) are directly employed by the organisation, taking on various roles such as sewing, cultural dancing, and basket weaving, which includes a cultural demonstration session dubbed "a day with a woman" as well as working in the guest house and restaurant. To enable different women benefit, they work in shifts of 25 women per day and are paid about UGX5000–9000 (approx. US$1.75–2.75) each day of work.

Case Study Two: Buhoma-Mukono Women's Group

Buhoma-Mukono women's group was founded in 1995 by Buhoma-Mukono Community Development Association (BMCDA), a community organisation that brings together over 7000 community members of Mukono Parish, the closest Parish to BINP. The organisation promotes improved livelihoods and conservation of natural resources, by enabling the locals to obtain reputable employment, and empowering them with the necessary tourism skills and careers. The organisation owns two lodges schools, a gravity water scheme, fruit, and vegetable project. It also owns a Cooperative Savings and Credit Society, offers village walk services, and manages a health insurance scheme for all its members. Its women's group focuses on handcrafts, cultural dances, and microfinance. The group brings together about eight women subgroups totalling about 200 members. In addition to membership fees of about UGX 50000 (US$13) and a monthly subscription of UGX2000 (less than half a dollar), each member contributes one basket to stock the group's craft shop that is strategically located near the park entrance. The group charges an interest of 10% per annum on borrowed funds from their microfinance savings.

BMCDA was first supported by the Uganda Wildlife Authority and the United States Peace Corp volunteer, John Dubois, in 1992. Other support organisations include the Uganda Community Tourism Association (UCOTA); the International Gorilla Conservation Programme (IGCP), Bwindi-Mgahinga Conservation Trust (BMCT), the United Nations

Development Programme (UNDP), Institute of Tropical Forest Conservation (ITFC), USAID and Mbarara University of Science and Technology for technical and financial support.

Findings

Benefits

This section provides an overview of the positive benefits being enjoyed by women tourism entrepreneurs in the Bwindi area. Women's engagement in tourism initiatives has enhanced their economic capabilities, including the opportunities to provide for themselves and their families, in terms of paying school fees for their children and attaining assets of their own such as land, livestock (goats, sheep, chicken), improved housing, and sewing machines (Boonabaana & Ochieng, 2022; Tucker & Boonabaana, 2012). The following are some of the expressions of women about these opportunities:

> *We are paying for family requirements like buying sugar, clothes but I have also bought land and built a house from craft business and my other small income sources like selling food and saving groups.*
>
> *(Member, Buhoma-Mukono women's group)*

> *I have bought mattresses, plates, cups and I am happy as a woman because I have my own money that I can spend on what I need without waiting upon my husband to provide. When there is money and I buy such things, my husband feels very happy.*
>
> *(Member, Buhoma-Mukono women's group)*

> *I used to dig but I no longer dig, I got a sowing machine and now I make items am called an officer and supply to people, I bought trees – which I got when I sold goats.*
>
> *(In-depth interview participant, Ride 4 a Woman group)*

By meeting some of their immediate needs at the household and community levels, women are gradually reducing their economic dependence on men. Women's narratives not only depict their emerging economic capabilities but also the growing disruptions to the traditional power relations and structures, related to access to productive resources, decision-making power, and mobility. There are also changes around reduced agricultural drudgery and unpaid labour due to women's tourism livelihoods. A study by Voumik et al. (2023) in South American and Caribbean countries shows that as women's participation in tourism increases, they tend to shift away from the traditional economic sectors, such as agriculture and the manufacturing industry, that have less productive potential for the women, compared to tourism.

Women's tourism entrepreneurship has also enabled them to gain self-confidence and self-worth, associated with collective power and agency (Boonabaana & Ochieng, 2022). A group leader of Ride 4 a Woman expressed that: *"the women were shy to speak but now can stand and speak in a congregation of 50 people."* A member of Buhoma-Mukono women's group mentioned that: *"Men respect us. For example, when I bring a mattress, bedsheets, blanket which he has not struggled for, he feels happy and is supportive."*

In addition, women have mobilised themselves to confront their "dark experiences" associated with gender-based violence (GBV). This has been through the acquisition of GBV preventive skills, peer learning and sharing of experiences, as well as putting in place a supportive infrastructure for the abused to find temporary refugee (Boonabaana & Ochieng, 2022). The following is what a group leader shared about this:

> *They have now learnt different forms of domestic violence for example when a man takes away your money, its domestic violence. Physical violence, a man forcing for sex, lack of decision making power are all forms of domestic violence.*
>
> *(Leader, Ride 4 a Woman)*

The earlier findings resonate with literature about the role of community-based tourism in facilitating women's empowerment (Boonabaana & Ochieng, 2022; Ferguson, 2011; Kimbu & Ngoasong, 2016; Scheyvens, 2000, 2002; Tucker, 2007; Tucker & Boonabaana, 2012). However, the contribution of tourism to GBV highlighted by this study is a relatively new insight into the growing evidence. While women in Latin America (Costa Rica and Honduras) have been able to save and invest in assets such as cars, fund their education, and travel (Ferguson, 2011), women in Buhoma mostly invest in their children's education, nutrition, and the wider family.

Gender Barriers

Despite the emerging positive tourism gains, most women in the study community are yet to be integrated into the tourism economy. They still grapple with various gender constraints ranging from the burden of domestic and care roles, limited mobility that involves negotiating permissions from their husbands, access to, and control over productive resources. As women take on tourism businesses, their share of domestic and care roles has largely remained the same. When women are unable to fulfil their traditional expected domestic "obligations," they are subjected to blame and judgement by their husbands and other community members as "marriage failures." A male community leader described such negative perceptions towards women as follows:

> *They leave house work unattended and when their husbands find it not done, they complain. Confidence of the public is negative. Giving women lead positions is sometimes perceived negative by the public. The perception is always that: Why does this person bring in women in key positions when there are men who can do the work....*
>
> *(Male, Chairperson, BMCDA)*

As they shoulder multiple domestic roles, women are increasingly taking on family provision roles, as their husbands either step back from these responsibilities or choose to spend on non-family items. Secondly, cultural expectations in many African contexts often confer power on men as key family decision-makers (Anunobi, 2002; Bassey & Bubu, 2019). This favours men to misuse these powers by restricting their wives from working within the public domain, including tourism spaces. Women often have to negotiate with their husbands for permission to work in tourism. Some men ask their wives to focus on the traditional household work, while others limit their work time claiming that they might be "hard to manage." Some women expressed having to push the gender boundaries through subtle resistances and negotiation by noting that: "*No, we are struggling. Some of us came by force. Like me, my husband had refused but I insisted. I didn't want to miss the opportunity, yet, he doesn't give me whatever I want.*" Another woman added that: "*Even me it was like that. He first refused but I insisted and came. He has now given up given that he sees me benefitting."* This limits women's freedom and choice to engage in the tourism labour market. This implies that women with lower negotiation power are unable to join the tourism work space. Other voices later depict the hustles that women have to go through to become part-time tourism entrepreneurs:

R3: Some husbands don't allow us to work and they really harass us when we insist

R4: Even when we work, they want us to give them the money.

In support for male control of women's labour, a male community leader expressed as follows:

Yes, they like dancing but I have heard their husbands complaining. Coming to lodges to dance is a burden. Maybe if they had one spot like their house because not all guests are interested in dancing for them, some want to rest. Also the issue of women of 60 years and 70 years running up and down to dance in lodges is not sustainable.
(Male, Chairperson, BMCDA)

In addition, managing a big group of women was considered a major challenge for the leaders. They expressed experiencing challenging group dynamics that tested their leadership abilities. One of the women leaders stated that: *"it is not easy, fights and quarrels are there. Others feel some are more liked than others."* In terms of access to resources, several women expressed having been restricted from accessing weaving materials from the national park, which renders them to move longer distances as far as Nteko and the Democratic Republic of Congo to collect specific weaving raw materials. In addition to walking long distance, the women have to pay high financial costs to secure the materials. A woman noted that: *"Given that the park officials refused us from accessing craft materials, we have a challenge of getting weaving materials. We buy them from Nteko and Congo and they are expensive...."*

Control over income is also problematic for some women entrepreneurs, whose husbands pressure into having control to their income. While this creates ground for family conflicts, women expressed determination and resilience to resist men control of their tourism income. Instead, they mentioned having to play out their "cards" in their favour, by giving out some small financial handouts to their husbands to "silence" them into continuing to grant them permission to work while controlling the bulk of their proceeds. All interviewed women were against men control of their income and supported "gifting" them as a negotiating mechanism. A woman from Ride 4 a Woman group expressed that: *Our husbands want us to give them the money we work for so that they spend it on alcohol, yet, they do not know even the price of salt!* Similarly, members of Buhoma women's group's focus group discussion expressed the following about income control:

R2: Sometimes I do, and sometimes I don't. At times when he insists, I pack my bags and leave home but later come back.

R1: I give him part of the money I earn to keep peace in the home.

R3: I don't give him my money. I know he will spend it on alcohol. But he always wants me to give it to him, so I can't accept.

Finally, the limited market and market competition, especially for craft products, were identified by the women as a key constraint. Women expressed the need to access more customers who could buy their products, as well as gaining better marketing skills and connections. The earlier depict the role of gender power relations in narrowing women's opportunities amidst tourism entrepreneurship potential in a rural and popular tourism destination. Notable challenges allude to the heavy burden of domestic and care work, the norms around patriarchal decision-making power, access to productive resources and control over income. While gender relations are context specific, some of these constraints have been cited in other destinations (Fergurson, 2011; Ferguson, 2007a; Tucker & Boonabaana, 2012; Voumik et al., 2023) except for the differences in the contextual dynamics and manifestations.

Women's Priority Needs

Beyond the gains and constraints experienced, women identified their pressing priority needs necessary for strengthening their entrepreneurial capacities. One of these was the need for capacity in the area of information technology, to enable them design their own websites for marketing purposes. The leader of Ride 4 a Womangroup noted that: "*I need more people on board to work with me to reduce stress. I need more volunteers who can design for us the website.*" Women unanimously expressed the need for engaging in exchange programmes that enable to learn from other community tourism destinations with successful women-run enterprises. One of the group leaders added to these views as follows:

> *...If other NGOs can come in the village and show women that women empowerment matters it would be helpful. Some of the celebrations should be put in the villages like women's day. Some women have never stepped in the bus or even moved beyond this community so I would want to take them out to see other places beyond this community.*
> *(Female, Founder and Director, Ride 4 a woman)*

In addition, women desire more skills in craft making and business management. These were deemed helpful in enabling women to produce better quality and marketable products. There was a common expression of the need for market expansion for women's craft businesses to address the current market competition that women are finding difficult to navigate. Women explained how other community groups like the youth and children were selling similar products, and creating high competition for customers. The following are examples of quotes by women from Buhoma women's group about this:

> *R4: ...At first we had market but now it has reduced. We had Prof William who would buy all our products and sale for us and bring us money, but now she no longer comes, so there is no morale. We no longer sell as before.*
>
> *R2: We lack of market for our products and this is our major challenge*
>
> *R7: We need more market. We can make the baskets but we need to be assured of the market like we used to enjoy in the past.*
> *(FGD members, Buhoma Women's Group)*

Previous studies in other tourism destinations in Latin America and the Caribbean have called for proper training and skills development plans for women, to enable them compete favourably in the tourism space. This has been attributed to the fact that they are "...newcomers to the tourism industry and previously worked in traditional economic sectors like agriculture and manufacturing" (Voumik et al., 2023, p. 14).

Lastly, women desire extra capital for their microfinance investment to enable them access more credit to expand both group and individual businesses. With this, they hope to acquire more sewing machines, expand their workshops, and invest in more lucrative craft materials. They also envisage this as an opportunity for diversifying tourism work with agribusiness that has great untapped potential. This would enable women to venture into organic food enterprises and supply the surrounding accommodation facilities. The following are examples that demonstrate women's desire for more credit opportunities:

I heard in Church that there was training so I wanted to join and I came over to inquire. I was immediately recruited with other 16 women for training and we are now training others. We now get some loans from the microfinance. One of the tourists donated the money to our saving group. We work and pay back with interest of 3%. Each woman gets a loan they can manage, and for me I took UGX700, 000 and bought my own sowing machine at UGX 600,000. For the second route, I got UGX 600,000 I bought materials for the machine and goats. Third route I got 500,000 and I paid fees. More support for our saving group would continue to help us a lot.
(Persona interview, Ride 4 a Woman)

R1: Microfinance has helped a lot so we need it stay and if possible, more funds injected in so that we borrow more money to support us.
(FGD member, Ride 4 a Woman)

One of the group leaders added a related view as follows:

...If our microfinance can grow from $14000 to $50000 it will bring in more money through interests. Ensuring that women are making their own businesses and leave less women at the centre... We intend to give each family 100 coffee seedlings for income generation. The goal is to have 100 women employed so that we can have 50 per day. We need to have a shop on the main road for marketing purposes.

Discussion

The earlier findings point to the importance of understanding the progress being made by women in tourism destinations, including constraints and opportunities. While women's priority desires and needs are connected to their immediate needs, they also depict everyday challenges that women tourism entrepreneurs have to circumvent to pursue tourism opportunities. Further, women's constraints are grounded in the traditional gender relations that define power, entitlements, and roles in this context.

The gender constraints around traditional roles, mobility, and capital identify with what other scholars have noted in other tourism destinations (Ferguson, 2007a; Handaragama & Kusakabe, 2021; Tucker & Boonabaana, 2012). However, women's responses and actions to these constraints provide new insights into how women in Bwindi have been able to circumvent the constraints to find space for themselves as tourism entrepreneurs. Handaragama and Kusakabe (2021) have argued that as women become tourism entrepreneurs, they still have to carry their domestic burdens that influence them to take up flexible tourism work that does not challenge their gender roles as domestic workers. This argument clearly resonates with this study findings.

Women further highlighted the need for capital, skills, marketing capabilities, and peer learning as critical factors for strengthening their tourism entrepreneurship opportunities. Related to this, in Sri Lanka, Handaragama and Kusakabe (2021) have noted how effective business networks, experienced mentors, access to financial, business information, and knowledge are crucial in promoting small-scale women entrepreneurs to overcome structural and institutional barriers to their tourism business growth.

However, women's entrepreneurship potential is less likely to be meaningfully realised if the underlying gender biases and inequalities remain unchallenged. This calls for a gender-transformative approach to tourism planning and development to enable genuine gender equality (Scheyvens & van der Watt, 2021). Jackman (2022) has explained the need for a gender-sensitive work force in tourism, which targets overcoming patriarchy. This calls for strong multi-stakeholder engagement (government, private sector/NGOs, community members and institutions, gender activities) to commit the requisite resources, including finances, political will, human resources, and time required to promote GTAs in rural tourism destinations with strong gender biases and injustice. The desired changes should target individual community members (women and men of different age groups), families, community institutions, tourism organisations, and policies.

The GTAs that include community conversations, journeys of transformation, engaging men as allies in women's economic empowerment, gender model family, and the gender household approach (FAO et al., 2020) will go a long way to challenge gender constraints that continue to limit women's engagement in tourism as entrepreneurs. Figure 14.1 provides a framework for GTAs towards equitable tourism entrepreneurship.

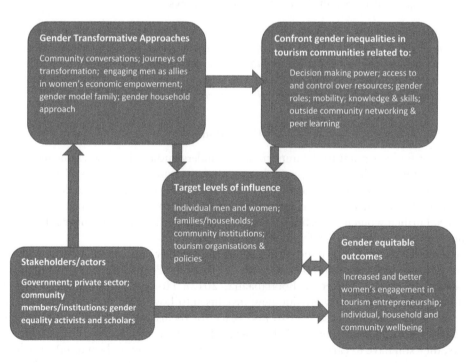

Figure 14.1 Framework for gender-transformative approaches towards equitable tourism entrepreneurship. (Source: Own contribution)

Conclusions and Recommendations

This chapter concludes that gender norms and inequalities constrain women's agency towards becoming tourism entrepreneurs. The chapter calls for GTAs (see Figure 14.1) that consider women as key players in the tourism development space. They also provide avenues for challenging the underlying gender constraints that limit women's entrepreneurship potential.

While the immediate needs proposed by women (capital, skills, and capacity building, peer learning) to strengthen their tourism entrepreneurship opportunities are very useful, their status might not improve much if the dominant gender relations around decision-making power, access to and control over resources, gender roles, mobility and knowledge, skills and network capital remain widely unchallenged. A multi-stakeholder approach comprised actors from government, private sector, community members, and activists will be required to move the transformative agenda in tourism destinations of the kind.

Drawing on one or more of the suggested GTAs illustrated in Figure 14.1 would enable changes in the local structures and systems that reinforce gender inequality in tourism destinations. This also allows paying intentional attention to the root cause of social injustice rather than treating the symptoms. This chapter has attempted to explain what might work better to strengthen women's tourism entrepreneurship in a rural, patriarchal, and tourism-dependent community of Uganda; and how such "strengthening" should entail to stimulate women's engagement and agency as key tourism entrepreneurs.

The chapter has implications for policy and programming around gender equality and tourism entrepreneurship and development and, further, contributes to ongoing debates on gender, women empowerment, and sustainable tourism. By centring the GTAs, the chapter bridges the knowledge gap in the field of tourism development and gender. While these approaches have been tested in the agricultural sector and proven to be successful (FAO et al., 2020) in communities with challenging gender biases and stereotypes, the tourism sector is ripe to integrate the same. While the GTAs are inherently gradual and long term in outlook, they are likely to offer a more sustainable and equitable pathway to meaningful gender equality and empowerment through tourism.

The author recommends the intentional integration of GTAs in the tourism sector by different stakeholders at all levels: local community, district, national, and global scales. In future, more studies would be required to understand the application, successes, and constraints to integrating GTAs in tourism destinations. In addition, given that the global COVID-19 pandemic severely disrupted women's gains in the tourism entrepreneurship space, it would be important for future research to pay attention to women's coping strategies, recovery mechanisms, and resilience to the pandemic-related shocks.

References

Al-Dajani, H., & Marlow, S. (2013). Empowerment and entrepreneurship: A theoretical framework. *International Journal of Entrepreneurial Behaviour & Research*, 19(5), 503–524.

Ampumuza, C. (2021). *Batwa, gorillas and the Ruhija road: A relational perspective on controversies at Bwindi Impenetrable National Park, Uganda* [Doctoral dissertation, Wageningen University and Research].

Ampumuza, C., Duineveld, M., & van der Duim, R. (2020). The most marginalized people in Uganda? Alternative realities of Batwa at Bwindi Impenetrable National Park. *World Development Perspectives*, 20, 100267.

Anunobi, F. (2002). Women and development in Africa: From marginalization to gender inequality. *African Social Science Review*, 2(2), 3.

Ayorekire, J., Obua, J., Mugizi, F., & Byaruhanga, B. M. (2020). Opportunities and challenges of religious tourism development in Uganda: Policy, planning and institutional perspectives. *International Journal of Religious Tourism and Pilgrimage, 8*(3), 12.

Bassey, S. A., & Bubu, N. G. (2019). Gender inequality in Africa: A re-examination of cultural values. *Cogito, 11*(3), 21–36.

Baum, T. (2013). *International perspectives on women and work in hotels, catering and tourism* (Bureau for Gender Equality, Working Paper No. 1/2013, Sectoral Activities Department, Working Paper No. 289). International Labour Organization: Geneva.

Boonabaana, B. (2012). *Community-based tourism development and gender relations in Uganda* [Doctoral dissertation, University of Otago].

Boonabaana, B. (2014). Negotiating gender and tourism work: Women's lived experiences in Uganda. *Tourism and Hospitality Research, 14*(1–2), 27–36.

Boonabaana, B., & Ochieng, A. (2022). Interrogating rural women's collective tourism entrepreneurship and social change in south western Uganda. In *Promoting social and cultural equity in the tourism sector* (pp. 85–104). IGI Global.

Buzinde, C., Shockley, G., Andereck, K., Dee, E., & Frank, P. (2017). Theorizing social entrepreneurship within tourism studies. In *Social entrepreneurship and tourism* (pp. 21–34). Springer.

Cole, S. M., Kantor, P., Sarapura, S., & Rajaratnam, S. (2014). *Gender-transformative approaches to address inequalities in food, nutrition and economic outcomes in aquatic agricultural systems* (Working Paper: AAS-2014-42). Penang, Malaysia: CGIAR Research Program on Aquatic Agricultural Systems.

Denzin, N. K., & Lincoln, Y. S. (2005). Introduction: The discipline and practice of qualitative research. In N. K. Denzin & Y. S. Lincoln (Eds.), *The Sage handbook of qualitative research* (3rd ed., pp. 1–32). Sage Publications, Inc.

FAO, IFAD, & WFP. (2020). *Gender transformative approaches for food security, improved nutrition and sustainable agriculture – A compendium of fifteen good practices*. Rome. https://doi.org/10.4060/cb1331en

Ferguson, L. (2011). Tourism, consumption and inequality in Central America. *New Political Economy, 16*(3), 347–371.

Ferguson, L. (2007a). Funding inequality: How socially conservative development projects limit the potential for gender equitable development. *Political Perspectives, 1*(1), 1–31.

Ferguson, L. (2007b). Reinforcing inequality: Service sector activities and the new entrepreneurial model of development in Central America. *Centre for International Politics Working Paper Series, 26*. The University of Manchester

Ferguson, L. (2010a). Interrogating 'gender' in development policy and practice: The World Bank, tourism and microenterprise in Honduras. *International Feminist Journal of Politics, 12*(1), 3–24.

Ferguson, L. (2010b). Tourism development and the restructuring of social reproduction in Central America. *Review of International Political Economy, 17*(5), 860–888.

Fossey, E., Harvey, C., McDermott, F., & Davidson, L. (2002). Understanding and evaluating qualitative research. *Australian and New Zealand Journal of Psychiatry, 36*, 717–732.

Gibson, H. J. (2001). Gender in Tourism: Theoretical perspectives. In Y. Apostolopoulos, S. Sönmez & D. J. Timothy (Eds.), *Women as producers and consumers of tourism in developing countries* (pp. 19-43). Westport: Praeger Publishers.

Guloba, M., Ssewanyana, S., & Birabwa, E. (2017). *Rural women entrepreneurship in Uganda: A synthesis report on policies, evidence, and stakeholders* (No. 677-2017-1124).

Gutierrez, E. L. M., & Vafadari, K. (2022) *Exploring the relationship between women's participation, empowerment, and community development in tourism: A literature review.*

Handaragama, S., & Kusakabe, K. (2021). Participation of women in business associations: A case of small-scale tourism enterprises in Sri Lanka. *Heliyon, 7*(11), e08303.

Harper, S. (2012). *Social determinants of health for Uganda's indigenous Batwa population.*

Hillenbrand, E., Karim, N., Mohanraj, P., & Wu, D. (2015). *Measuring gender-transformative change. A review of literature and promising practices* (Working Paper developed for: World Fish and the CGIAR Research Program on Aquatic Agricultural Systems). https://core.ac.uk/download/pdf/132685604.pdf

International Finance Corporation. (2017). Women and tourism: Designing for inclusion. In *Tourism for development knowledge series*. World Bank Group.

Jackman, M. (2022). The effect of tourism on gender equality in the labour market: Help or hindrance? *Women's Studies International Forum, 90*, 102554. Pergamon.

Jennings, R. G. (2005). Interviewing: A focus on qualitative techniques. In B. W. Ritchie, Burns, P., & Palmer, C (Ed.), *Tourism research methods: Integrating theory with practice*: CAB International.

Kabeer, N. (1999). Resources, agency, achievements: Reflections on the measurement of women's empowerment. *Development and Change, 30*(3), 435–464.

Kabeer, N. (2001). Conflicts over credit: Re-evaluating the empowerment potential of loans to women in rural Bangladesh. *World Development, 29*(1), 63–84.

Kagumba, A. K. (2021). The Batwa Trail: Developing agency and cultural self-determination in Uganda through Indigenous tourism and cultural performance. *AlterNative: An International Journal of Indigenous Peoples, 17*(4), 514–523.

Kimbu, A. N., & Ngoasong, M. Z. (2016). Women as vectors of social entrepreneurship. *Annals of Tourism Research, 60*, 63–79.

Kituyi, M. (2018). *Economic development in Africa report 2017. Tourism for transformative and inclusive growth*. United Nations Conference on Trade and Development (UNCTAD).

Malapit, H., Quisumbing, A., Meinzen-Dick, R., Seymour, G., Martinez, E. M., Heckert, J. Rubin, D., Vaz, A., Yount, K.M. & Gender Agriculture Assets Project Phase 2 (GAAP2) Study Team (2019). Development of the project-level Women's Empowerment in Agriculture Index (pro-WEAI). *World Development, 122*, 675–692.

Meethan, K. (2001). *Tourism in global society: Place, culture, consumption*. Palgrave.

Ministry of Gender, Labour and Social Development (MGLSD). (2007). *Gender policy*. Government of Uganda.

Ministry of Tourism, Wildlife, and Antiquities (MTWA). (2020). *The impact of COVID-19 on the tourism sector in Uganda*. Government of Uganda.

Ministry of Tourism, Wildlife, and Antiquities (MTWA). (2014). *Uganda tourism development master plan (2014–2024)*. Government of Uganda.

Mukasa, N. (2014). The Batwa indigenous people of Uganda and their traditional forest land: Eviction, non-collaboration and unfulfilled needs. *Indigenous Policy Journal*, Vol. XXIV, No. 4 (Spring).

National Planning Authority (NPA). (2020). *Third National Development Plan (NDPIII), 2020/2021-2024-2025*. Government of Uganda.

Nomnga, V. J. (2017). Unlocking the potential of women entrepreneurs in the tourism and hospitality industry in the eastern Cape Province, South Africa. *Journal of Economics and Behavioral Studies, 9*(4 (J)), 6–13.

Parashar, S. (2014). Marginalized by race and place: A multilevel analysis of occupational sex segregation in post-apartheid South Africa. *International Journal of Sociology and Social Policy, 34*(11/12), 747–770.

Phillimore, J., & Goodson, L. (2004). Progress in qualitative research in tourism: Epistemology, ontology and methodology. In J. Phillimore & L. Goodson (Eds.), *Qualitative research in tourism: Ontologies, epistemologies and methodologies* (pp. 3–29). Routledge.

Poulsen, L. (2018). *Implementing gender-transformative approaches (GTAs) in agricultural initiatives*. Integrating gender and nutrition within agricultural extension services. Discussion paper June 2018.

Rowlands, J. (1995). Empowerment examined. *Development in Practice, 5*(2), 101–107.

Rowlands, J. (1997). *Questioning empowerment: Working with women in Honduras*. Oxfam.

Scheyvens, R. (2000). Promoting women's empowerment through involvement in ecotourism: Experiences from the Third World. *Journal of Sustainable Tourism, 8*(3), 232–249.

Scheyvens, R. (2002). *Tourism for development: Empowering communities*. Pearson Education.

Scheyvens, R., & van der Watt, H. (2021). Tourism, empowerment and sustainable development: A new framework for analysis. *Sustainability, 13*(22), 12606.

Sinclair, M. T. (1997a). Gendered work in tourism: Comparative perspectives. In M. T. Sinclair (Ed.), *Gender, Work and Tourism*. London: Routledge.

Sinclair, M. T. (1997b). Issues and theories of gender and work in tourism. In M. T. Sinclair (Ed.), *Gender, work and tourism* (pp. 1–15). Routledge.

Staritz, C., & Reis, J. G. (2013). *Global value chains, economic upgrading, and gender: Case studies of the horticulture, tourism, and call center industries*.

Tshabalala, S. P., & Ezeuduji, I. O. (2016). Women tourism entrepreneurs in KwaZulu-Natal, South Africa: Any way forward? *Acta Universitatis Danubius. Œconomica, 12*(5), 19–32.

Tucker, H. (2007). Undoing shame: Tourism and women's work in Turkey. *Journal of Tourism and Cultural Change, 5*(2), 87–105.

Tucker, H., & Boonabaana, B. (2012). A critical analysis of tourism, gender and poverty reduction. *Journal of Sustainable Tourism, 20*(3), 437–455.

Tumusiime, D. M., & Vedeld, P. (2015). Can biodiversity conservation benefit local people? Costs and benefits at a strict protected area in Uganda. *Journal of Sustainable Forestry, 34*(8), 761–786.

Van Eerdewijk, A. H. J. M., Wong, F., Vaast, C., Newton, J., Tyszler, M., & Pennington, A. (2017). *White paper: A conceptual model on women and girls' empowerment.*

Voumik, L. C., Nafi, S. M., Majumder, S. C., & Islam, M. A. (2023). The impact of tourism on the women employment in South American and Caribbean countries. *International Journal of Contemporary Hospitality Management 35* (9), 3095–3112

World Tourism Organization. (2011). *Global report on women in tourism.* United Nations World Tourism Organization.

World Tourism Organization. (2019). *Global report on women in tourism* (2nd ed.). United Nations World Tourism Organization.

15

GENDER AND ETHNICITY INTERPLAYS IN MIGRANT TOURISM ENTREPRENEURSHIP

Evidence from Ghana

Ewoenam Afua Afenyo-Agbe, Issahaku Adam,
Albert Nsom Kimbu, and Frederick Dayour

Abstract

While migrant entrepreneurship has received considerable research attention within tourism research over the years, the interplay of gender and ethnicity on the migrant tourism entrepreneurship phenomenon has rarely been discussed. To aid our understanding and appreciation of the migrant tourism entrepreneurship phenomenon in developing country contexts, this chapter focuses on Ghana and explores the role of gender and ethnicity in shaping migrant tourism entrepreneurs' opportunities, inequalities, and constraints. Using a qualitative methodology, the chapter unpacks the experiences of 16 internal migrant entrepreneurs in coastal Ghana. The chapter contributes to the body of knowledge on migrant tourism entrepreneurship by providing insights into the migrant tourism entrepreneurs' mobility decision-making process as well as their experiences of integration in and relationships with their host communities.

Keywords

Ethnicity, Ghana, Identities, Internal Migrants, Tourism Entrepreneurship

Introduction

Tourism is often presented as an easy-to-enter sector for migrant entrepreneurs (MEs) due to its low entry requirement in terms of start-up capital, technological requirements, access to market and networks (Iversen & Jacobsen, 2016; Lugosi & Allis, 2019). The literature is replete with studies on how tourism has provided an entrepreneurial pathway for migrants (Bosworth & Farrell, 2011; Iversen & Jacobsen, 2016) and accelerated the economic and socio-cultural integration of migrants (Bosworth & Farrell, 2011; Iversen & Jacobsen, 2016). However, migrants' ability to construct successful entrepreneurial experiences in tourism is often shaped by their multiple identities such as ethnicity, race, and other defining markers of power and access to resources and opportunities in the host country or community (Iversen & Jacobsen, 2016; Lugosi & Allis, 2019). The gender and ethnicity of migrants are important identity pillars that do not only determine migrants' entrepreneurial opportunities, but they also define their ability to access funding, entry into tourism entrepreneurial space, access to inputs and markets, and networks (Iversen & Jacobsen, 2016; Lugosi & Allis, 2019). They further shape business restrictions on migrants' entrepreneurial ventures as well as ease of business registration and licensing and other political and socio-cultural expectations that collectively shape migrants' ability to construct meaningful and successful entrepreneurial experiences (Barberis & Solano, 2018; Iversen & Jacobsen, 2016; Paniagua, 2002; Su & Chen, 2017).

Despite these multiple identity nuances that shape migrants' entrepreneurial engagements in the tourism sector, little attention has been dedicated to understanding how multiple identities affect the entrepreneurial engagements of migrants in tourism. Previous studies have mostly focused on examining the intersections between single-identity variables such as race, and ethnicity and migrants' entrepreneurial engagements in tourism (e.g., Barberis & Solano, 2018; Iversen & Jacobsen, 2016; Paniagua, 2002; Su & Chen, 2017). Thus, previous examination of the subject has assumed a single-identity intersection with migrant entrepreneurship without recourse to how multiple identities simultaneously shape migrants' entrepreneurial engagements in tourism. Even so, the single-identities focused on have left out gender despite it being an important determinant in entrepreneurial activities pursuit and growth of such businesses (e.g., Barberis & Solano, 2018; Iversen & Jacobsen, 2016; Su & Chen, 2017), especially in sub-Saharan African countries where conservative views of gender still shape entrepreneurial opportunities, and access to critical entrepreneurial success factors (e.g., Barberis & Solano, 2018; Iversen & Jacobsen, 2016; Tucker & Boonabaana, 2012). Consequently, there is limited understanding of how the multiple migrant identities of ethnicity and gender shape the entrepreneurial engagements of migrants from the under-researched context of Africa. Gender and ethnicity are two significant migrant identities that intersect to shape their ability to construct meaningful economic and socio-cultural lives in their host countries or communities (Ribeiro et al., 2021); hence, there is need to understand the interplay of ethnicity and gender in the migrant entrepreneurial space in tourism in Africa. This chapter seeks to explain the mechanisms through which the identities of ethnicity and gender of internal migrants intersect with their tourism entrepreneurial experiences in Ghana. Internal migrant tourism entrepreneurs within the context of this chapter are defined as persons who have established and manage tourism businesses in a geographical locality different from that in which they were born or were raised. Given the acknowledgement in the extant literature of the role and importance of migrant tourism entrepreneurship in employment creation as well as the development of sustainable tourist destinations (Iversen & Jacobsen, 2016; Lugosi & Allis, 2019), this chapter's practical implications contribute to knowledge that would aid relevant tourism industry stakeholders in the facilitation of this form of entrepreneurship. The next section of this chapter provides a synopsis of the literature on gender, ethnicity, migration, and tourism entrepreneurship, followed by a brief description of the method and study context. The findings and discussion are then presented after which there is a reflection of the study's findings and their implications for theory and practice.

Literature Review

Linking Migration and Tourism Entrepreneurship

Human mobilities have been of interest to researchers over the years. People's movements vary by space, time, motivation with diverse levels of permanency (Choe & Lugosi, 2022). Migration has been conceptualised as "movement across the boundary of an aerial unit" (Boyle et al., 1998, p. 34) with some level of permanence (Williams & Hall, 2000). The emphasis on tourism as 'a temporal movement' of persons situates it within the wider migration discourse. Williams and Hall (2000) and Salazar (2020) both emphasise the interconnectedness between migration and tourism as analogous but different forms of human mobility. Migration and tourism involve interactions which are complex, and scholarly works have posited the former as a product of the latter and vice versa (Choe & Lugosi, 2022). Both phenomena arise out of a combination of social, economic and/or political factors, fuelled by hopeful expectations, fears, and uncertainties (Zhang & Su, 2020). In aiding further understanding of the relationship between tourism and migration, Williams and Hall (2000) recognised five forms of tourism-related mobilities; labour migration, entrepreneurial migration, return (labour) migration, consumption-led economically active migration, and retirement migration. According to O'Reilly (2003), these forms have different structures with accompanying behaviours, attitudes, and identities for both migrants and hosts.

Tourism entrepreneurial migration, which is of interest to this chapter, has equally assumed many forms with varied dynamics which are reflected in the literature. While much attention has been given to entrepreneurial migration across international borders (e.g., Iversen & Jacobsen, 2016), notice has also been made of in-migration or internal migration tourism entrepreneurial activities (e.g., Xiong et al., 2020; York et al., 2021). Themes such as ethnic migrant entrepreneurship (e.g., Abd Hamid et al., 2019) and refugee entrepreneurship in tourism (e.g., Alrawadieh et al., 2019) have also been explored. Distinctions have also been made between the categories of tourism MEs with labels such as lifestyle (Sun et al., 2020) and amenity seekers (Matarrita-Cascante & Suess, 2020; Snepenger et al., 1995). Nevertheless, Adams (2021) argues that the current literature on migration and tourism connections is often obscured by western tourism concepts and is for this reason that emphasis on geographic context in the discussions is very critical. As alluded to by Kondo (1990), entrepreneurship is deeply shaped by the historical, political, economic, socio-cultural context within which it occurs.

Gender, Ethnicity, and Migrant Tourism Entrepreneurship

The subject of gender and ethnicity within migrant entrepreneurial activities has been addressed but often in isolation. Except for studies such Essers and Benschop (2007) and Essers (2009) which attempted to address the intersectionality of the two concepts in entrepreneurship, there is a lack of clarity regarding how the intersection of these two factors shapes the experiences of MEs. The social constructivist perspective illustrates how the constructed meanings of both gender and ethnicity shape migrant entrepreneurship experiences. Gender is defined by systems and people within those systems reflect and sustain it accordingly (Essers & Benschop, 2007), thus what is gender with its accompanying expectations could differ from context to context. Raffaelli and Ontai (2004) found that communities and families are very instrumental in defining and implementing appropriate behaviours and actions as defined by their conceptualisation of gender. The concept of ethnicity, on the other hand, may be seen as the process by which individuals allude to a sense of belonging to groups with similar socio-cultural traits and normative behaviour (Coles & Timothy, 2004). Depending on people's ethnicity, gender may be experienced differently, and vice versa, and therein lies opportunities as well as challenges (Wekker & Lutz, 2001). Gender, ethnicity, and migrant status could serve as an obstacle to MEs' integration into host communities and meeting their economic expectations. Trupp and Sunanta (2017) in their study of ethnic tourism in Thailand observed that the existence of a gender hierarchy in Thai, underlined by issues of class, ethnicity, and rural-urban dynamics, makes the Akha ethnic minority women a double marginalised group and that also places them in the lowest part of the Thai tourism industry. In another study of Asian female entrepreneurs in Australia, Collins and Low (2010) noted how female entrepreneurs with Asian educational qualifications and skills were less regarded in the labour market or in the starting up of business as compared to other immigrants. Broadly speaking, the extant research (cf. Figueroa-Domecq et al., 2022; Ngoasong & Kimbu, 2016, 2019) clearly indicates the critical role of gender in determining entrepreneurs' access to resources especially finance for start-up and expansion capital, training and capacity building opportunities and general information about opportunities in both developed and emerging destinations. To negotiate these hurdles, MEs have been found to adopt diverse strategies to renegotiate their identities. They achieve this through creation and reliance on networks and being very flexible in their entrepreneurial activities all in the drive to access resources to start and sustain their businesses. Some female MEs have been found to constantly renegotiate and/or even denounce their gender (Kimbu et al., 2019) and ethnic identities within the context in which they were operating as a strategy to continue pursuing their business activities (Essers & Benschop, 2007).

Study Setting and Methods

Ghana has over the years established itself as a key tourism destination within the West African subregion both in terms of volume and value. It is blessed with diverse tourism offerings including but not limited to flora and fauna, landscapes, culture, and heritage (special emphasis on Trans-Atlantic Slave Trade relics). From its pre-colonial era, the country witnessed initial efforts by the colonial leaders and associate expatriates to develop leisure facilities to meet their personal recreational needs. Most of these initiatives were centred around the natural resources found along the coastal and middle sections (Akyeampong & Asiedu, 2008). Post-independence, the state took active interest in the development of tourism which saw some level of institutionalisation of the phenomenon, provision of overnight accommodations (e.g., Continental Hotel, Meridian Hotel, Ambassador Hotel, Volta Hotel), and entertainment centres popularly referred to as catering rest house across the country more consciously targeting the lower class and public servants (Akyeampong & Asiedu, 2008). The liberalisation of the country's economy from the late 1980s coupled with the country's return to democratic rule encouraged private sector participation in the tourism sector. Consequently, the country can now boast of the presence of multinational businesses in all sectors of tourism, including Marriot, Holiday Inn, KFC, Burger King, British Airways, KLM, Emirates Airlines, and Avis Car Rentals. In 1990, international arrivals stood at 145,780 with receipts of about US$81 million. This increased to 456,275 visitors in 2000 with an accompanying receipt of US$289.5 million (Ministry of Tourism, 2013). In 2019, the country launched the "Year of Return" tourism campaign. According to the Minister of Tourism, Arts and Culture, the initiative brought in about 1.1 million international tourist arrivals, total airport arrivals increased by 45% over previous years and over 500 visas on arrivals were processed and over 200 African American and Afro Caribbeans who applied for Ghanaian citizenship were granted. It was also purported that a total of US$1.9 billion has been receipted because of the year of return through direct and indirect activities.

Cape Coast and Elmina are two coastal tourism communities in the Central region of Ghana. Cape Coast is the capital of the region and has a land area of about 122 km sq. According to the 2021 population and housing census, Cape Coast has a total population of about 189,925 with 92,790 males and 97,135 females. Elmina also has an estimated land area of about 372.45 km sq. with a population of 166,017 (80,570 males and 85,447 females). Both communities have a history with the Trans-Atlantic Slave Trade and are home to three world heritage sites: Cape Coast Castle, Elmina Castle, and Fort St Jago. The Cape Coast castle was built in 1630 by the Dutch. The Castle played a very significant role during the infamous Transatlantic slave trade. The Elmina Castle was built in 1482 by the Portuguese. It is the oldest European building in sub-Sahara Africa and was also intensively used during the slave trade.

The popularity of the Forts and Castles in these communities as tourism attractions has led to the emergence of tourism-related economic activities such as tour guiding services, accommodation services, souvenir production and sale and food and beverage operations in the communities. Other economic activities in the areas include fishing and related activities, retailing, petty trading, and sand mining. Reflective of the general population in Ghana, Cape and Elmina have a slightly higher female population; nevertheless due to high illiteracy rates among women, the majority of the economically active women are within the informal sector (agriculture, wholesale and retail trade, and tourism/hospitality) and their economic activities are defined along cultural and biological lines.

Data Collection and Analysis

The study adopted a qualitative methodology as it sought to explore the experiences of migrant tourism entrepreneurs. The choice of methodology was consistent with previous studies which have explored people's perceptions, experiences, meanings, and emotions (Silverman, 2013). In consonance with the research methodology, a structured in-depth interview guide was designed and employed for the data collection. The in-depth interview guide was divided into two sections. The first section captured information on the migration journeys of the MEs, while the second section focused on the influence of gender and ethnicity on their migrant tourism entrepreneurship experiences. A total of 16 internal migrant tourism entrepreneurs who had stayed in the study areas for at least two years were purposively selected and interviewed. The breakdown of the sample (Table 15.1) was ten MEs engaged in souvenir/craft production and sale and six MEs in the accommodation, food, and beverage subsector. These categories of businesses were selected because of the dominance of MEs in those subsectors. Ten females and six males were interviewed. With respect to age, the youngest respondent was 24 years old and the oldest was 52 years old. Educational levels of the respondents were generally low. Majority (14) of the respondents had not attended school beyond the Junior High School (JHS) level.

Table 15.1 Profile of study participants

Pseudo name of interviewees	Sex	Age	Active years in tourism business in the area	Educational level	Marital status	Service sector
Akua	F	24	1	JHS	Single	Souvenir/Craft
Kwame	M	52	13	JHS	Married	Souvenir/Craft
Mansa	F	48	10	Vocational	Married	Souvenir/Craft
Kofi	M	21	3	JHS	Single	Souvenir/Craft
Elizabeth	F	43	9	JHS	Married	Souvenir/Craft
Amina	F	35	5	Vocational	Single	Food/Beverage
Sammy	M	49	6	JHS	Married	Souvenir/Craft
Serwaa	F	31	4	JHS	Married	Food/Beverage
Nicholas	M	25	4	Vocational	Single	Food/Beverage
Anitha	F	29	5	SHS	Single	Souvenir/Craft
Paa Kwesi	M	38	5	Technical	Married	Souvenir/Craft
Efe	F	29	`6	Vocational	Married	Accommodation
Akosua	F	27	3	JHS	Single	Souvenir/Craft
Nimo	M	48	7	JHS	Married	Accommodation
Asabea	F	40	6	Vocational	Married	Souvenir/Craft
Irene	F	37	5	JHS	Single	Food/Beverage

However, some of them had undertaken vocational and technical training in various professions (including dress making, catering, hairdressing, woodwork). The respondents' average length of stay in the study areas was five years.

The lowest level of education attained by the respondents was primary school and the highest was tertiary. At least all the respondents had some form of formal education. The interviews were conducted in the local dialect Akan language and were audio-recorded for later transcription. The interviews averaged about 45 minutes and were conducted in July 2022. After each interview, the tape recordings were replayed and reviewed to look out for new emergent issues. Reflections and observations on each interview were also written down in the field notebook. These emergent issues and reflections were further explored in subsequent interviews. This was done to ensure the depth of data collected. After the data collection process, the recorded interviews were transcribed verbatim. This was followed by a content analysis of the transcripts. A non-directed content analysis approach was employed in the data analysis as it enables the themes to emerge from the data without any bias to theory (Hsieh & Shannon, 2005). Coding of the data was done inductively. The inductive coding was done based on the key issues which emerged from the transcripts. Specifically, the following steps were followed in the conduct of the content analysis as opined by Strauss (1987). The first step focused on first level of coding on a sentence-by-sentence basis and then on a paragraph-by-paragraph basis. At this stage, recurrent words as well as their possible meanings or associated emotions evoked were noted. This was followed by the labelling of the data/codes into categories as well as their properties and dimensions. The last step critically compared the data to look out for patterns, trends, and contradictions between the codes. A narrative approach (cf. Kimbu et al., 2021) was adopted in the presentation of the results and discussion. Direct quotations from the transcripts which were relevant to the study were used to support the findings.

Results and Discussion

Tourism Entrepreneurial Migration

In many developing countries, people migrate to seek employment and entrepreneurial opportunities that can offer them a high income, improved standard of living, and enhanced social status (Zhou et al., 2017). It was evident similar motives especially economic reasons influenced the migration of the MEs and caused them to establish small businesses in accommodation, food and beverage, and the souvenir trade. Specifically, the MEs were motivated to relocate, start a business, make, and save money to transform their livelihoods and well-being and that of their families. However, when it came to the choice of economic activity or economic sector to participate in, not all the MEs made their mobility decisions around tourism and its related activities. Consequently, while for some respondents, their desire to undertake an entrepreneurial activity in the tourism sector initiated the migration journeys to their current location; others chanced upon tourism while trying to pursue entrepreneurial activities in their new settlements. For the MEs whose migration decisions centred around pursuing tourism-related opportunities, they were found to have had prior contact with the sector by visiting and experiencing places where tourism activities thrived. For example, a 31-year-old female pub owner stated that:

> When I was working in a pub in Takoradi [coastal city in ...Ghana], I realised most of our patrons were foreigners and tourists. So, I decided to find out where they mostly visit so I can set up a business there. That was why I moved down to Cape Coast and targeted the vicinity of the Cape Coast Castle.
>
> *(Serwaa, 31-year-old married female, Food/Beverage)*

From the earlier submission of the MEs, it was obvious that some prior awareness of tourism's business stimulation potentials impacted on their decision to migrate to towns where tourism activities were vibrant. This supports Gossling and Schulz (2005)'s assertion that the decisions of migrants regarding tourism entrepreneurship can be determined by specific knowledge they possess about the phenomenon.

Most respondents who leaned towards this thinking transferred their existing experiences, skills, and knowledge in starting the new ventures in their new settlements. They further demonstrated more understanding of the nature of tourism and its possible effect on their businesses. As commented by a 40-year-old female craft trader:

> I was already in the Kente and adinkra artefact business in my hometown so when I decided to follow the tourists to Cape Coast, I said I will do the same business... this our job, you need patience. When the season is right, you will make good sales.
>
> *(Asabea, 40-year-old married female, Souvenir/Craft)*

For MEs who later got involved in tourism-related entrepreneurial activities, their initial preoccupation was the need to find any economic opportunity to improve upon their livelihoods as well as that of their dependents. Hence, prior to venturing into tourism, they had either taken up jobs or established businesses in other sectors. For a 35-year-old female pub owner, she initially relocated from Kumasi to Accra and engaged in petty trading, odd jobs before working in a pub. Her interaction with the pub business led her to open her own pub in her new settlement.

> *I also wanted to hustle and make money like my friends, so I decided to Accra to look for a job. When I arrived in Accra, I had to do menial jobs like carrying load in the market, selling things before I finally ended up working in a drinking spot. I saw the money that was made while I was working there so I also decided to start this business.*
>
> *(Amina, 35-year-old single female, Food/Beverage)*

Another craft dealer also started working as a clearing agent in a fishing and shipping harbour before venturing into the craft trade.

> *I was working in the shipping harbour and the job was paying well. Initially I didn't want to relocate but I was having marital issues, so I decided to quite the job and join my husband here in Elmina. It took me over a year after I arrived to get something to do because the economic options in this town are very limited. Because I am industrious and I have already acquired skills in sewing, I decided to venture into this Africa wear business. I have seen the foreigners patronize my products and business can be good.*
>
> *(Mansa, 48-year-old married female, Souvenir/Craft)*

It is therefore plausible that for this category of MEs, if they had found themselves in non-tourism destinations, their choice of economic activity might have been different. Regardless, it was evident that all the MEs irrespective of their gender were motivated by the perceived tourism revenue or profits and their prolonged continuance of running tourism-related enterprises may be indicative of their expectations being met.

Ethnicity and Tourism Migrant Entrepreneurship

The issue of ethnicity was found to have had varied influences on the MEs experiences in the study areas in terms of choice of migration destination and integration into host community. Ethnicity was not a determinant of MEs choice of migration destination. The main consideration rather was on the economic opportunities available at the destination. As reiterated by two souvenir traders:

> *I never think of my ethnicity or that of the place before moving there. I was more concerned about what businesses I could do while there.... Yes, I could not speak the Fante language initially but that did not deter me from moving here since I knew that my business will be more with the foreigners.*
>
> *(Kofi, 21-year-old single male, Souvenir/Craft)*

This is more likely to be due to the metropolitan nature of the communities under study. Unlike some rural communities in the country where native status and ethnicity were necessary requirements for engaging in tourism-related activities (Afenyo, 2011), same cannot be said of the urban areas where cultural issues are quite blurred due to their diverse and multi-cultural make-up. Further, communal living is still present in most communities across the country, therefore making it relatively easier for persons from different ethnicities to undertake business anywhere across the country:

> *This is Ghana and we are all Ghanaians so I don't think it will be possible that I will move somewhere in Ghana, and I would not be accepted...I don't think Ghanaians are like that...the business I am doing well is legitimate so there is nothing I must worry about.*
>
> *(Nimo, 48-year-old married male, Accommodation)*

Again, the experiences shared are not unexpected, especially when considering the cultural distance between the MEs and their hosts. The Akan language for example is widely spoken across Ghana. This makes it relatively easier for many people in the country to communicate with others when they move outside their localities or places of residence. This plausibly acts as a unifier and aids in making people regardless of their ethnicity feel included when they travel within the country. In contexts where the cultures are similar, MEs feel less socially excluded (Hamid et al., 2018) and that was the observed case in this study where the MEs and the hosts shared the same national culture.

In terms of MEs integration into the host community, issues of discrimination were rare. Except for one respondent who indicated she had challenges securing land from a private landowner for her shop because she was considered a foreigner, others did not encounter same.

> *When I wanted a land to rent to put up a shop, the landowner I approached refused to give the land to me. The excuse was that I was not from the community. I believe If I had made a native to rent the land for me, it would have been easier. I later got a space from another landowner.*
>
> *(Elizabeth, 43-year-old married female, Souvenir/Craft)*

Rogers et al. (2009) have maintained that the presence of MEs in communities can become a source of community tension, particularly if they (MEs) represent competition for jobs. This has implications for the MEs integration into the host community. This was observed from the narratives as there seemed to be some sense of tension though subtle between the MEs and the host residents. This was emanating from the host residents' perception of the MEs taking up their jobs and making money. For instance, some respondents recounted that:

> *The local people say that it is only the migrants who get money when they come and work here than they the locals...*
>
> *(Anitha, 29-year-old married female, Souvenir/Craft)*

> *...they (host residents claim we (MEs) are using juju (mystical powers) to sale that is why we are in good business...sometimes too you will hear them say that we (ME) have come to take over their jobs.*
>
> *(Paa Kwesi, 38-year-old married male, Souvenir/Craft)*

The entrepreneurship literature has posited higher entrepreneurial propensity among migrants as compared to their hosts (Clark et al., 2017). Guerrero et al. (2021) have further argued that migrants are more likely to choose the entrepreneurial pathway due to the discrimination they are likely to face in the host labour market. This may explain the presence of migrants in tourism entrepreneurship in the study areas; a situation the hosts may not be pleased with. Tourism destinations that seek growth cannot wish away the role of MEs. Through concerted efforts, valuable lessons and skills can be exchanged between the hosts and the MEs in order to build a formidable tourism sector that benefits all.

While having local friends has been found to be helpful to MEs in negotiating their stay in the host communities (Ward & Rana-Deuba, 1999), this was not the case in the study areas as the MEs felt that the host residents were "jealous" of them. The word "jealous" was used by all the respondents, and they used it to position the host residents as being against them. Thus, they expressed a lot of hesitation in attempting to build friendship or form synergies with the host residents. Consequently, the nature of the relationship that existed between the hosts and the MEs was largely superficial in nature but not conflictual; a situation that was also noted by McAreavey and Argent (2018). As such, majority of the MEs preferred to isolate themselves as much as possible from the host residents and limited the non-economic encounters that occurred among them:

When I close my shop, I go home. I don't have friends and it is better not to have friends. These people will be smiling with you, but they are jealous of you. I try to greet those around me but nothing beyond that.
(Anitha, 29-year-old single female, Souvenir/Craft)

As for the community, when they are doing programmes, I attend like the upcoming festival, I will fully participate but beyond that I just keep to myself.
(Serwaa, 31-year-old married female, Food/Beverage)

Secondly, it was observed that the MEs used their own ethnicities as a lens to see and form an impression about their hosts. In all the interviews, the MEs used words and phrases such as "lazy," "gossips" "unwilling to work" "always begging for alms" to describe the host residents. They found these attributes as unacceptable. A 52-year-old male ME who owned an African wear shop saw it as an exploitative attitude when he observed that:

So far as they ask you for food or money and you give them, the friendship between you will be ok but the day they ask and you are unable to give, then you will begin to have issues with them. We don't do that from where I come from...we are known to be hard working people so I find it odd that here everyone beg instead of them finding some work to do.
(Kwame, 52-year-old married male, Souvenir/Craft)

Drawing from the interviews conducted, the host residents' perception of the MEs as their competitors may be due to that fact that the MEs may have better capacity and knowledge to better engage in such tourism-related activities because of their past experiences, interactions, and understanding of the tourism phenomenon (Pernecky, 2012), so will make better gains from it.

Gender and Tourism Migrant Entrepreneurship

Gender shapes various aspects of the tourism phenomenon, including entrepreneurial migration (Aitchison, 2001; Tucker & Boonabaana, 2012). The gender influences observed within the context of this study were on the motivation for migration, entrepreneurial risk taking, family support, as well as integration into the host community. The propensity to take risk as an entrepreneur was exhibited more by the male MEs than the female MEs. This is supported by the extant literature which posits female entrepreneurs as more conservative in risk taking compared to their male counterparts (Kozubíková et al., 2017) even though this assumption is now increasingly being questioned (see, e.g., Figueroa-Domecq et al., 2020). At the inception of their enterprises, majority of the male MEs did not have family or social contact prior to their relocation to their current settlement, while their female counterparts had some contact (husbands, fiancé, former school mates). Thus, the female MEs found it easier to settle into the new community and establish networks compared to their male counterparts. Additionally, the majority of the female MEs mentioned receiving both financial and non-financial support from their families to aid their ventures but such was not mentioned by the male MEs.

My boyfriend really helped me to settle in this community. He comes from here and knows his way around. He gave me some money and went around with me to set up my shop.

(Amina, 35-year-old single female, Food/Beverage)

My husband and family had to agree to me moving to Cape Coast before I moved down. When I was facing challenges at the initial stages of my business, I went back to them do discuss the situation and they were very supportive.

(Asabea, 40-year-old married female, Souvenir/Craft)

The above findings concur with similar observations made by Kimbu and Ngoasong (2016) in their study on the transformational role of women through tourism social entrepreneurship in emerging destinations. This support for the females perhaps lessened the risk associated with the operation of their tourism ventures in the new locations.

Family considerations also underscored the migration motivation of the female respondents and influence their destination choice. Female MEs especially were noted to be motivated to relocate to join their spouses after they got married or were in the process of getting married. These female entrepreneurs revealed that it was in the scarcity of employment in their new locations that motivated them to search for alternative economic activities and their decision to go into tourism-related activities since it was in existence there. In further buttressing this position, a 48-year-old African wear dealer commented that:

I was working in the shipping harbour and the job was paying well. Initially I didn't want to relocate but I was having marital issues, so I decided to quite the job and join my husband here in Elmina. It took me over a year after I arrived to get something to do because the economic options in this town are very limited. Because I am industrious and I have already acquired skills in sewing, I decided to venture into this Africa wear business. I have seen the foreigners patronize my products and business can be good.

(Mansa, 48-year-old married female, Souvenir/Craft)

Another female Kente trader noted that she had to seek the consent of her husband and family before migrating to her new settlement to undertake her new trade:

> *My husband and family had to agree to me moving to Cape Coast before I moved down. When I was facing challenges at the initial stages of my business, I went back to them do discuss the situation and they were very supportive.*
>
> *(Asabea, 40-year-old married female, Souvenir/Craft)*

The above quote clearly indicates the role of culture and tradition in determining the power relationships within families in emerging destinations of Africa and Asia where women are still viewed as, and expected to be, subservient to men/husbands and families when important decisions are to be made (Zhang et al., 2020).

Regarding the type of activity to engage in, gender was found to have played a deciding role. A female ME indicated that her choice of tourism-related activity was informed by the flexibility inherent in the operation of the venture as compared to working in the formal government sector or taking up employment elsewhere. In all the discussions, the MEs indicated that they believe it was easier getting assistance as a female than male even though they could not substantiate this claim. Nevertheless, this is very necessary for the success of female businesses given that family support has been seen to be vital in entrepreneurial journeys of females, especially in economies where the social status of women is low relative to men (Kimbu & Ngoasong, 2016; Lindvert et al., 2017).

Reflections

This chapter explored the experiences of internal migrant tourism entrepreneurs in the context of Ghana. By focusing on internal migrants, this chapter responded to the challenge of providing insight into movements within a particular country by its residence (either by citizenship or birth) as compared to international migrant tourism entrepreneurship which has dominated the literature. It further sought to aid our understanding of how multiple identities and their interplays particularly that of gender and ethnicity affect the entrepreneurial engagements of migrants in tourism.

First, the evidence obtained confirmed the assertion that economic motives drive entrepreneurial migration decisions. Given the fact that in resource-scarce contexts like Ghana where economic options are very limited, people are compelled to seek economic opportunities wherever they exist. The chapter evidenced how entrepreneurial migrations are borne out of necessity not opportunity or passion as the extant literature on the migrant tourism entrepreneurship phenomenon has posited. Hence MEs will be attracted to areas, places, or destinations that will enable them to achieve the economic objectives of earning income and transforming their livelihoods. Even though tourism was not central to the mobility decisions of all the MEs, it was established that tourism is relatively easy-to-enter sector (cf. Figueroa-Domecq et al., 2020) for MEs especially in places where gender and ethnicity restrictions are minimal.

Secondly, gender and ethnicity have been identified as having some influence on MEs migration decision-making as well as integration into host communities. Marriage was found to have primarily underscored the mobility decisions of female MEs, a reflection of the perpetuation of patriarchy in the study context. This is irrespective of the calls for more gender equality in tourism by (inter)national civil society organisations and important strides made by the Ghanaian government advancing gender equity through the introduction of gender-aware policies and regulations. While the male tourism MEs demonstrated more entrepreneurial migration risk-taking propensity as compared to their female counterparts, the entrepreneurial experiences of both genders were not significantly different. Narrations regarding start-up of the tourism enterprises, integration in the host communities, and host-migrant relationships were similar. We recognise the possible intervening influence that the urban context with its multi-cultural characteristics might have on the constructions as well as the interplays of these identities, i.e., gender and ethnicity on the subject matter. This is a contribution to the on-going debate regarding the multiplicities of identities of migrant tourism entrepreneurs and the interplays of these identities on their entrepreneurial experiences.

Thirdly, in exploring the nature of the relationship between the MEs and their hosts, the relationship was identified as being superficial and stemmed out of how both the MEs and the hosts perceived each other and acted accordingly. Given that MEs have been found to be more entrepreneurial than their hosts, they will profit more from the tourism establishments in the communities. This may make them more economically advantageous than the hosts and continuously place them against each other. But beyond the perceived economic gains, will the hosts be interested and willing to make the same or necessary investments in tourism enterprises as the MEs? This question needs to be further explored in subsequent studies on migrant tourism entrepreneurship in Ghana. A major limitation of this chapter which is duly acknowledged is the limited focus on the experiences of MEs located in the urban areas of Cape Coast and Elmina. Moving beyond and exploring ME experiences from an urban-rural context perspective would arguably present much more complex and varied findings. Hence the findings presented in this chapter cannot be said to be reflective of the situation across the entire country, but they provide directions for further research. This chapter has therefore brought attention to internal migrant entrepreneurship in Ghana's tourism development.

References

Abd Hamid, H., O'Kane, C., & Everett, A. M. (2019). Conforming to the host country versus being distinct to our home countries: Ethnic migrant entrepreneurs' identity work in cross-cultural settings. *International Journal of Entrepreneurial Behaviour & Research*, 25(5), 919–935.

Adams, K. M. (2021). What western tourism concepts obscure: Intersections of migration and tourism in Indonesia. *Tourism Geographies*, 23(4), 678–703.

Afenyo, E. A. (2011). *Community-based ecotourism in Ghana. An evaluation of the Tafi Atome Monkey Sanctuary Project* [Unpublished master's dissertation, Department of Hospitality and Tourism Management, University of Cape Coast, Ghana].

Aitchison, C. C. (2001). Theorising other discourses of tourism, gender, and culture: Can the subaltern speak (in tourism)? *Tourist Studies*, 1(2), 133–147.

Akyeampong, O., & Asiedu, A. B. (Eds.). (2008). *Tourism in Ghana: A modern synthesis*. Assemblies of God Literature Centre.

Alrawadieh, Z., Karayilan, E., & Cetin, G. (2019). Understanding the challenges of refugee entrepreneurship in tourism and hospitality. *The Service Industries Journal*, 39(9–10), 717–740.

Barberis, E., & Solano, G. (2018). Mixed embeddedness and migrant entrepreneurship: Hints on past and future directions. An introduction. *Sociologica*, 12(2), 1–22.

Bosworth, G., & Farrell, H. (2011). Tourism entrepreneurs in Northumberland. *Annals of Tourism Research*, 38(4), 1474–1494.

Boyle, P., Halfacree, K., & Robinson, V. (1998). *Exploring contemporary migration*. Longman.

Choe, J., & Lugosi, P. (2022). Migration, tourism and social sustainability. *Tourism Geographies*, 24(1), 1–8.

Clark, K., Drinkwater, S., & Robinson, C. (2017). Self-employment amongst migrant groups: New evidence from England and Wales. *Small Business Economics*, 48(4), 1047–1069.

Coles, T., & Timothy, D. J. (2004). "My field is the world": Conceptualizing diasporas, travel and tourism: Conceptualizing diasporas, travel and tourism tourism, migration and mobility: A missing piece of the jigsaw? In T. Coles & D. J. Timothy (Eds.), *Tourism, diasporas and space* (Vol. 6, pp. 15–44). Routledge.

Collins, J., & Low, A. (2010). Asian female immigrant entrepreneurs in small and medium-sized businesses in Australia. *Entrepreneurship and Regional Development*, 22(1), 97–111.

Essers, C. (2009). *New directions in postheroic entrepreneurship: Narratives of gender and ethnicity* (Vol. 25). Copenhagen Business School Press DK.

Essers, C., & Benschop, Y. (2007). Enterprising identities: Female entrepreneurs of Moroccan or Turkish origin in the Netherlands. *Organization Studies*, 28(1), 49–69.

Figueroa-Domecq, C., de Jong, A., Kimbu, A. N., & Williams, A. M. (2022). Financing tourism entrepreneurship: A gender perspective on the reproduction of inequalities, *Journal of Sustainable Tourism*, https://doi.org/10.1080/09669582.2022.2130338.

Figueroa-Domecq, C., Kimbu, A. N., de Jong, A., & Williams, A. M. (2020). Sustainability through the tourism entrepreneurship journey: A gender perspective. *Journal of Sustainable Tourism*, 30(7), 1562–1585.

Gossling, S., & Schulz, U. (2005). Tourism-related migration in Zanzibar, Tanzania. *Tourism Geographies*, 7(1), 43–62.

Guerrero, M., Mandakovic, V., Apablaza, M., & Arriagada, V. (2021). Are migrants in/from emerging economies more entrepreneurial than natives? *International Entrepreneurship and Management Journal*, 17(2), 527–548.

Hamid, H. A., Everett, A. M., & O'Kane, C. (2018). Ethnic migrant entrepreneurs' opportunity exploitation and cultural distance: A classification through a matrix of opportunities. *Asian Academy of Management Journal*, 23(1), 151–169.

Hsieh, H. F., & Shannon, S. E. (2005). Three approaches to qualitative content analysis. *Qualitative Health Research*, 15(9), 1277–1288.

Iversen, I., & Jacobsen, J. K. S. (2016). Migrant tourism entrepreneurs in rural Norway. *Scandinavian Journal of Hospitality and Tourism*, 16(4), 484–499.

Kimbu, A. N., de Jong, A., Adam, I., Ribeiro, A. M., Adeola, O., Afenyo-Agbe, E., & Figueroa-Domecq, C. (2021). Recontextualising gender in entrepreneurial leadership. *Annals of Tourism Research*, 88. https://doi.org/10.1016/j.annals.2021.103176

Kimbu, A. N., & Ngoasong, M. Z. (2016). Women as vectors of social entrepreneurship. *Annals of Tourism Research*, 60, 63–79.

Kimbu, A. N., Ngoasong, M. Z., Adeola, O., & Afenyo-Agbe, E. (2019). Collaborative networks for sustainable human capital management in women's tourism entrepreneurship: The role of tourism policy. *Tourism Planning & Development*, 16(2), 161–178.

Kondo, D. K. (1990). *Crafting selves: Power, gender, and discourses of identity in a Japanese workplace.* University of Chicago Press.

Kozubíková, L., Dvorský, J., Cepel, M., & Balcerzak, A. P. (2017). Important characteristics of an entrepreneur in relation to risk taking: Czech Republic case study. *Journal of International Studies*, 10(3), 220–233.

Lindvert, M., Patel, P. C., & Wincent, J. (2017). Struggling with social capital: Pakistani women micro entrepreneurs' challenges in acquiring resources. *Entrepreneurship & Regional Development*, 29(7–8), 759–790.

Lugosi, P., & Allis, T. (2019). Migrant entrepreneurship, value-creation practices and urban transformation in São Paulo, Brazil. *Revista Brasileira de Pesquisa em Turismo*, 13, 141–163.

Matarrita-Cascante, D., & Suess, C. (2020). Natural amenities-driven migration and tourism entrepreneurship: Within business social dynamics conducive to positive social change. *Tourism Management*, 81, 104140.

McAreavey, R., & Argent, N. (2018). New immigration destinations (NID) unravelling the challenges and opportunities for migrants and for host communities. *Journal of Rural Studies*, 64, 148–152.

Ministry of Tourism (2013). *National tourism development plan (2013–2027)*. Retrieved from http://www.ghana.travel/wp-content/uploads/2016/11/Ghana-Tourism-Development-plan.pdf. Accessed on 3/8/2015.

Ngoasong, M. Z., & Kimbu, A. N. (2016). Informal microfinance institutions and development-led tourism entrepreneurship. *Tourism Management*, 52, 430–439.

Ngoasong, M. Z., & Kimbu, A. N. (2019). Why hurry? The slow process of high growth in women-owned businesses in a resource-scarce context. *Journal of Small Business Management*, 57(1), 40–58.

O'Reilly, K. (2003). When is a tourist? The articulation of tourism and migration in Spain's costa del sol. *Tourist Studies*, 3(3), 301–317.

Paniagua, Á. (2002). Urban-rural migration, tourism entrepreneurs and rural restructuring in Spain. *Tourism Geographies*, 4(4), 349–371.

Pernecky, T. (2012). Constructionism: Critical pointers for tourism studies. *Annals of Tourism Research*, 39(2), 1116–1137.

Raffaelli, M., & Ontai, L. L. (2004). Gender socialization in Latino/a families: Results from two retrospective studies. *Sex Roles*, 50(5), 287–299.

Ribeiro, M. A., Adam, I., Kimbu, A. N., Afenyo-Agbe, E., Adeola, O., Figueroa-Domecq, C., & de Jong, A. (2021). Women entrepreneurship orientation, networks and firm performance in the tourism industry in resource-scarce contexts. *Tourism Management*, 86, 104343.

Rogers, A., Anderson, B., & Clark, N. (2009). *Recession, vulnerable workers and immigration*. Retrieved October 12, 2022 from https://www.compas.ox.ac.uk/wp-content/uploads/PR-2009-Recession_Vulnerable_Workers.pdf

Salazar, N. B. (2020). Labour migration and tourism mobilities: Time to bring sustainability into the debate. *Tourism Geographies*, 24(1), 141–151.

Silverman, D. (2013). *Doing qualitative research: A practical handbook*. Sage.

Snepenger, D. J., Johnson, J. D., & Rasker, R. (1995). Travel-stimulated entrepreneurial migration. *Journal of Travel Research*, 34(1), 40–44.

Strauss, A. L. (1987). *Qualitative analysis for social scientists*. Cambridge University Press.

Su, X., & Chen, Z. (2017). Embeddedness and migrant tourism entrepreneurs: A Polanyian perspective. *Environment and Planning A: Economy and Space*, 49(3), 652–669.

Sun, X., Xu, H., Köseoglu, M. A., & Okumus, F. (2020). How do lifestyle hospitality and tourism entrepreneurs manage their work-life balance? *International Journal of Hospitality Management*, 85, 102359.

Trupp, A., & Sunanta, S. (2017). Gendered practices in urban ethnic tourism in Thailand. *Annals of Tourism Research*, 64, 76–86.

Tucker, H., & Boonabaana, B. (2012). A critical analysis of tourism, gender and poverty reduction. *Journal of Sustainable Tourism*, 20(3), 437–455.

Ward, C., & Rana-Deuba, A. (1999). Acculturation and adaptation revisited. *Journal of Cross-Cultural Psychology*, 30, 372–392.

Wekker, G., & Lutz, H. (2001). A wind-swept plain: the history of ideas on gender and ethnicity in the Netherlands" translations from "Een hoogvlakte met koude winden: de geschiedenis van her gender-en etniciteitsdenken in Nederland" in M. *Caleidoscopische visies: de zwarte, migranten en vluchtelingenvrouwenbeweging in Nederland, Koningklijk Instituut voor de Tropen, Amsterdam*, 25–49.

Williams, A. M., & Hall, C. M. (2000). Tourism and migration: New relationships between production and consumption. *Tourism Geographies*, 2(1), 5–27.

Xiong, Y., Zhang, Y., & Lee, T. J. (2020). The rural creative class: An analysis of in-migration tourism entrepreneurship. *International Journal of Tourism Research*, 22(1), 42–53.

York, Q. Y., Yan, L., & Ben, H. Y. (2021). My life matters here: Assessing the adjusted identity of domestic migrant workers at intangible cultural heritage tourism businesses in China. *Tourism Management Perspectives*, 39, 100856.

Zhang, C. X., Kimbu, N. A., Lin, P., & Ngoasong, M. Z. (2020). Guanxi influences on women intrapreneurship. *Tourism Management*, 81. https://doi.org/10.1016/j.tourman.2020.104137

Zhang, H., & Su, X. (2020). Lifestyle migration and the (un) making of ideal home. *Geoforum*, 115, 111–119.

Zhou, L., Chan, E., & Song, H. (2017). Social capital and entrepreneurial mobility in early-stage tourism development: A case from rural China. *Tourism Management*, 63, 338–350.

16

THE EFFECT OF FAMILY SUPPORT ON RURAL WOMEN'S TOURISM ENTREPRENEURIAL INTENTION IN CHINA

Xiangli Fan, Haili Qin, Jiamei Zhang, and Fan Zhong

Abstract

Women entrepreneurs play a crucial role in rural revitalisation in China; however, Women entrepreneurs in rural areas have not been given adequate attention in research arena. This chapter explored the internal influence mechanism of family support on rural women's tourism entrepreneurial intention with mixed methods. A paradigm between family support and rural women's tourism entrepreneurial intention has been established. It concluded that gender perception is an important regulatory variable that affects the willingness of rural women to start a business; family supports the tourism entrepreneurial intentions of rural women through multiple pathways of perceptual aspiration and perceptual feasibility; the influence of family emotional support on perceptual aspiration and perceptual feasibility is more significant.

Keywords

Family support; rural; women; entrepreneurial intentions; tourism enterprise

Introduction

The recession of rural areas is increasingly evident in China, which has been obviously related to the irresistible urbanisation trend, thus curbing rural declining has become an urgent mission of this era. In 2017, the rural vitalisation strategy was proposed as a key move at the 19th National Congress of the Communist Party of China (CPC), and the CPC Central Committee and the State Council released a package of policies in 2018 charting the roadmap for rural vitalisation, which identified rural tourism as the main path for achieving rural revitalisation (Zhang et al., 2022).

However, in recent years, the phenomenon of "rural hollowing" in China has become serious, and retaining people in rural areas has proved to be the basis for the realisation

DOI: 10.4324/9781003286721-23

of the rural revitalisation strategy (Abdullah et al., 2019). Several studies showed that rural recession resulted from more and more women leaving the village for cities (Jia et al., 2018; Khanna & Fellow, 2010). In 2017, the Women's Tourism Committee of the China Tourism Association released the "Report on Rural Women's Tourism Entrepreneurship Problems and Countermeasures," affirming the role played by rural women in tourism entrepreneurship. Rural revitalisation can probably be achieved, only if women, the main group living in the countryside, can work or start businesses in the countryside they live. Whereas rural tourism business is one of their most preferable business directions (Akpinar et al., 2005). However, rural women's entrepreneurial activities are different from men's and are influenced by more factors. On the one hand, due to the educational background of rural women themselves, gender role cognition and other reasons, rural women's willingness to start a business may not be as strong as men's (Madanaguli et al., 2021); on the other hand, due to the traditional concept of Chinese, women's main field of activity is the family, and external resources available to women, such as funds and contacts, are also relatively scarce(Shi, 2001; Arunachalam, 2014). But in some rural tourist destinations, there still emerged some outstanding rural female entrepreneurs (Mcgehee et al., 2007), such as Brkez Saymu helping nearly 200 rural women earn a living through making and selling embroidery to tourists in Kashgar prefecture, Northwest China's Xinjiang Uygur autonomous region, Vlogger Li Ziqi from Sichuan Province who amazes world with China's countryside life earning more than 100 million followers. As the main group of rural tourism entrepreneurship, women's entrepreneurial intention is the beginning of entrepreneurial activities, which is directly related to the retention of rural women, thus affecting the realisation of the rural revitalisation strategy. This study attempts to explore the mechanism among family support and entrepreneurship intention within China's rural context, in the hope to understand how to encourage women's entrepreneurship will, expecting to contribute to China's rural revitalisation strategy.

Literature Review

Entrepreneurship and Gender

Women entrepreneurship has an important impact on the country's economic growth, and it is conducive to raising the level of gender equality and reducing poverty (Enemuo, Udeze, & Ugwu, 2019). Some scholars argued entrepreneurship was never gender-neutral, with women showing a lower propensity for entrepreneurial behaviour compared to men (Muñoz-Fernández et al., 2016; Obschonka et al., 2014). Studies also showed that women's entrepreneurial intention, as a valid predictor of entrepreneurial behaviour, was also lower than that of men. Rural women were a marginalised group, and the factors affecting their entrepreneurship include rural women's limited education, lack of entrepreneurial opportunity recognition, lower social capital, less prior experience, lack of professional skills and knowledge in tourism (Yang & Yang, 2017). In addition, Luo Mingzhong et al. believe that the biggest difference between rural female entrepreneurs and men is that they need more support, understanding and encouragement from family members (Luo, 2011), otherwise female entrepreneurs will face greater role conflicts.

Rural Tourism Entrepreneurship and Women

Rural tourism research appeared China in the 1990s, and then it got attention by numerous scholars worldwide. Existing research mainly focuses on the definition of concepts and connotations (Liu, 2005; Xiao et al., 2001), developing mode and characteristics (Cantallops et al., 2015; Xue, 2008), driving mechanisms and influencing factors (Shen et al., 2020; Wang et al., 2011), community participation and stakeholder research in rural tourism development (Johnson, 2010), etc. Getz and Carlsen focused on entrepreneurial couples in rural areas and examined entrepreneurship and operation of the accommodations run by families.(Getz & Carlsen, 2000). The research showed that the couples had a strong motivation to live and work in rural areas where they lived, and this was proved to be true by other researchers (Dong et al., 2017). In the study of rural tourism, gender issue has always been a very important component. The rural tourism improved women's economic status in the family, studies suggested that women play a quite central and dominant role in the development of rural tourism (Nilsson, 2002). However, there were also some research that suggest the proportion of women in rural households who were actually engaged in tourism was not high (Gasson & Winter, 1992).

The Chinese Context

That women's holding up half the sky is beyond doubt. Since the founding of the People's Republic of China, women have been liberated and their productive forces have also been liberated. They actively participated in the socialist modernisation drive, played a vital role in all walks of life, and made great contributions to the socialist construction (Croll, 1985). In the vast countryside, women are an important force in building a new socialist countryside (Gallin & Bossen, 2004). Due to urbanisation and a large number of young and middle-aged men working in cities, women have become the main part of the rural labour force, reaching 60% in some places, and even higher in some places (Shui & Liu, 2020). In order to mobilise the enthusiasm of women to participate in rural construction, in February 2018, the All-China Women's Federation issued the Implementation Opinions on the "Rural Revitalization Action for Women," which fully affirmed the important role of women in the implementation of rural revitalisation strategy and pointed out that "women are the important force to promote agricultural and rural modernization, the beneficiaries, the promoters and builders of rural revitalization," the specific action plan to mobilise rural women to actively participate in rural revitalisation was put forward, including the implementation of the "Rural Women's Quality Improvement Plan," the implementation of the "Beautiful Home" construction activity, the expansion of the "Most Beautiful Family" activity, and the continuous deepening of the "Women's Poverty Alleviation Action." Under the organisation of the All-China Women's Federation, women's enthusiasm for participating in rural revitalisation has been greatly enhanced.

Research Hypotheses and Theoretical Frame

Theoretical Model Construction

Entrepreneurial Intentions

Shapero and Sokol (1982)'s model of entrepreneurial event theory suggested that convergence of attitudes and situational factors led to business start-ups. In Shapero and Sokol's model, willingness is based on perceived feasibility and perceived desirability (Shapero &

Sokol, 1982). Perceived desirability measures the attractiveness of a particular behaviour (such as becoming an entrepreneur) for an individual. Perceived feasibility is defined as awareness of one's ability to perform a particular action (Liñán & Santos, 2007). The individual's perceived desirability and perceptual feasibility are influenced by external circumstances, such as social and cultural environments. The perception of desirability depends not only on the value system of the individual entrepreneur but also on the social system. The feasibility perception depends on the social capital that the entrepreneur has at that time, such as the support of entrepreneurial funds and the support of entrepreneurial partners (Matriano & Kiyumi, 2021). Krueger et al. (2000) found that the entrepreneurial event model was effective in explaining actual entrepreneurial intentions and is statistically more reasonable. Some scholars improved it, like Krueger and Brazeal (1994) added the variable "credibility" to the theory of entrepreneurial events, and argued that both perceptual aspiration and perceptual feasibility had an impact on entrepreneurial intentions through "trustworthiness." Fitzsimmons and Douglas (2011) suggested that perceptual desirability, perceptual feasibility, and their interactions had an impact on entrepreneurial intentions. Schlaegel and Koenig (2014) integrated entrepreneurial event theory and planned behaviour theory, arguing that attitudes, subjective norms, entrepreneurial self-efficacy, and perceptual behaviour control influenced entrepreneurial intentions through perceptual desirability, and entrepreneurial self-efficacy and perceptual behaviour control influenced entrepreneurial intentions through perceived feasibility. Most scholars considered the influence of the two variables of perceptual desirability and perceptual feasibility on entrepreneurial intentions when studying the theory of entrepreneurial events, and this study also focused on the two variables of perceptual desirability and perceptual feasibility when constructing a theoretical model based on entrepreneurial event theory.

Family Support

Family has always been one of the important factors influencing the willingness to start a business. Conflicts between family and business can affect an entrepreneur's intention to give up or continue to run the business. Dan et al. (2016) argued that entrepreneurs might shorten their time at home for their careers or change their business participation for their family members (Spivack, & Desai, 2016). Henderson and Robertson (2000) argued that family was the second largest factor influencing individual career choices, second only to personal experience. Family members and friends had a crucial influence on an individual's career choices, which were often considered to provide funds and role models. Pruett and his colleagues conducted a study on college students in the United States, Spain, and China, taking social, cultural, and psychological factors as predictors of entrepreneurial willingness. They found strong influences of psychological factors, while also highlighting the important relationship between family support and entrepreneurial willingness. Having family support or lack of family support is related to individual entrepreneurial willingness (Pruett et al., 2009). Turker (2009) established the entrepreneurial support model (ESM), in which entrepreneurial willingness was influenced by three factors: education, relationships, and structural support. In his study, relationship support was primarily expressed as emotional and financial support from family and friends. Perception of this support encouraged individuals to choose entrepreneurship and affected their entrepreneurial willingness (Turker & Sonmez Selcuk, 2009).

Social Gender Theory

Joan Scott (1988), an American feminist historian, proposed that gender is a representative form of power relations, a product of the combined action of culture, economy, and society. Since the 1990s, gender theory has emphasised the cross-examination and analysis of gender dimensions with political, ethnic, and family factors, focusing on the multiplicity and diversity of people's identities. From the concept of gender as an analytical tool to the advancement of theorisation, gender theory has been continuously enriched. The social gender theory emphasises the comprehensive analysis of multiple identities of people and pays attention to the gender relations shaped by social culture (Figueroa-Domecq et al., 2020), such as the changing process of gender relations in different fields, the gender division of labour in economic activities, gender order, and the understanding of their own gender roles. The family, as an important field, is closely associated with women. In the family domain, the roles between men and women had obvious gender differences, which is the result of the constant evolution of social culture and the changing concept of gender in different eras. The traditional concept of gender mainly characterised by "men dominate the outside and women dominate the inside" has dominated the thinking of the masses for a long time. Based on the concept and theory of gender, it can be seen that: first, gender is constructed by socio-culture, in other words, with the change of gender culture, it may change various problems caused by gender system, such as gender inequality. Second, gender perception is dynamic and changeable. Third, the essential characteristics of gender concepts such as social construction and changeability make it possible to promote the reform and liberation of gender concept through external environments and means. Exploring the influencing factors and influencing relationships of gender concepts can provide ideas for solving problems related to gender concept. Therefore, based on the theory of entrepreneurial events and gender theory, this study introduced family support into the model, using gender concept as a regulatory variable and tool to analyse and discuss the influencing factors and mechanisms of rural women's entrepreneurship willingness. The theoretical model included family emotional support, family instrumental support, perceptual desirability, perceptual feasibility, entrepreneurial intentions, and gender concepts, as shown in Figure 16.1.

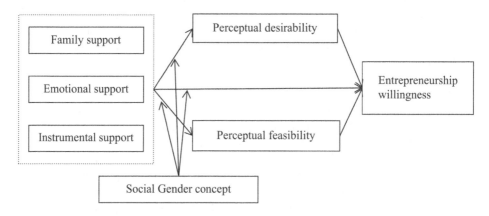

Figure 16.1   Theoretical model.

Hypothesis

Family Support Positively Affects Rural Women's Entrepreneurship Willingness

Rural women's entrepreneurship is distinct because of their "rural" and "female" identities. First of all, China is a traditional relation-oriented society. Compared with urban areas, rural social interactions are more family-oriented, and traditional concepts such as clan and local culture are strong; thus, family's emotional support for entrepreneurs is more important, which can provide strong spiritual support for entrepreneurs and motivate them to develop the will to start businesses and carry out entrepreneurial activities. Second, rural women's entrepreneurship often needs more financial support from their families. The regional constraints of living in rural areas make the source of income of rural women mainly from farming, which is susceptible to natural conditions and is unstable. The identity of rural women has a certain impact on their entrepreneurship willingness, and the idea of "son preference (prefer sons to daughters, which is rooted from the ancient notion that women are inferior to men)" in rural China is relatively serious (Li. & Cooney, 1993), and women are not supported to show up (the old role distribution "men outside, and women inside"). Third, women face gender discrimination in entrepreneurship. They are considered to be weak in risk-bearing ability, physical strength, human capital, and social capital are weak, so they are often considered unsuitable for entrepreneurship and not supported.

Dong and Zhao (2019) based on the sample of 4531 farmer entrepreneurs obtained from China's "Thousand Villages Survey," starting from family economic support and emotional support, studied the impact of family support on farmer entrepreneurs, and the research results showed that the economic support and family emotional support of families had an impact on farmers' entrepreneurship, and family economic support is an important financial support for farmers in the early stage of entrepreneurship(Dong & Zhao, 2019). Due to the higher levels of financial constraints in rural areas, farmers are more dependent on informal financing channels for entrepreneurship, and family members are a central component of this channel. Jiang and Guo (2012) argued that the experience of successful entrepreneurship among acquaintances will have a positive impact on farmers' entrepreneurship willingness, and the social relationships perceived by farmers could enhance their confidence in entrepreneurship and thus enhance their entrepreneurship willingness.

It can be seen from the literature that family support is not only reflected in economic funds but also includes emotional and relationship resources, which has an important impact on rural women's willingness to start a business. Therefore, the following assumptions are drawn:

H1a: Family emotional support positively affects rural women's willingness to start a tourism business.
H1b: Family instrumental support is positively influencing rural women's willingness to start a tourism business.

The Mediating Role of Perceptual Desirability and Perceptual Feasibility

Research have demonstrated the impact of family support on perceptual aspiration and perceptual feasibility. Krueger (1993) found that positive entrepreneurial experience sharing among family members had a significant impact on respondents' perceived eagerness. Edelman et al. (2016a) and Jaskiewicz et al. (2015) also showed that a cohesive family could become a source of emotional support for family members, and the exchange and sharing of entrepreneurial experience and information in a cohesive family could promote the accumulation of experience in different fields, which had an impact on the individual's perceived desire, thereby affecting their willingness to start a business (Edelman et al., 2016b; Jaskiewicz et al., 2015). Drennan et al. (2005) proved that parental entrepreneurship had a significant impact on children's perceived desire, and when children perceived positive or negative entrepreneurial experiences, they had a significant impact on children's perceived desire and perception feasibility. The findings of Linan and Santos (2007) showed that the presence of entrepreneurs in the family has a significant impact on perceived desire, and experienced entrepreneurial environment has a significant impact on perceived feasibility. Family support provides entrepreneurs with the space to make their own choices, gives them the freedom to take the initiative, and recognises their entrepreneurial views, thus promoting the passion for entrepreneurship. When there is a lack of emotional support such as encouragement, understanding, and attention from the family, it often leads to a decrease in entrepreneurial satisfaction and increased life pressure, so that it is difficult to identify the value of entrepreneurial activities from the heart (Parasuraman et al., 1996). Another research showed that family members' understanding, comfort, and love for entrepreneurs, as well as interest in entrepreneurial projects, and expressed their relational support for entrepreneurs (Tao & Li, 2018). With this support, encouragement and positive feedback from family members can enhance entrepreneurs' confidence in achieving their entrepreneurial goals. Therefore, the following assumptions are drawn:

H2a: Family emotional support positively affects the perceived desirability of rural women's tourism entrepreneurship.

H2b: Family emotional support positively affects the perceived feasibility of rural women's tourism entrepreneurship.

H2c: Family instrumental support positively affects the perceived desirability of rural women's tourism entrepreneurship.

H2d: Family instrumental support positively affects the perceived feasibility of rural women's tourism entrepreneurship.

It can be seen from past research that the mediating role of perceptual desirability and perceptual feasibility is not simple, and its role is different under different conditions. On the one hand, the entrepreneurship information and experience in family support can enrich the theoretical knowledge of rural women's entrepreneurship, establish entrepreneurial awareness, stimulate women's motivation for achievement, and make women more eager for entrepreneurship. On the other hand, the part of family support on financial and interpersonal resources can increase the confidence of rural women in entrepreneurship and sense the stronger feasibility. The desire for entrepreneurship and confidence in entrepreneurship can also stimulate women's entrepreneurial will. With the greater family support, the more it can induce their desire for entrepreneurship, need for achievement, and the confidence in entrepreneurship, thus generating a stronger entrepreneurial will. Accordingly, the following assumptions are drawn:

H3a: Perceptual desirability plays an intermediary role between family emotional support and rural women's willingness to start a tourism business.

H3b: Perceived feasibility mediates between family emotional support and rural women's willingness to start a tourism business.

H3c: Perceptual desirability plays an intermediary role between institutional support in the family and rural women's willingness to start a tourism business.

H3d: Perceived feasibility mediates between family instrumental support and rural women's willingness to start a tourism business.

The Regulatory Role of Gender Concepts

Gender concepts are used by sociologists to describe the group characteristics of gender roles, social responsibilities, division of activities, and other group characteristics formed by the socio-cultural environment in a specific society, which refers to people's specific views and views on the existence of gender in both sexes. From the perspective of traditional society and modern society, the concept of gender can be divided into two categories: first, the traditional gender concept constructed by in accordance with the patriarchal culture, the core of which is mainly embodied in "men are superior to women." Men are considered to be strong, independent, adventurous and rational, and women should be gentle, considerate, vulnerable, and emotional. It is the responsibility of men to work and support their families, and it is the duty of women to manage housework; the second is the modern social gender concept formed with the advancement of the modernisation, and its core view is "gender equality." There are still certain stereotypes in social and cultural traditions about the role of women, and they tend to position female roles in the family rather than in the shopping mall; especially in traditional Chinese culture, there has always been a subjective impression of "male outside, female inside" (Li, Wu, & Shi, 2023). Basco (2019) argued that families provided many unique resources for entrepreneurship and that families should be embedded in the entire process of women's entrepreneurship. Family support means family role sharing, spiritual and financial support, etc. When women's gender concepts are more traditional, they are more subject to the stereotype of women, that women should take care of children and complete housework at home. Under such conditions, family support plays an important role in the generation of women's entrepreneurial intentions, and the degree of family support is different. The degree of attractiveness of and confidence in self-entrepreneurship will vary with the degree of family support, that is, perceptual desirability and perceptual feasibility will change accordingly. The more traditional women's gender concept is, the more perceptual desirability and perceptual feasibility are influenced by family support. Vice versa. From this, the following assumptions are drawn:

H4a: Gender perceptions regulate family emotional support with perceived desirability. That is, the more traditional the concept of gender, the more the role between family emotional support and perceptual desirability is strengthened.

H4b: Gender perceptions mediate between emotional support in the family and perceived feasibility. That is, the more traditional the gender concept, the more the role between family emotional support and perceived feasibility is strengthened.

H4c: Gender perceptions mediate between instrumental support in the family and perceptual desirability. That is, the more traditional the concept of gender, the more strengthened the role between instrumental support of the family and perceptual desirability.

H4d: Gender perspectives mediate between instrumental support of the family and perceived feasibility. That is, the more traditional the gender concept, the more the role between institutional support of the family and perceived feasibility is strengthened.

Data Analysis

Research Tool and variable Measurement

In this study, instrumental support from family in entrepreneurship was divided into financial as well as social capital, mainly referring to the scale by Edelman et al. (2016). The measures of the emotional support were mainly based on the work-family support scale developed by Li and Zhao (2009). Finally, based on the work-family support scale of King et al. (2010), this study developed a family support scale in China (Table 16.1).

The measures of perceived desirability and perceived feasibility were mainly referred to Norris and Krueger's study (Krueger, 1993), and the specific measurement items are shown in Table 16.2.

The measurement of entrepreneurial intentions was mainly referred to in the study by Liñán et al. (2007). In their study, Liñán pointed out that, considering that it is also common for previous scholars to measure entrepreneurial intentions with a single question, that is, directly by asking respondents whether they were ready to start a business, among the seven questions set, one directly answered whether they have entrepreneurial intentions. However, the authors also pointed out that this question was mainly used for comparison

Table 16.1 Family support scale

Variable	Item	Main source
Emotional support	EFS1 Members of my family often provide a different way of looking at my work-related problems	Li and Zhao (2009)
	EFS2 When something at work is bothering me, members of my family show that they understand how I'm feeling	
	EFS3 When I'm having a difficult week at my job, my family members try to do more of the work around the house	
	EFS4 When I have a tough day at work, family members try to cheer me up	
	EFS5 If I have a problem at work, I usually share it with my family members	
	EFS6 When I have a problem at work, members of my family express concern	
Instrumental support	IFS1 My parents/family provide me with debt capital	Edelman et al. (2016a)
	IFS2 My parents/family provide me with equity capital	
	IFS3 the capital provided by my parents/family has favourable and flexible conditions	
	IFS4 My parents/family provide me with contracts to people that might help me with pursuing an entrepreneurial career	
	IFS5 My parents/family introduce me to business networks, providing contacts to potential business partners and/or customers	

Table 16.2 The perceived desirability and perceived feasibility scale

Variable	Item	Main source
Perceived desirability	PD1 I would love do it PD2 I would be tense PD3 I would be enthusiastic	Norris and Krueger et al. (Krueger, 1993)
Perceived feasibility	PF1 I think it would be hard PF2 I'm very certain of success PF3 I'm very overworked PF4 I know enough to start a business PF5 I'm very sure of myself	Norris and Krueger et al. (Krueger, 1993)

and not for validation of the questionnaire. Therefore, this item was removed from the questionnaire for this study. The remaining six questions were used to measure entrepreneurial intentions (see Table 16.3).

Furthermore, this study also used some scales from the third Survey of Chinese Women's Social Status. The survey is conducted every ten years and is a comprehensive social survey devoted to gender equality concepts, gender role norms, and attitudes (Jia & Ma, 2015). Table 16.4. provides detailed measurement items.

Table 16.3 The entrepreneurial intentions scale

Variable	Item	Main source
Entrepreneurial intentions	EI1 My professional goal is becoming an entrepreneur EI2 I will make every effort to start and run my own firm EI3 I have very seriously thought about starting a firm EI4 I'm determined to create a firm in the future EI5 I'm ready to make anything to be an entrepreneur EI6 I've got the firm intention to start a firm some day	Liñán et al. (2007)

Table 16.4 Gender equality perceptions scale

Variable	Item	Main source
Gender equality perceptions	XBGN1 The ability of women is not worse than men XBGN2 The proportion of men and women in leadership positions should be approximately equal XBGN3 The men should also take the initiative to undertake household chores XBGN4 The development of the husband is more important than the development of the wife XBGN5 It is better to marry up than to do a good job XBGN6 Men should give priority to society and women should give priority to their families XBGN7 Earn money to support the family is mainly a matter of men	The third survey of Chinese Women's Social Status

Sample

In the formal research of this study, the demographic variables of the sample are shown in Table 16.5, mainly including age, occupation, marital status, income, and education.

Hypothesis Verification

Overall Structural Equation Model Testing

AMOS24.0 was used to draw the model and test the data. First, the fitting degree of the initial model should be verified. When the mean square and square root of progressive residual RMSEA are less than 0.08, it is acceptable, and less than 0.05 is good; the smaller the CMIN, the better; SMRM (Standardised root mean square residual) is better than 0.5. When the values of GFI, NFI, RFI, IFI, TLI, and CFI are generally greater than 0.9, the fitting degree of the model is good. It can be seen from Table 16.6 that the SRMR value of the modified model is 0.0439, less than 0.5; RMSEA value is 0.070, less than 0.08, good; the values of GFI, NFI, RFI, IFI, TLI, and CFI are all greater than 0.9, indicating that the modified model has a good fit. The model built in this study is shown in Figure 16.2.

Table 16.5 Demographic variables of the sample

Demographic variables	Classification	Frequencies	Percent	demographic variables	Classification	Frequencies	Percent
Age	18 and below	37	9.1	Marital status	Married and child-bearing	222	54.4
	19–24	78	19.1		Married and non-fertility	55	13.5
	25–34	105	25.7		Unmarried	121	29.7
	35–44	143	35.0		Divorced	4	1.0
	45–54	36	8.8		Others	6	1.5
	55–64	8	2.0	Income level	≤2,500	187	45.8
	64 and above	1	0.2		2,501–5,000	153	37.5
Occupation	Civil servants	21	5.1		5,001–10,000	52	12.7
	Technology professionals	20	4.9		10,001–20,000	11	2.7
	Business and service personnel	126	30.9		≥20,001	5	1.2
	Teachers	22	5.4	Education level	Middle school and below	131	32.1
	Soldier	3	.7		High school or technical secondary school	111	27.2
	Students	96	23.5		Junior college	87	21.3
	Farmers	17	4.2		Undergraduate	69	16.9
	Fishermen	8	2.0		Graduate degree or above	10	2.5
	Others	95	23.3				

Table 16.6 Total model fit index

Model fit index	CMIN	CMIN/DF	GFI	NFI	RFI	IFI	TLI	CFI	SRMR	RMSEA	P
Initial model	729.100	5.099	.852	.879	.855	.900	.880	.900	.0917	.100	0
Modified model	412.453	2.967	.903	.931	.916	.953	.942	.953	.0439	.070	0
Advice standard		[1,5]	>0.9	>0.9	>0.9	>0.9	>0.9	>0.9	<0.05	<0.08	<0.05

Hypothesis Verification of the Relationship between Family Support and Perceived Desirability, Perceived Feasibility, Entrepreneurial Intentions

The results showed that family emotional support had effect on perceived desirability (p < 0.001), in support of H2a. Family instrumental support had effect on perceived feasibility (p < 0.001), in support of H2d. Similarly, family emotional support had effect on perceived feasibility(p < 0.001), in support of H2b. Family instrumental support had effect on perceived desirability(p < 0.001), in support of H2c. Furthermore, p-value is less than 0.001, perceived desirability had an impact on entrepreneurial intentions; p-value is less than 0.001, perceived feasibility had an impact on entrepreneurial intentions. The T-values are all positive, indicating that the effects are all positive (see Table 16.7).

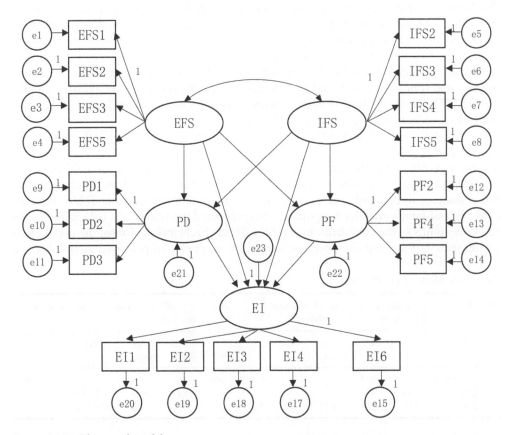

Figure 16.2 The tested model.

Table 16.7 Hypothesis verification of the relationship between family support and perceived desirability, perceived feasibility, entrepreneurial intentions

Research path		Path coefficient β	S.E.	T-value	P
Family emotional support	Perceived desirability	0.462	0.092	5.029	***
Family instrumental support	Perceived feasibility	0.252	0.059	4.303	***
Family emotional support	Perceived feasibility	0.433	0.081	5.377	***
Family instrumental support	Perceived desirability	0.392	0.069	5.663	***
Perceived desirability	Entrepreneurial intentions	0.485	0.087	5.58	***
Perceived feasibility	Entrepreneurial intentions	0.755	0.109	6.933	***

Verification of Mediating Effect

To test the mediating effect of perceived desirability and perceived feasibility on relationship between family support and entrepreneurial intentions, it is first necessary to test the direct effect of family support on entrepreneurial intentions based on two dimensions: emotional support and instrumental support, i.e., the existence of main effects. As shown in Table 16.7, the p-value was lower than 0.05 and the 95% confidence intervals (CI) did not include zero. Thus, H1a and H1b were supported. The standardised coefficient of family instrumental support was 0.4, and the standardised coefficient of family affective support was 0.22. The larger the standardised coefficient was, the greater the influence of the independent variable on the dependent variable was.

Model 14 in the PROCESS in SPSS 23.0 was conducted to verify whether there is a mediating role of perceived desirability and perceived feasibility on relationship between family support and entrepreneurial intentions. The upper and lower limits of the confidence interval did not include zero, indicating that p-value is less than 0.05 and the mediating effect was considered to be significant, and vice versa. The results are shown in Table 16.8., the total effect of family affective support on entrepreneurial intentions was 0.289. The upper and lower limits of the confidence interval for direct effects included zero, the p-value was greater than 0.05 and the effect was not significant, meaning that perceived desirability and

Table 16.8 Main effect regression

DV	IV	Unstandardised coefficients		Standardisation coefficient	t	sig.	B 95.0% confidence intervals		Collinearity	R²
		B	SE	Beta			Upper limits	Lower limits	VIF	
EI	(constant)	1.120	.277		4.047	.000	.576	1.664		.310
	EFS	.289	.065	.222	4.435	.000	.161	.418	1.475	
	IFS	.428	.054	.400	7.979	.000	.322	.533	1.475	

Table 16.9 Verification of the mediating role of perceived desirability and perceived feasibility

Path				β	SE	T	P	BootLLCI	BootULCI
EFS	→	EI	Total effect	0.289	0.065	4.435	0	0.161	0.418
			Direct effect	0.004	0.045	0.087	0.931	−0.084	0.092
			PD	0.120	0.032			0.066	0.189
			PF	0.165	0.043			0.085	0.254
IFS	→	EI	Total effect	0.428	0.054	7.979	0	0.322	0.533
			Direct effect	0.107	0.038	2.838	0.005	0.033	0.181
			PD	0.114	0.027			0.069	0.177
			PF	0.207	0.046			0.122	0.302

perceived feasibility are completely mediating role between family affective support and entrepreneurial intention; thus, H3a and H3b were supported. The total effect of family instrumental support on entrepreneurial intentions was 0.428. The upper and lower limits of the confidence interval for direct effects did not include zero, the p-value was less than 0.05, and the effect was significant, indicating that perceived desirability and perceived feasibility had partial mediation effects on the relationship between family instrumental support and entrepreneurial intentions. The path coefficients of perceived desirability and perceived feasibility are 0.114 and 0.207, respectively (see Table 16.9). These findings provided support for hypotheses H3c and H3d.

Verification of the Regulatory Role of Social Gender Concept

The regulatory role of social gender concept was tested through model 11 in the PROCESS in SPSS 23.0. The results were as shown in Table 16.10. The upper and lower limits of the confidence interval did not include zero and p-value is less than 0.05 in inspecting the regulating effect of social gender concept, indicating that social gender concept played a positive regulating role between family emotional support and perceived desirability, perceived feasibility, and entrepreneurial intentions, respectively. H4a, H4b, and H5a were supported. Social gender concept regulates the relationship between family instrumental support and perceived desirability, perceived feasibility, and entrepreneurial intentions, in support of H4c, H4d, and H5b.

As can be seen from Figure 16.3, the fitting curves for family emotional support and perceived feasibility were flatter when gender perceptions were at low levels, and steeper when gender perceptions were at high levels. This indicates that when rural women have traditional gender perceptions and are more conservative in their thinking, the more obvious the influence

Table 16.10 Verification of the regulatory role of social gender concept

Path	Interaction term	coeff	se	t	p	LLCI	ULCI
EFS PD	XBGN EFS	0.098	0.028	3.502	0.001	0.043	0.152
EFS PF	XBGN EFS	0.104	0.029	3.581	0.000	0.047	0.160
EFS EI	XBGN EFS	0.069	0.033	2.054	0.041	0.003	0.134
IFS PD	XBGN IFS	0.121	0.022	5.465	0.000	0.077	0.164
IFS PF	XBGN IFS	0.132	0.030	4.486	0.000	0.074	0.191
IFS EI	XBGN IFS	0.070	0.027	2.611	0.009	0.017	0.123

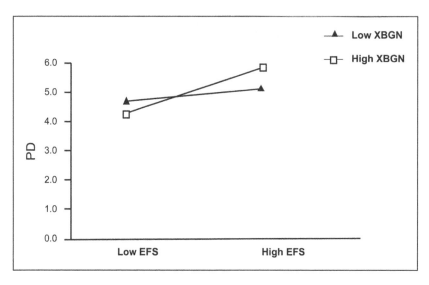

Figure 16.3 The regulatory role of social gender concept on the relationship between family emotional support and perceived desirability.

of family emotional support on their perceived feasibility, and vice versa. Thus, it can be seen that gender perceptions have a positive role in regulating the relationship between family emotional support and perceived feasibility, i.e., the more traditional gender perceptions are, the more the positive effect of family emotional support on perceived feasibility is enhanced.

As shown in Figure 16.4, the fitting curve for family emotional support and entrepreneurial intentions were flatter when gender perception was at lower levels, and steeper

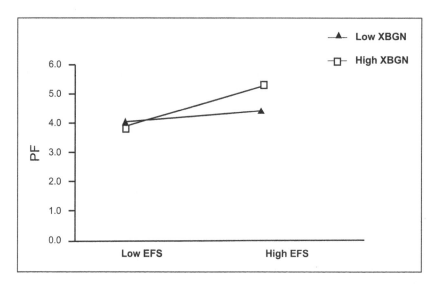

Figure 16.4 The regulatory role of social gender concept on the relationship between family emotional support and perceived feasibility.

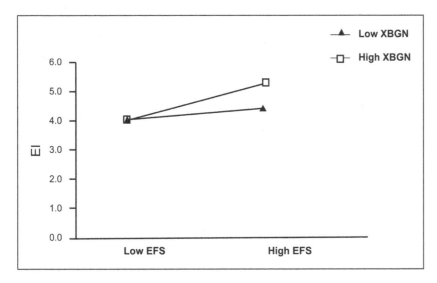

Figure 16.5   The regulatory role of social gender concept on the relationship between family emotional support and entrepreneurial intentions.

when gender perceptions at higher levels. This indicates that when rural women have more traditional social gender concept and conservative ideological concept, the impact of family emotional support on their entrepreneurial intention is more obvious, and vice versa. It can be concluded that gender concept has a moderating effect on the positive impact of family affective support on entrepreneurial intention. The more traditional gender concept, the more positive impact of family affective support on entrepreneurial intention.

As can be seen from Figure 16.5, when the social gender concept is at a low level, the fitting curve of family instrumental support and perceived desirability is relatively gentle, while when the social gender concept is at a high level, the fitting curve of family instrumental support and perceived desirability is relatively steep. This indicates that when rural women have more traditional and conservative social gender concepts, the impact of family instrumental support on their perceived desirability is more obvious, and vice versa. It can be seen that social gender concept has a certain moderating effect on the positive impact of family instrumental support on perceived desirability, that is, the more traditional social gender concept, the more positive impact of family instrumental support on perceived desirability.

As can be seen in Figure 16.6, the fitted curves for instrumental family support and perceived feasibility are flatter when gender perceptions are at lower levels and steeper when gender perceptions are at higher levels. This suggests that when rural women have a more traditional gender perspective and a more conservative mindset, the more obvious the influence of instrumental family support on their perceived viability, and vice versa. Thus, it is clear that gender perception moderates the positive effect of family instrumental support on perceived feasibility, i.e., the more traditional the gender concept, the more it can strengthen the positive effect of family instrumental support on perceived feasibility.

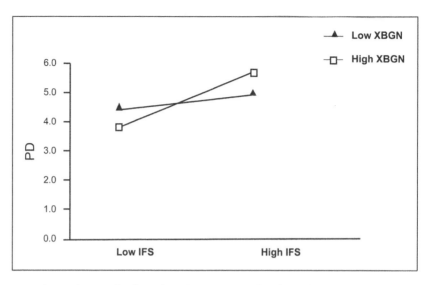

Figure 16.6 The regulatory role of social gender concept on the relationship between family instrumental support and perceived desirability.

Findings

Combined with Table 16.11, we can see that the main findings of this study are as follows:

Table 16.11 Hypothesis testing

Path		Results
Main effect	Family emotional support entrepreneurial intentions	H1a was supported
	Family instrumental support entrepreneurial intentions	H1b was supported
	Family emotional support perceived desirability	H2a was supported
	Family emotional support perceived feasibility	H2b was supported
	Family instrumental support perceived desirability	H2c was supported
	Family instrumental support perceived feasibility	H2d was supported
Mediating effect	Family emotional support perceived desirability entrepreneurial intentions	H3a was supported, full mediation
	Family emotional support perceived feasibility entrepreneurial intentions	H3b was supported, full mediation
	Family instrumental support perceived desirability entrepreneurial intentions	H3c was supported, partial mediation
	Family instrumental support perceived feasibility entrepreneurial intentions	H3d was supported, partial mediation
Regulating effect	Family emotional support perceived desirability	H4a was supported
	Family emotional support perceived feasibility	H4b was supported
	Family instrumental support perceived desirability	H4c was supported
	Family instrumental support perceived feasibility	H4d was supported
	Family emotional support entrepreneurial intentions	H5a was supported
	Family instrumental support entrepreneurial intentions	H5b was supported

The Influence of Family Emotional Support on Perceived Desirability and Perceived Feasibility Is More Significant

In the process of entrepreneurial intention generating the emotional support and encouragement provided by the family is more important for rural women. The financial and network support provided by the family also affects rural women's entrepreneurial intention to a lesser extent than the emotional and attitudinal support. With the emotional encouragement of family and the support of money and contacts, etc., the individual's perception of the feasibility and attractiveness of tourism entrepreneurship will be enhanced, followed by entrepreneurship desire. Therefore, with women becoming the main force in tourism in rural areas, it is not only necessary to pay attention to the entrepreneurs themselves, i.e. rural women, but also to their family environment, like guiding and encouraging families to give more understanding and support to women entrepreneurs etc.

The Family Support Impact on Entrepreneurial Intention through the Intermediary Effect of Multipath of Perceived Desirability and Perceived Feasibility

Family affective support positively influences entrepreneurial intentions, and perceived desirability and perceived feasibility have a complete mediating effect, consistent with the findings of Edelman et al. (2016b), Jaskiewicz (2015), and Tao and Li (2018). The results indicate that family affective support have an effect on tourism entrepreneurial intentions relying on individual perception processes of desirability and feasibility in the model. Family instrumental support positively influences entrepreneurial intentions through partial mediation of perceived desirability and perceived feasibility. In the process of family support influencing entrepreneurial intentions, it is more difficult to establish a direct link between family emotional support and entrepreneurial intentions; perceptions of feasibility and desirability are needed before entrepreneurial intentions can be influenced. In contrast, instrumental support such as family capital and connections can act directly on rural women's entrepreneurial intentions, as well as through perceptions of feasibility and desirability.

Gender Perception Is an Important Moderating variable Affecting Rural Women's Entrepreneurial Intentions

This study introduced gender perceptions into the research model of tourism entrepreneurial intentions, taking into account the specificity of rural women's family environment. The results verified there was a positive and significant moderating effect of gender perceptions on the process of the effect of family support on perceived feasibility, perceived desirability, and entrepreneurial intentions. The study shows the more traditional the gender concept of rural women is, the more obvious is the impact of family support on perceived desirability, perceived feasibility, and entrepreneurial intention. It indicates that in the process of generating rural women's entrepreneurial intention, when the more traditional their social gender perceptions are and the more backward their perceptions of the gender roles are, the more they rely on their families' emotional understanding, attitudinal encouragement, and financial and networking support, etc., so that they will have a stronger entrepreneurial intention. On the contrary, the more modern the gender concept of rural women is and the more advanced the perception of gender roles is, the stronger the individual's sense of independence is, and the more the creation of entrepreneurs will depend on their own considerations and decisions. Therefore, in the process of generating rural women's willingness to start a business, it is necessary to pay attention to the content of gender concepts and gender role perceptions and to focus on the cultivation of modern social gender concepts.

Discussion and Conclusion

Theoretical Implications

This study contributes to women's entrepreneurship and rural leadership literature in several significant and meaningful ways.

First, this study highlights rural women's dual disadvantaged identity (namely rural identity and gender identity), which limited their entrepreneurship knowledge, experience, and interest. Previous studies predominantly approached rural tourism entrepreneurship from rural environment in the context of Western developed or developing countries, whereas the dual structure of urban and rural China is very different from that of foreign countries (Liu, 2016). To be specific, the urban-rural disposable income gap in China is one of the highest gaps in the world (Sicular et al., 2010). Furthermore, the rural-urban migration and *Hukou* system (the official form of identity based primarily on place of birth) in China made children, wives, and elderly parents composing the bulk of the left-behind population, while husbands moving into cities for employment (Li, 2010). China rural women are considered under double burdens of production work and housework, double dependencies on men for economic and emotional support (Zhao, & Guo, 2020). All these reasons mentioned earlier plus rural existing gender ideologies and the rigidity of prescribed gender roles limited rural women's social capital, resource, and self-efficacy to be entrepreneurs (Zhao et al., 2011). Therefore, it is important to acknowledge that family support plays a much more significant role in rural women entrepreneurship than in their Western counterparts (Bensemann & Hall, 2010; Hisrich & Fan, 1991). Their gendered and cultured identity has much relevance to their entrepreneurial intention, particularly in cultures with a strong Confucian influence.

Second, previous studies have mainly focused on the conflict and balance between the family and work roles of female entrepreneurs, with less discussion on the positive impact of families on women's entrepreneurship (Neneh, 2021). This study reveals that women's family role can also gain with respect to their entrepreneurship. To be specific, female entrepreneurs build social networks through family activities to obtain specific channels of funding, resources, and information support (Constantinidis et al., 2019), which is beneficial for them to better overcome the difficulties in the early stages of enterprise survival and achieve entrepreneurial opportunities. Female entrepreneurs share the hardships and joys of entrepreneurship through family activities, win the attention and encouragement of family members, and enhance their confidence in entrepreneurship. Moreover, family members participating in their entrepreneurial discussions and providing them with spiritual support are beneficial for female entrepreneurs to overcome the uncertainty of the external environment; the professional skills of family members can also help female entrepreneurs improve the overall efficiency of entrepreneurial opportunity development to a certain extent (Hannah, 2020). In addition, female entrepreneurs have more work time in the division of family roles due to the sharing of household chores by family members, which is beneficial for them to search for entrepreneurial information and discover hidden business opportunities in the context. These findings might contribute to work-family relation research of female entrepreneurs from a positive perspective.

Third, this research combines ESM by Turker (2009) and theory of the entrepreneurial event model by Shapero and Sokol (1982). On the one hand, this study verifies the multi-path mesomeric effect between family support and entrepreneurial willingness. Based on the entrepreneurial event model, the process of family support influencing entrepreneurial willingness through perceived desirability and perceived feasibility was verified. The research results show that the mediating degree of family support in the two dimensions is different, and in future research, attention should be paid to the distinction and degree of influence between the two dimensions. On the other hand, this study introduces the variable of family support in the context of rural women's tourism entrepreneurship and analyses the factors that affect women's entrepreneurial willingness in the context of rural tourism development at a more micro level. To some extent, it has expanded the entrepreneurial event model and enriched relevant theoretical systems such as rural tourism and entrepreneurial willingness. Studying the role of family support as a separate variable in the generation of women's willingness to travel and entrepreneurship has also enriched relevant research on family support. Finally, this study verified the moderating effect of gender perspectives on the process of rural women's willingness to engage in tourism entrepreneurship. The results of this study indicate that in the context of rural tourism entrepreneurship, the generation of entrepreneurial willingness among rural women with modern gender perspectives is more dependent on individual planning and decision-making, rather than relying on family. Rural women with more traditional gender perspectives have a stronger willingness to start a business when they receive more family support. The more modern the gender concept, the stronger the independent consciousness women exhibit, and their personal entrepreneurial willingness is more influenced by personal intentions and decisions, rather than relying on support from various aspects of the family to drive it.

Practical Implications

In the current employment guidance work for women, most of the focus is on the entrepreneurial subject itself, while ignoring the impact of the potential entrepreneurial subject's family support environment on their entrepreneurial willingness. From the results of this study, it can be seen that whether in the practical work of tourism entrepreneurship or in promoting ideological liberation, the family has special significance for women, especially rural women. When encountering disagreements, troubles, and fatigue during the entrepreneurial process, emotional support, such as family sharing, encouragement, and understanding, as well as possible financial and social support, all ensure the smooth progress of entrepreneurial activities to a certain extent and become important factors affecting individuals' willingness to start a business. Therefore, to stimulate the entrepreneurial willingness of rural women, it is necessary not only to pay attention to rural women themselves but also to instil ideas, consciousness, and ideas into the family members of potential female entrepreneurs, encourage them to share family affairs, support them to create new chances for their own career, and give them some emotional understanding and financial support.

Rural women generally have low economic income due to their low education level, lack of skills, limited job opportunities, and limited geographical environment. The lack of entrepreneurial funds has become a driving force in the cradle that stifles their entrepreneurial willingness. From the empirical results of this study, it can also be seen that instrumental support such as household funds can affect entrepreneurial willingness through multiple pathways. Therefore, providing certain financial and policy support is the key to stimulating rural women's willingness to start tourism businesses.

The transformation of gender concepts from traditional to modern requires the joint efforts of women themselves and the overall environment, in order to truly promote women's ideological liberation work in place. According to the traditional and modern gender perspectives of rural women, targeted guidance should be provided for their entrepreneurship. Faced with women with modern gender perspectives, due to their more independent self-awareness, they can provide more entrepreneurial information and opportunities and more active entrepreneurial guidance in a timely manner, enabling them to perceive the needs and possibilities of entrepreneurship from the outside to the inside, thus stimulating stronger entrepreneurial willingness. In the face of women with traditional social values, the first thing to do is to cultivate their sense of independence and autonomy, and women can also go out of their homes and have their own careers. Through the popularisation of knowledge, we guide them to abandon outdated gender concepts, encourage women to go out of their homes, cultivate independent personalities and protagonist awareness, and regard themselves as the main body of entrepreneurship, in order to encourage everyone to invest in the cause of entrepreneurship.

Limitations

Firstly, although the sample collection of research data meets the required standards, considering that the larger the sample size, the better the accuracy of the study, it can be achieved by increasing the sample size to get better research effectiveness and more accurate research conclusions. Secondly, this study only introduced an antecedent variable of family support in the entrepreneurial event model, and family support was only divided into two dimensions. More influencing factors and antecedent variables can be considered to explore other possible influencing factors. Finally, this study only considers family support as the antecedent variable to study the impact of entrepreneurial intention. It can be attempted to use family support as a moderating variable to study the moderating effect of other antecedent variables on entrepreneurial intention.

Declaration

This article is a phased achievement of the Fujian Social Science Foundation project (FJ2022B053) and Philosophy and Social Sciences Project at Huaqiao University (22SKGC-QG02).

References

Abdullah, N. B., Zainudin, A. Z., & Idris, N. H. (2019). Review on the manifestations of rural hollowing. *Asia Proceedings of Social Sciences*, 4(2), 143–145. DOI:10.31580/apss.v4i2.767

Akpinar, N., lkden Talay, Ceylan, C., & Gündüz, S. (2005). Rural women and agrotourism in the context of sustainable rural development: A case study from Turkey. *Environment Development & Sustainability*, 6(4), 473–486. doi: 10.1007/s10668-005-5633-y

Arunachalam, P. (2014). Women empowerment and entrepreneurship through self-help groups. *International Journal of Physical & Social Sciences*, 12(4), 187–195. https://www.indianjournals.com/ijor.aspx?target=ijor:ijpss&volume=4&issue=12&article=013

Basco, R. (2019). What kind of firm do you owner-manage? An institutional logics perspective of individuals' reasons for becoming an entrepreneur. *Journal of Family Business Management*, 9(3), 297–318. https://doi.org/10.1108/JFBM-09-2018-0032

Bensemann, J., & Hall, C. M. (2010). Copreneurship in rural tourism: Exploring women's experiences. *International Journal of Gender and Entrepreneurship*, 2(3), 228–244. doi: 10.1108/17566261011079224

Cantallops, A. S., Cardona, J. R., & Muntaner, R. E. (2015). Characteristics and peculiarities of rural tourism in the Balearic Islands. *Tourism Planning & Development*, 12(2), 125–144. doi: 10.1080/21568316.2014.933121

Constantinidis, C., Lebègue, T., Abboubi, M. E., & Salman, N. (2019). How families shape women's entrepreneurial success in morocco: An intersectional study. *International Journal of Entrepreneurial Behaviour & Research*, 25(8), 1786–1808. doi: 10.1108/IJEBR-12-2017-0501

Croll, E. (1985). Women and rural development in China: production and reproduction. *International Labour Office*. doi: 10.1080/02255189.1995.9669594

Dan, K. H., Wiklund, J., Anderson, S. E., & Coffey, B. S. (2016). Entrepreneurial exit intentions and the business-family interface. *Journal of Business Venturing*, 31(6), 613–627. doi: 10.1016/j.jbusvent.2016.08.001

Dong, L., Huang, P., & Luo, J. (2017). Demands and supporting policies of rural tourism enterprises in rural tourism destinations: Based on the investigation of Zhouwu village in Dayu country. *Journal of Central South University of Forestry & Technology (Social Sciences)*, 11, 76–83. doi: 10.14067/j.cnki.1673-9272.2017.05.013

Dong, J., & Zhao, C. (2019). Influence of family support on Farmers' Entrepreneurial motivation: A substitution of interpersonal relationship. *Chinese Journal of Population Science*, 1, 61–75. CNKI:SUN:ZKRK.0.2019-01-006

Drennan, J., Kennedy, J., & Renfrow, P. (2005). Impact of childhood experiences on the development of entrepreneurial intentions. *International Journal of Entrepreneurship and Innovation*, 6(4), 231–238.

Edelman, L. F., Manolova, T. S., Shirokova, G., & Tsukanova, T. (2016a). Student entrepreneurship in emerging markets: Can family help overcome the institutional voids? *Academy of Management Annual Meeting Proceedings*, 2(1), 11966.

Edelman, L. F., Manolova, T., Shirokova, G., & Tsukanova, T. (2016b). The impact of family support on young entrepreneurs' start-up activities. *Journal of Business Venturing*, 31, 428–448. doi: 10.1016/j.jbusvent.2016.04.003

Enemuo, J., Udeze, C., & Ugwu, J. N. (2019). Rural women empowerment and entrepreneurship development: Leveraging the gender factor. *International Journal of Finance & Economics*, 14(3), 43–59.

Figueroa-Domecq, C., de Jong, A., & Williams, A. M. (2020). Gender, tourism & entrepreneurship: A critical review. *Annals of Tourism Research*, 84, 1–13.doi:10.1016/j.annals.2020.102980

Fitzsimmons, J. R., & Douglas, E. J. (2011). Interaction between feasibility and desirability in the formation of entrepreneurial intentions. *Journal of Business Venturing*, 26, 431–440.doi:10.1016/j.jbusvent.2010.01.001

Gallin, R. S., & Bossen, L. (2004). Review of Chinese women and rural development: Sixty years of change in lu village, Yunnan. *The China Journal*, 51(51), 147–149. doi: 10.2307/3182153

Gasson, R., & Winter, M. (1992). Gender relations and farm household pluriactivity. *Journal of Rural Studies*, 8(4), 387–397. doi: 10.1016/0743-0167(92)90052-8

Getz, D., & Carlsen, J. (2000). Characteristics and goals of family and owner-operated businesses in the rural tourism and hospitality sectors. *Tourism Management*, 21(6), 547–560. doi: 10.1016/S0261-5177(00)00004-2

Hannah, C. A. (2020). *Making of a women entrepreneur: Understanding the fundamental role of family* [Doctoral dissertation, Tata Institute of Social Sciences].doi:10.13140/RG.2.2.17458.84162

Henderson, R., & Robertson, M. (2000). Who wants to be an entrepreneur? young adult attitudes to entrepreneurship as a career. *Career Development International*, 5(6), 279–287.

Hisrich, R. D., & Fan, Z. (1991). Women entrepreneurs in the People's Republic of China: An exploratory study. *Journal of Managerial Psychology*, 6(3), 3–12. doi: 10.1108/02683949110144855

Jaskiewicz, P., Combs, J. G., & Rau, S. B. (2015). Entrepreneurial legacy: Toward a theory of how some family firms nurture transgenerational entrepreneurship. *Journal of Business Venturing*, 30(1), 29–49. doi: 10.1016/j.jbusvent.2014.07.001

Jaskiewicz, P., Luchak, A. A., Oh, I. S., & Chlosta, S. (2015). Paid employee or entrepreneur? how approach and avoidance career goal orientations motivate individual career choice decisions. *Social Science Electronic Publishing*, 43(4), 349–367.

Jia, L., Liu, Z. J., Zhang, L. X., Fang, Z. Q., & Qin, M. N. (2018). The experiences of Japan's rural revitalization and its implications to China. *Research of Agricultural Modernization*, 39(3), 359–368. doi: 10.13872/j.1000-0275.2018.0032

Jia, Y., & Ma, D. (2015). Changes in a gender perspective from multifaceted perspective: The case with men dominating the outside while women dominating the inside of households. *Journal of Chinese Women's Studies*, 5, 29–36.

Jiang, J., & Guo, H. (2012). Entrepreneurship atmosphere, social networks, and farmers' entrepreneurial intentions. *China Rural Survey*, 2, 20–27.

Johnson, P. A. (2010). Realizing rural community-based tourism development: Prospects for social economy enterprises. *Journal of Rural & Community Development* (1): 150–162. https://journals.brandonu.ca/jrcd/article/download/349/81

Khanna, P., & Fellow, F. N. (2010). How secure is national rural employment guarantee as a safety net (working paper). https://global-labour-university.org/wp-content/uploads/fileadmin/GLU_conference_2010/papers/52._How_Secure_is_National_Rural_Employement_Guarantee_as_a_Safety_Net.pdf

King, L. A., Mattimore, L. K., & Adams, K. (2010). Family support inventory for workers: A new measure of perceived social support from family members. *Journal of Organizational Behavior*, 16(3), 235–258. doi: 10.1002/job.4030160306

Krueger, N. (1993). The impact of prior entrepreneurial exposure on perceptions of new venture feasibility and desirability. *Entrepreneurship Theory and Practice*, 18, 5–21. doi: 10.1177/104225879301800101

Krueger, N. F. Jr, & Brazeal, D. V. (1994). Entrepreneurial potential and potential entrepreneurs. *Entrepreneurship Theory and Practice*, 18(3), 91–104. doi: 10.1177/104225879401800307

Krueger, N., Reilly, M. D., & Carsrud, A. L. (2000). Competing models of entrepreneurial intentions. *Journal of Business Venturing*, 15(5), 411–432. doi: 10.1016/S0883-9026(98)00033-0

Li., S. (2001). Rural women: Employment and income – An empirical analysis based on the data from sample villages. *Social Sciences in China*, 3, 56–71. doi: 10.3969/j.issn.1001-862X.2019.01.009

Li., J., & Cooney, L. (1993). Son preference and the one child policy in China: 1979–1988. *Population Research & Policy Review*, 12, 277–296. doi: 10.1007/BF01074389

Li, W., & Li, S. (2010). The institutional cause of the problem of left behind women – The registered residence system and the dual division of urban and rural areas. *Journal of Hubei University of Economics(Humanities and Social Sciences)*, 7(4), 21–22.

Li, X., Wu, H., & Shi, Y. (2023). Gender roles and the household consumption structure: A study based on the China family panel studies data, *Journal of Central University of Finance and Economics*, 8, 73–90.

Li, Y., & Zhao, N. (2009). Structure and Measurement of work-family support and its moderation effect. *Acta Psychologica Sinica*, 14(9), 863–874.

Liñán F., & Chen Y. W. (2006). Testing the entrepreneurial intention model on a two-country Sample. *Working Paper*, 7(06).

Liñán, F., & Santos, F. (2007). Does social capital affect entrepreneurial intentions? *International Advances in Economic Research*, 13, 443–453. https://doi.org/10.1007/s11294-007-9109-8

Liu, H. (2005). Thinking about the connotation of rural tourism. *Journal of Sichuan Teachers College (Social Science)*, 2, 15–18. doi: 10.16246/j.cnki.51-1674/c.2005.02.008

Liu, T. (2016). A study on the psychological obstacles of female college students returning to their hometowns for entrepreneurship from a gender perspective. *Shandong Higher Education*, 4(09), 20–26.

Luo, M. (2011). A review of research on factors influencing farmers' entrepreneurship. *Economic Dynamics*, 8, 133–136.

Madanaguli, A. T., Kaur, P., Bresciani, S., & Dhir, A. (2021). Entrepreneurship in rural hospitality and tourism. A systematic literature review of past achievements and future promises. *International Journal of Contemporary Hospitality Management*, 33(8), 2521–2558. doi: 10.1108/IJCHM-09-2020-1121

Matriano, M. T., & Kiyumi, H. (2021). A critical study on entrepreneur's perception on sme's opportunities and challenges: Case of Oman. *Advances in Social Sciences Research Journal*, 8(1), 421–455. doi: 10.14738/assrj.81.9630

Mcgehee, N. G., Kim, K., & Jennings, G. R. (2007). Gender and motivation for agri-tourism entrepreneurship. *Tourism Management*, 28(1), 280–289. doi: 10.1016/j.tourman.2005.12.022

Muñoz-Fernández, G. A., Rodríguez-Gutiérrez, P., & Santos-Roldán, L. (2016). Entrepreneurship in higher education in tourism, gender issue? *Electronic Journal of Research in Education Psychology, 14*, 45–66. doi: 10.14204/ejrep.38.15040

Neneh, B. N. (2021). Role salience and the growth intention of women entrepreneurs: Does work-life balance make a difference? *The Spanish Journal of Psychology, 24*(4), 1–16. doi: 10.1017/SJP.2021.9

Nilsson, P. Å. (2002). Staying on farms: An ideological background. *Annals of Tourism Research, 29*(1), 7–24. doi: 10.1016/S0160-7383(00)00081-5

Obschonka, M., Schmitt-Rodermund, E., & Terracciano, A. (2014). Personality and the gender gap in self-employment: A multi-nation study. *PLoS One, 9*(8), e103805. doi: 10.1371/journal.pone.0103805

Parasuraman, S., Purohit, Y. S., Godshalk, V. M., & Beutell, N. J. (1996). Work and family variables, entrepreneurial career success, and psychological well-being. *Journal of Vocational Behavior, 48*(3), 275–300. doi: 10.1006/jvbe.1996.0025

Pruett, M., Shinnar, R., Toney, B., Llopis, F., & Fox, J. (2009). Explaining entrepreneurial intentions of university students: A cross-cultural study. *International Journal of Entrepreneurial Behavior & Research, 15*(6), 571–594. doi: 10.1108/13552550910995443

Shapero, A. & Sokol, L. (1982). The social dimensions of entrepreneurship. In C.A. Kent, D.L. Sexton, & K.H. Vesper (Eds.), *Encyclopedia of Entrepreneurship. Englewood Cliffs*, 72–90. NJ: Pretice-Hall.

Schlaegel, C., & Koenig, M. (2014). Determinants of entrepreneurial intent: A meta–analytic test and integration of competing models. *Entrepreneurship Theory and Practice, 38*(2), 291–332. doi: 10.1111/etap.12087

Scott, J.(1988). *Gender and the politics of history.* (Gender and culture). Columbia University Press.

Shen, H., Zhang, Y., & Zhang, J. (2020). The ecological environment optimization mechanism of rural tourism community based on "Two Mountains Theory". *3rd International Conference on Advances in Management Science and Engineering (IC-AMSE 2020).* doi: 10.2991/aebmr.k.200402.026

Shui, G., & Liu, Y. (2020). Rational looking at the ecological environment of female talent on the context of rural. *Chongqing Social Sciences, 8*, 136–144. doi: 10.19631/j.cnki.css.2020.008.012

Sicular, T., Yue, X., Gustafsson, B., & Li, S. (2010). The urban–rural income gap and inequality in China. *Review of Income and Wealth, 53*(1), 93–126.

Spivack, A. J., & Desai, A. (2016). Women entrepreneurs' work-family management strategies: A structuration theory study. *International Journal of Entrepreneurship and Small Business, 27*(2–3), 169–192.

Tao, Y., & Li., Y. (2018). A family embeddedness and gross-domain perspective on development mechanism of entrepreneurial passion. *Chinese Journal of Management, 15*, 1810–1818. doi: 10.3969/j.issn.1672-884x.2018.12.008

Turker, C. (2009). *Evidential support, reliability, and Hume's problem of induction.* Blackwell Publishing Ltd.

Turker, D., & Sonmez Selcuk, S. (2009). Which factors affect entrepreneurial intention of university students? *Journal of European Industrial Training, 33*, 142–159. doi: 10.1108/03090590910939049

Wang, D., Wang, J., Chen, T., & Zhang, Y. (2011). Influence model and mechanism of the rural residents for tourism support: A comparison of rural destinations of Suzhou in different life cycle stages. *Acta Geographica Sinica, 66*(10), 1413–1426. doi: 10.3724/SP.J.1011.2011.00415

Xiao, Y., Ming, Q., & Li, S. (2001). On the concept and types of rural tourism. *Tourism Science.* doi: 10.16323/j.cnki.lykx.2001.03.003

Xue, Q. (2008). Distribution pattern and regulation of Chinese rural tourism. *Research of Agricultural Modernization, 29*(6), 715–718. doi: 10.1016/S1002-0721(08)60105-2

Yang, X., & Yang, P. (2017). Empirical research of entrepreneurial opportunity identification on rural tourism. *Tourism Tribune, 32*(2), 89–103. doi: 10.3969/j.issn.1002-5006.2017.02.014

Zhang, Y., Guo, Y., & Ji, L. (2022, August). Going somewhere or for someone? The sense of human place scale (shps) in Chinese rural tourism. *Tourism Management, 91*, 104530. https://doi.org/10.1016/j.tourman.2022.104530

Zhao, D., Guo,W.(2020). Compromising in the Gap: A case study of Rural Women's Erotic Practice Model. *Journal of China Agricultural University (Social Sciences Edition), 37*(2), 55–64.

17

WOMEN'S ENTREPRENEURSHIP IN THE HOSPITALITY AND TOURISM INDUSTRY

A Systematic Literature Review
and Future Research Directions

Magdalena Petronella (Nellie) Swart,
Vanessa S. Bernauer, and K Thirumaran

Abstract

Research often portrays women as weaker entrepreneurs in the hospitality and tourism (H&T) industry. Our systematic literature review sheds light on women's entrepreneurial success in H&T by examining four determining factors to success (or failure) for women's entrepreneurship: (1) motivations that lead women to become entrepreneurs, (2) barriers they face, (3) non-government strategies, and (4) governmental policies. Our findings suggest that financial and legal support, training, political, social, and cultural backing are among the consistent success factors examined in previous studies. However, scholarship is silent on how the four key themes can be aligned to support women to succeed in H&T entrepreneurship. As a key contribution, this systematic literature review provides a template for future research that starts from the identifiable success factors. In addition, this work provides practitioners such as policymakers and industry leaders with a better understanding of how to reduce barriers to women's entrepreneurship.

Keywords

Women entrepreneurs, Tourism and Hospitality, Motivations, Barriers, Strategies, Policies

Introduction

Women's contribution to all levels of economic activity through entrepreneurship in hospitality and tourism (H&T) has been acknowledged in the United Nations Sustainable Development Goals (UN, 2015), the Global Report on Women in Tourism (UNTWO, 2019), and the World Economic Forum Global Gender Gap (WEF, 2019). Women's entrepreneurship in H&T has been well researched, especially since women represent between 60% and 70%

DOI: 10.4324/9781003286721-24

of the general workforce (ILO, 2020) and further account for about 54% of employment in the H&T industry (UNTWO, 2019; WTTC, 2020). In their critical literature review on gender, tourism, and entrepreneurship, Figueroa-Domecq et al. (2020a) as well as Wilson-Youlden and Bosworth (2019), highlight that more feminist approaches are needed. On the one hand, they contend that gender biases still prevail, and women are often portrayed as weaker entrepreneurs. But, for instance, in his study on gender, customer satisfaction, and service quality in H&T, Koc (2020) highlights women's strengths, their potential for entrepreneurship, and their prospects for being employed in the industry.

The nature of the H&T industry requires service personnel and business owners to focus on customer service, product innovation, and sustainable growth (Lapan et al., 2016; Morgan & Winkler, 2020; Wilson-Youlden & Bosworth, 2019), especially now while the world anticipates the recovery of the industry. Ideally, men and women should be equal in competency and capability. This ideology, as the various literature suggests, is not a reality, we need to rethink whether women entrepreneurs are allowed to make their mark in the H&T business (Costa et al., 2016). A consensus that women's entrepreneurship should be promoted seems clear. However, since business models have changed, it is necessary to understand how women's entrepreneurship in H&T can be supported and what hinders women to be successful entrepreneurs. We provide a systematic literature review with a comprehensive synthesis of existing knowledge of determining factors to success (or failure) for women's entrepreneurship and create a basis for transferring relevant insights to those women, scholars, practitioners, and policymakers. This chapter aims to present a critical analysis of the current state of research on women's entrepreneurship in H&T. We carve out the following: (1) which factors motivate women to become entrepreneurs; (2) what are the barriers they face; (3) which strategies will support; and (4) which governmental policies need to be aligned to support women's success in entrepreneurship.

The next section outlines the method of data collection and analysis. This is followed by the findings section, wherein we present the types of businesses and geographical locations women engage in and highlight the determining factors to success (or failure) for women's entrepreneurship identified in the reviewed articles. Finally, we discuss our findings and conclude with takeaway points.

Methodology

This study aimed to review the current body of knowledge on women's entrepreneurship in H&T. We conducted a systematic literature review, adhering to a set of core principles outlined by Briner and Denyer (2012). The review is (1) organised, by following an explicit, systematic, (2) transparent, and (3) replicable process. Furthermore, to develop a conceptual map of research, we (4) combine and synthesise the findings of various relevant studies. The review is based on rigorous search criteria, as illustrated in Table 17.1. Automated searches were performed in January 2021 and updated in March 2022, in five databases (Ebsco, Scopus, Web of Science, One Search, and Pro Quest) with Boolean operators. Furthermore, inclusion criteria were articles written in English and published in peer-reviewed journals between 2010 and 2021, which focused on women's entrepreneurship (not merely women employees nor management), and which fell in an H&T context.

The screening of articles was performed by all authors using the software Rayyan and EndNote for systematic reviews. Inclusion-exclusion appraisals were extensively discussed

Table 17.1 Inclusion and exclusion criteria and their explanations

I/E	Criteria	Criteria explanation
Exclusion	Search engine reason	Articles
		• which do not fit Boolean operators: (Wom*n OR female OR gender) AND (entrepreneur*) AND (hospitality OR touris*)
		• not in English, including articles with titles, abstracts, and keywords in English but not full text
		• published before 2010 and after 2021
	Non-related	Articles not published in peer-reviewed journals, e.g., book chapters, conference reviews, and others
	Loosely related	Articles focusing on women employees or managers in H&T
	Partially related	Articles focusing on women's entrepreneurship in other industries or entrepreneurship in H&T business schools
Inclusion	Closely related	Articles explicitly focusing women's entrepreneurship on H&T

Note: "Wom*n" can either be "women" or "woman".

in four steps: identification, screening, eligibility, and inclusion (see Figure 17.1 for a PRISMA diagram of the process). Out of 523 articles, we finally identified a total of 73 relevant articles. We could identify determining factors to success (or failure) for women's entrepreneurship by analysing the full texts: (1) motivations, (2) barriers, (3) non-government strategies, and (4) government policies. The content analysis was done by all authors with extensive discussions.

Findings

In this section, we first present the types of businesses women entrepreneurs are mainly involved in according to the reviewed articles to understand the trends. Drawing on the literature, we further analyse the study locations on the world map to identify prospects for future studies. Thereafter, we deep-dive into the determining factors to success (or failure) for women's entrepreneurship, namely motivating factors, barriers, non-government support strategies, and government policies that (dis)empower women entrepreneurs in H&T.

Types of Businesses and Geographical Foci

Studies about women's entrepreneurship in tourism mainly address small and medium-sized enterprises of which the majority focus on rural areas and hospitality services. Table 17.2 displays an overview of the various sectors of the H&T women entrepreneurs are engaged in as reported by the selected studies.

After a close evaluation of the various studies and the locations at which these studies were conducted, a strong concentration of study contexts emerges in the United States of America, Scandinavia, the Baltic region, Spain, South and West Africa, India, and Southeast Asia (see Figure 17.2). On the other hand, North Africa, South and Central America, Central Europe, Russia, and Central Asia are the least represented in the studies on women's entrepreneurship in the H&T sector.

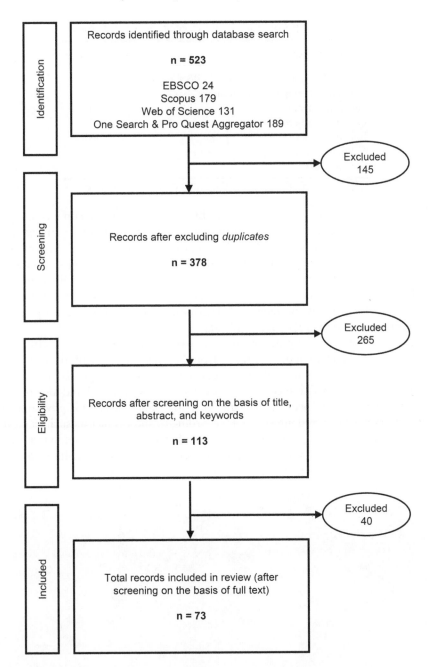

Figure 17.1 The PRISMA flow diagram showing the data collection steps.

Determining Factors to Success (or Failure) for Women's Entrepreneurship

After an extensive systematic literature review on women's entrepreneurship in the H&T industry, we identified four determining factors for the success (or failure) of women's entrepreneurship. First, we focused on what motivates women to become entrepreneurs;

Table 17.2 Types of business addressed in the selected articles

Type of business	Studies
Accommodations (e.g., homestay, B&B etc.)	20
Food and beverage (e.g., restaurant, bar etc.)	15
Rural and farm tourism	14
Souvenirs and crafts	13
Tour operators	13
Wellness and spa	3
Not specified	26

Note: Some studies addressed multiple types of businesses

second, which barriers they face in this field; third, we analysed the support strategies of non-governmental organisations, and fourth, we looked at government policies that support women's entrepreneurship in H&T. Through textual analysis and supported by the analysis of prominent words, we identified education and training, family, cultural, social, economic and financial, legal, and environmental circumstances as well as the business and gender mainstreaming in policies or strategies as the main areas of investigation. The matrix in Table 17.3 shows the identified determining factors to success (or failure) for women's entrepreneurship in the H&T industry.

Our findings show that many studies focus on barriers that hinder women to contribute as entrepreneurs in the H&T industry. Apart from business engagement, the characteristics that lead to women's entrepreneurial success depend on their family, social, economic, and financial circumstances. These success factors seem to be widespread and also relate to issues to motivate women to become or being entrepreneurs. Considering government policies, the authors have targeted intervention according to success factors such as economics and finance, business, gender mainstreaming, and education and training. In the discussion later, we present how the identified determining factors of success (or failure) for women's entrepreneurship are addressed in the literature of the last decade.

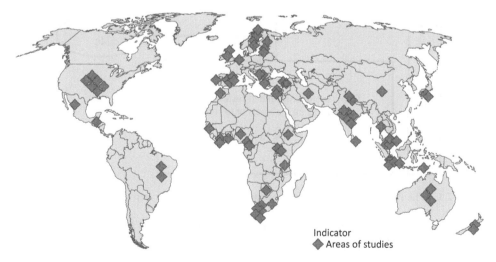

Indicator
◆ Areas of studies

Figure 17.2   Areas and frequency of research on women's entrepreneurship in hospitality and tourism in the selected articles.

Table 17.3 Determining factors to success (or failure) for women's entrepreneurship in hospitality and tourism

Areas of investigation	Determining factors to success (or failure) for women's entrepreneurship in H&T			
	Motivation	Barriers	Non-governmental support strategies	Government intervention policies
Education and training		Entrepreneurial training and work experience barriers (10*)	Strategic training and technological capacitation (7*)	Policies must accommodate entrepreneurship training programs (6*)
Family	Economic independence from husbands (5*)	Family-centric decision-making and dual role barriers (14*)	Strategies to balance woman's work demands and family responsibilities are needed (5*)	
Cultural		Traditional norms and social sanctions (8*)		
Social	Lifestyle and economic survival (*4) Combine house and business responsibilities (9*)	Discrimination, inequity, and inequality (20*)	Networking (8*)	
Economic and financial	Supporting family and society (15*)	Gendered economic disadvantages (6*) Systemic economic barriers (15*)	Investment opportunities (5*)	Financial inclusion and financing policies (8*)
Legal		Legal entrepreneurial hurdles (5*)		Formalisation and regulation policies (3*)
Business	Explore new business concepts (2*)	Challenges and constraints include competition, seasonality, work intensity, managerial hurdles, infrastructure, corruption, and information access (13*)	Leadership competence (10*) and Innovation (*4)	Capacity building through business incubator centres (5*)
Environmental	Climate change concerns (3*)		Eco-tourism entrepreneur strategies (1*)	

Note: *Number of articles referring to the sub-factor, supported by the most prominent example

Women's Entrepreneurship Motivations in Hospitality and Tourism

Several success factors emerged from the selected literature for women to become an entrepreneur in H&T, one of which is motivations. The following section addresses women's motivations to engage in entrepreneurial activities.

It is not uncommon that women's circumstances drive them to become H&T entrepreneurs (Wilson-Youlden & Bosworth, 2019), and it is evident that various family, social, economic and financial, and environmental factors contribute to their motivations to become business owners. In a family context, many women strive to become economically independent from their husbands (Kimbu & Ngoasong, 2016; Maliva et al., 2018; Möller, 2012; Movono & Dahles, 2017; Tajeddini et al., 2017b), or to explore new business concepts (Carvalho et al., 2018). For example, some bed and breakfast owners in Mthatha, South Africa, are content with the time flexibility to attend to personal matters and network for professional gains which their business affords them to do (Hlanyane & Acheampong, 2017).

Social circumstances (Cikic et al., 2018; Iwu & Nxopo, 2015; Wilson-Youlden & Bosworth, 2019) are often influenced by women's lifestyles and economic (survival) livelihood conditions. This can be partly explained by where in rural areas, women and their family members often face challenges to find employment (Maksimovic et al., 2019; Movono & Dahles, 2017; Pallarès-Blanch et al., 2015; Wilson-Youlden & Bosworth, 2019). Hence, to make ends meet, women may play a crucial role in becoming an entrepreneur. Some studies indicate instances where women favour the opportunity to combine their house responsibilities with the running of a business. Such a role allows for taking care of their family (Amrein, 2013; Arbarini et al., 2019; Kimbu & Ngoasong, 2016; Möller, 2012), and contributing to the household's finances (Kimbu & Ngoasong, 2016; Möller, 2012). When women feel valued by their families and supported by spouses (Surangi, 2016), they can become highly motivated to invest in a business. Hence, the respect gained from family members (Tajeddini et al., 2017b) can have a positive psychological impact to become an entrepreneur (Hlanyane & Acheampong, 2017). Women entrepreneurs are extremely resilient and successful in overcoming their challenges, despite intimidations from male counterparts in the business world, primarily because of their positive attitudes which are manifested by strong personalities (Hlanyane & Acheampong, 2017).

Complex economic circumstances motivate women to become entrepreneurs in H&T (Kwaramba et al., 2012; Maliva et al., 2018; Wilson-Youlden & Bosworth, 2019) to support their families and society. These motivations include the need to financially support their family (Möller, 2012; Wilson-Youlden & Bosworth, 2019; Zapalska et al., 2015); to supplement their household income (Kimbu & Ngoasong, 2016; Suminar et al., 2019); to alleviate poverty (Hillman & Radel, 2021); to feel economically empowered (Movono & Dahles, 2017) and to create jobs (Zapalska & Brozik, 2014). Several studies indicate that women possess a natural maternal instinct which creates a desire to contribute to their family's needs (Carvalho et al., 2018; Maliva et al., 2018; Möller, 2012; Wilson-Youlden & Bosworth, 2019).

A concern for the environment, which is rooted in sustainability trends, as well as the looming threats of climate change, are catalysts for women's H&T entrepreneurial motives (Zapalska & Brozik, 2017), especially when this allows them to live close to nature (Pallarès-Blanch et al., 2015).

Women's Entrepreneurship Barriers in Hospitality and Tourism

Several socio-economic challenges continue to impede women's entrepreneurship in H&T. These barriers primarily manifest in the way women are "required" to balance their life and livelihood with socially and economically imposed conditions. Evidence suggests women entrepreneurs in H&T experience barriers in education and training, family life, cultural-, social-, economic & financial-, legal-, and business environments, as postulated in the discussions later.

Even with education and training in modern society, women experience entrepreneurial training and work experience barriers (Ali, 2018; Iwu & Nxopo, 2015; Kurtege Sefer, 2020; Rao et al., 2021; Tshabalala & Ezeuduji, 2016), especially when the man is the sole breadwinner (Amrein, 2013). Personal development is hampered by limited managerial and technological skills (Ali, 2018). We learned that in some instances, where women were trained in digital technology and enterprise, it has not had the desired outcomes after four years of intervention, as is the case in the Swiss valley of Val d'Anniviers (Amrein, 2013). Reasons for this failure stem from women's bound to their families and home living in a patriarchal environment despite training provisions. On the other hand, governments, like Ghana, have been unable to develop facilities for training interventions in H&T (Ali, 2018; Handaragama & Kusakabe, 2021; Ribeiro et al., 2021), and these have met with some success. In this milieu of differences in societal and cultural values, urban women with training in H&T are employed in hotels, such as the case in Bali, however, rural women are often exploited as cheap labourers in the industry (Tajeddini et al., 2017b).

Family barriers associated with family-centric decision-making and dual role barriers are highlighted as among the biggest challenge for the development of women's entrepreneurship in H&T (Carvalho et al., 2018; Chong & Velez, 2020; Costa et al., 2016; Iwu & Nxopo, 2015; Katongole et al., 2013; Kurtege Sefer, 2020; Maksimovic et al., 2019; Morgan & Winkler, 2020; Priyadarshini & Ramakrishnan, 2016; Savage et al., 2020). In related literature, women are often depicted as the secondary breadwinners, who need to consult with their family before they can make a decision (Tajeddini et al., 2017a). In a study in Epirus, Greece, the women see themselves as entrepreneurs working to contribute to their families as opposed to making a profit. Women, the few who are successful in Epirus, are adept at navigating the dual needs of business and family (Bakas, 2017). Bensemann and Hall (2010) differentiate between gender roles where housecleaning responsibilities resonated with women and the business aspects with men, while the barriers associated with the combining of business and household tasks are highlighted by Kimbu and Ngoasong (2016).

Cultural barriers limit women entrepreneurs' business relationships and networking possibilities (del Mar Alonso-Almeida, 2012; Hillman & Radel, 2021; Iwu & Nxopo, 2015; Priyadarshini & Ramakrishnan, 2016) as they need to conform to traditional norms and constrained by social sanctions. Traditional norms restrict women in conservative communities from having limited contact with men, as this is against the traditional norms and values of some cultures (Maliva et al., 2018). Consequently, women tend to have less support in these societies than men (Kimbu & Ngoasong, 2016). Cultural traditions associated with Guanxi, Blat, and Ubuntu have consistently challenged the dominant Western ideologies of business relationships, especially where Western H&T firms operate in non-Western locations (Zhang et al., 2020). Women are often subject to Machismo and social sanctions (Morgan & Winkler, 2020) which limit their opportunities to excel as H&T entrepreneurs.

Social barriers further highlight the discrimination, inequality, and inequity women in society are confronted with (Cikic et al., 2018; Hillman & Radel, 2021; Iwu & Nxopo, 2015; Kimbu & Ngoasong, 2016; Lindberg et al., 2014; Maksimovic et al., 2019; Priyadarshini & Ramakrishnan, 2016; Savage et al., 2020; Tshabalala & Ezeuduji, 2016). For example, female Akha ethnic minorities face discrimination and are looked down upon by mainstream Thais in Bangkok. With no citizenship, these women do not have access to education or business ownership (Trupp, 2015). Another challenge is that women may also experience difficulty in traversing between collaborations and managing their own businesses, especially as communities may perceive their businesses as "controversial" and "irrelevant" (Lindberg et al., 2014, p. 106). Women working in H&T are subject to perceptions in society (Surangi, 2018) and are confronted by gender racism (Ali, 2018; Hikido, 2018), gender inequity, and inequality (Chong & Velez, 2020; Hillman & Radel, 2021), stereotyping, and discrimination (Albors-Garrigos et al., 2021; Carvalho et al., 2018; Chong & Velez, 2020; Hillman & Radel, 2021; Kimbu & Ngoasong, 2016; Kurtege Sefer, 2020), as well as patriarchy (Hillman & Radel, 2021; Surangi, 2018). Women also suffer from a lack of respect and trust from the public or authorities (Bernhard & Olsson, 2020), which is evident in the relationship between financial barriers and gender barriers (Morgan & Winkler, 2020) in H&T businesses. As a result, some of these types of discrimination and gender-related challenges could make women feel psychologically disempowered and have a lack of self-efficacy (Morgan & Winkler, 2020; Yoopetch, 2020).

Gendered economic disadvantages and systematic economic barriers are amongst the biggest economic and financial stumbling blocks for women to become H&T entrepreneurs (Ali, 2018; Chong & Velez, 2020; Cikic et al., 2018; Handaragama & Kusakabe, 2021; Hillman & Radel, 2021; Hlanyane & Acheampong, 2017; Iwu & Nxopo, 2015; Kimbu & Ngoasong, 2016; Kurtege Sefer, 2020; Maksimovic et al., 2019; Priyadarshini & Ramakrishnan, 2016; Zapalska & Brozik, 2014, 2015). Systemic economic barriers for women entrepreneurs are synonymous with macro-environmental factors, such as when an economic turndown necessitates entrepreneurs to implement strategies to sustain their businesses (Perez & Bui, 2010; Vukovic et al., 2021). Hlanyane and Acheampong (2017) report on related barriers experienced by women in Mthatha, South Africa, who are operating accommodation businesses. The seasonality of tourists and access to financial resources, coupled with the practice of corruption and bribery, means that women entrepreneurs are often challenged by circumstances prevalent in the South African economy.

Other barriers include limited access to business networks (Costa et al., 2016; Handaragama & Kusakabe, 2021; Surangi, 2018), fair remuneration (Favre, 2017), limited decision-making powers (Carvalho et al., 2018; Morgan & Winkler, 2020), access to markets and suppliers (Song-Naba, 2020), economic participation (Ashrafi & Hadi, 2019), and economic gratification (Morgan & Winkler, 2020). While political circumstances often polarise (Hillman & Radel, 2021) a lack of investment by the government and the private sector (Ashrafi & Hadi, 2019), where the bureaucracy (Kimbu & Ngoasong, 2016) hampers investments in women H&T businesses.

Many women in H&T experience gendered economic disadvantages. A study conducted among women-owned small and medium enterprises (SMEs) in Turkey revealed that a mere 15% of the participants had the opportunity to receive governmental financial support (Kurtege Sefer, 2020). Furthermore, access to financial capital and community credits (Ali, 2018; Maksimovic et al., 2019; Song-Naba, 2020; Zapalska & Brozik, 2014), as well as unreasonable loan conditions and the repayment terms thereof (Chong & Velez, 2020), continue to marginalise women.

Legal barriers represent challenges which women in H&T experience (Cikic et al., 2018; Kimbu & Ngoasong, 2016; Kurtege Sefer, 2020; Lindberg et al., 2014; Priyadarshini & Ramakrishnan, 2016). Entrepreneurial hurdles are associated with informal governance structures and local legitimacy practices (Lindberg et al., 2014), bureaucracy and taxes (Kimbu & Ngoasong, 2016), as well as gendered and class-based legal barriers (Kurtege Sefer, 2020) when they embark on entrepreneurship.

Business environment barriers are another impediment to women's entrepreneurship in H&T (Maksimovic et al., 2019; Maliva et al., 2018). Challenges and constraints in the entrepreneurial landscape include women experience an increase in competition (Costa et al., 2014), seasonality (Cikic et al., 2018; Costa et al., 2016), demanding working hours, work intensity (Costa et al., 2016), challenges as an owner or manager, business operations (Kimbu & Ngoasong, 2016), insufficient infrastructure (accessible roads) (Ashrafi & Hadi, 2019), lack of business rules (Cikic et al., 2018; Kurtege Sefer, 2020), lack of business skills (Kurtege Sefer, 2020), and high levels of corruption in male networks (Maksimovic et al., 2019), as well as the perception that women are less competent than men (Carvalho et al., 2018) to name a few. In addition, product innovations do not immediately, or easily, transfer to a commercial value (Lindberg et al., 2014), as women may not understand the needs of tourists (Ashrafi & Hadi, 2019), know how to promote their tourism business services (Ali, 2018), or know how to use social media (Halim et al., 2020). Hence, H&T entrepreneurship in rural areas is an independent and commercial form of livelihood (Möller, 2012). Furthermore, women entrepreneurs in H&T crave to have access to information (Kimbu & Ngoasong, 2016; Maksimovic et al., 2019). Qureshi and Ahmed (2012) found that women who are taking up independent businesses had limited access to business information. As a result, very few women succeed in the handloom business, especially for tourism products.

Non-Government Support Strategies for Women's Entrepreneurship in Hospitality and Tourism

The literature surveyed for this review suggests that, within the private sector, several strategies associated with education and training, family, social, economic and financial, business, and environmental exist to support women's entrepreneurship in H&T.

Strategic training and technological capacitation are key to the empowerment of women's entrepreneurship in H&T (Abou-Shouk et al., 2021; Vukovic et al., 2021). These strategic training programs (Handaragama & Kusakabe, 2021; Zapalska et al., 2015) can foster family entrepreneurship, as evident in the Samin Village (Suminar et al., 2019). Capacitating women's entrepreneurship through technological skills (Figueroa-Domecq et al., 2020b) is essential, as H&T business women in Kuala Lumpur have mastered the skill in using the Airbnb platform to such an extent that they are willing to assist others who are finding the platform a challenge to use (Lim & Bouchon, 2020).

Strategies to foster family support are indicative (Möller, 2012). Women living in Semarang Regency, Indonesia, use H&T businesses to support their families (Arbarini et al., 2019). The latter's success was attributed to family members, such as the husbands and children, who encourage these women to sell their crafts to tourists. Gender mainstreaming in this latter case appears to be beneficial to women being an entrepreneur and contributing to their families' well-being. Furthermore, in a study by Moswete and Lacey (2015), women participating as entrepreneurs and leaders in a government cultural tourism program in southern Botswana found that men were supportive of their wives as they pursue H&T

business opportunities. Therefore, based on these success, stories strategies are needed to support cooperative tourism in micro-entrepreneurship to further assist women entrepreneurs to balance their work demands and family responsibilities (Lapan et al., 2016; Morgan & Winkler, 2020).

Social strategies are informed by structural (Hillman & Radel, 2021), psychological (Ayala & Manzano, 2014), and feminist (del Mar Alonso-Almeida, 2012) strategies. Ecofeminism aims to support conservation (del Mar Alonso-Almeida, 2012), while feminine sensibilities (care) (Costa et al., 2016) related to women's entrepreneurship in H&T. It is known that feminist strategies also address other aspects of gender inequality such as economic, social, structural, environmental strategies, but this was not evident in the consulted literature. Gomez-Perez and Jourde (2021), in their study, found that women in the Hajj pilgrimage business are highly successful and skilful as entrepreneurs, as they can mobilise their resources with help from their family members, community connections, and political leaders which ensures permits are secured to access markets. Their ability to travel frequently to the Hajj destination in Saudi Arabia also means that they can develop relationships and network with key contacts despite the travel business being a men-dominated field. Business networking strategies are a way to overcome gender stereotyping (Hillman & Radel, 2021) and remain a sensitive topic in many patriarchal communities. To overcome patriarchy, women entrepreneurs in rural areas are socially supported by their peers through business networking and collaborations to increase their family income (Suminar et al., 2019). This can be a powerful strategy for negotiating patriarchal norms within a family setting. Psychological strategies (as a social science discipline) focus on the resilience of women's entrepreneurship in H&T, as they have the skills to adapt to different situations (Ayala & Manzano, 2014) and these women become transformational leaders (Zapalska et al., 2015). Othman et al. (2015) found that women's entrepreneurship in the Islamic Spa sector is inspired by the success of other women business leaders and the increasing demand by Muslim women for spa services. This has led to a rise in the number of women entrepreneurs in the Muslim spa sector. Networking is thus a key strategy as it is used to address the structural barrier and brings psychological benefits to women entrepreneurs.

Economic strategies (del Mar Alonso-Almeida, 2012; Hillman & Radel, 2021) supported by investment opportunities are needed to support women's entrepreneurship, such as the women from Kuala Lumpur, who invested in Airbnb and are highly motivated by the monetary gains of their businesses. These businesswomen are willing to take the risk of meeting strangers as potential investors in their H&T business (Lim & Bouchon, 2020). Once H&T women entrepreneurs receive loans, they supplement the loan with their investments (Çiçek et al., 2017). Therefore, strategies in support of economic growth and employment are needed (Hillman & Radel, 2021) to enable women to generate higher levels of per capita income for their businesses (del Mar Alonso-Almeida, 2012), which are donor-driven from local economic development strategies (Kwaramba et al., 2012) and supported by political strategies (del Mar Alonso-Almeida, 2012).

Business strategies support women's entrepreneurship in H&T in developing a supportive business environment when they adjust their attitude towards business (Ertac & Tanova, 2020; Vukovic et al., 2021). It instils leadership competence (Zapalska & Brozik, 2014) and inform them how to communicate their business goals and their innovative tourism product designs to have a sustainable competitive advantage (Zapalska & Brozik, 2015). From these studies, it is evident that women are agents of economic development and growth where a culture of self-employment enhances the family economy (Ashrafi

& Hadi, 2019; Vukovic et al., 2021). This leads to intrapreneurship as women are more flexible, trustworthy, and harmonious, as well as risk and result-orientated (Zhang et al., 2020). According to Bakas (2017), women with entrepreneurial instincts, tend to achieve family stability and welfare before their entrepreneurial instincts take a more intense direction towards profitability.

Business strategies are informed by innovation. Lindberg and colleagues (2014) used the Quadruple Helix model to identify ways in which NGOs can ameliorate gender differences when addressing innovation and entrepreneurship. These women facilitate collaborations by bridging the gap between gender roles that are socially constructed by societies. Furthermore, Akha ethnic migrants from Thailand's highlands seek opportunities in the capital city of Bangkok where women entrepreneurs primarily sell souvenirs. As microbusiness entrepreneurs, their ingenuity is seen in their innovations in transforming cultural items into palatable souvenirs and adding colour and vibrancy to the street vendor ship in the capital city (Adelaar, 2015). Polish women also capitalise on natural resources, culture, and heritage to develop innovative tourism products sustainably (Zapalska et al., 2015), while women in Sri Lanka invest in business associations to create a network of business support (Handaragama & Kusakabe, 2021).

Environmental strategies are supported by eco-tourism entrepreneurs. Women in Nepal's lowlands have benefited enormously from eco-tourism, especially in the hospitality sector (accommodation). Their active participation is due to easy access to loans, information, and government policies which are based on merits and needs rather than gender. These resulted in more self-confident women entrepreneurs, which were economically independent and self-driven in a culturally male-dominated society. The authors call for more effort in easing traditional practices that enforce women in a subservient role (Panta & Thapa, 2018) as they contribute to the overall preservation of the environment.

Government Intervention Policies to Support Women's Entrepreneurship in Hospitality and Tourism

Governments also play a crucial role in enabling and empowering women's entrepreneurship and aspirations. The impact and need for policies are outlined in this subsection.

Educational and training intervention policies are proposed by several scholars (Iwu & Nxopo, 2015; Jaafar et al., 2015; Zapalska & Brozik, 2017). Moswete and Lacey (2015) studied the Botswana tourism policy which entrusted leadership and community-based cultural tourism to women. The study surveyed the women in Southern Botswana who were involved in the project and found that many women felt confident and empowered to be independent without the harbingers of men and family traditions. Those with formal education and who have engaged in vocational training became more active and ventured into cultural tourism opportunities. Therefore, policies should accommodate entrepreneurship training programs that prepare women with managerial and leadership skills (Zapalska & Brozik, 2017), together with financial and technology literacy (Yoopetch, 2020) through action learning (Rao et al., 2021).

As the provision of securities, such as an asset, to obtain formal financing from lending agencies remains a challenge, the relevant economic and financing support policies need to be in place (Iwu & Nxopo, 2015; Jaafar et al., 2015; Ribeiro et al., 2021; Zapalska & Brozik, 2017). Financial inclusion and financing policies are central to the support of women's H&T entrepreneurship development. Microfinance opportunities to support women in rural areas must be supported through the government, as well as non-government organisations (Ertac & Tanova, 2020; Maliva et al., 2018; Zapalska & Brozik, 2017). Government policies need to support indigenous economic growth and freedom to become self-sufficient (Zapalska & Brozik, 2017), provide social security relief, stimulate infrastructure investment (Jaafar et al., 2015), support women's entrepreneurship with a transport allowance (Kimbu & Ngoasong, 2016) and develop initiatives to integrate women into the tourism economy (Kwaramba et al., 2012).

Legal support policies (Iwu & Nxopo, 2015) that strengthen the formalisation and regulations of women entrepreneurs in H&T are a priority. Scholars call for policies to empower women to formalise their businesses (Hikido, 2018), and to regulate H&T activities (Jaafar et al., 2015) to protect both the entrepreneurs and the customers.

Kimbu and colleagues (2019) studied national policies and academic literature related to human capital management in Ghana, Nigeria and Cameroon. Based on the caveat that women are often in a unique position in society that circumstantially deprives them of opportunities, Capacity building through business incubator centres can foster women's entrepreneurship in H&T, especially amongst marginalised women. Therefore, relevant business support policies need to be aligned to support women's entrepreneurship on how to develop micro-enterprises through incubator centres (Zapalska & Brozik, 2017), the creation of H&T incentive programs, support for marketing (Jaafar et al., 2015), human resources to retain employees (Costa et al., 2014), or how to define the core purpose of the H&T business (Hallak et al., 2015), and to capacitate participation by women in the decision making of the business (Movono & Dahles, 2017; Zapalska & Brozik, 2015). These initiatives can be facilitated through university, government and private-sector partnerships (Untari & Suharto, 2021).

Discussion and Conclusion

Our study sheds light on the determining factors for women's entrepreneurial success or failure in the H&T industry, which are highlighted in Table 17.3. Therewith, we contribute to the literature on women's entrepreneurship in the specific context of the H&T industry by providing a conceptual summary of the last decades' existing research. Through a systematic review of the literature, we identified success factors that motivate women to become entrepreneurs and which barriers they face. Furthermore, we point out non-government strategies and government intervention policies that support and promote women's entrepreneurship in the H&T industry.

Women's entrepreneurship could be a fruitful way for women to find their position in the economy and succeed as entrepreneurs, rethinking traditional family and economic models. Through our findings, it is evident that especially social as well as economic and financial circumstances motivate women to become entrepreneurs or run their businesses in H&T. Women are indeed eager to contribute to the family income and hence attempt to create a delicate balance between business and family life (Amrein, 2013; Kimbu & Ngoasong, 2016; Möller, 2012). However, the latter can unfortunately also have the opposite effect, when women need to combine business and family life (Kimbu & Ngoasong, 2016) and are taken seriously neither by their family nor by society or authorities (Bernhard & Olsson, 2020). In addition to these family-related barriers, other obstacles are often of a social or economic nature, such as the lack of information (Maksimovic et al., 2019) or access to government financial support (Kurtege Sefer, 2020).

Among the most discussed non-government support strategies in women's entrepreneurship in H&T research are business-related aspects, such as developing women's business environment (Ertac & Tanova, 2020; Vukovic et al., 2021) with leadership competencies or communication skills (Zapalska & Brozik, 2015). Government policies that are targeted at women empowerment in H&T can be enablers of women's entrepreneurship are also evident. Scholars mainly address economic- and financial-related policies, such as offering a safety net through the implementation of social security or investments in infrastructure (Jaafar et al., 2015). Overall, in our analysis of women's entrepreneurship in H&T literature, it becomes clear that social, business, economic and financial factors determine the success of women entrepreneurs the most, these are the levers with which the women can be supported by all involved actors. Practitioners, such as policymakers and industry leaders in the H&T industry, should provide targeted entrepreneurial education and training, economic, financial, legal and business support specifically for women or gender mainstreaming policies to promote their entrepreneurial success.

The H&T industry can provide women with business opportunities, economic and social empowerment and escape from precarious circumstances. However, inequalities based on gendered practices, substructures, subtexts or logics, such as gendered roles in agriculture, still exist and are constantly (re)produced. We are aware of a possible reification through the study of women-specific success factors for entrepreneurship. Moreover, some scholars in the field contribute in (re)producing stereotypical attributions and prejudices of women's entrepreneurship, for instance by highlighting a woman's natural maternal instinct and a desire to contribute to their family's needs (e.g., Carvalho et al., 2018), their feeling of being disempowered and lack of self-efficacy (Morgan & Winkler, 2020; Yoopetch, 2020) or referring to women being more flexible, trustworthy, and harmonious, as well as risk and result orientated (Zhang et al., 2020). It is necessary to point out and make visible the existing strengths and weaknesses to create future paths. Societal structures and resources for women need to be supported as they are still disadvantaged in society.

The reviewed literature generally supports policies that promote and safeguard women's entrepreneurship in the H&T sector. However, an exception exists as some studies found no significant disparities in the business performance of men and women operating accommodation establishments (García-Machado et al., 2020). Further research is essential to understand how H&T policies can effectively protect all genders, particularly those within marginalised groups.

Following the extensive research conducted by Figueroa-Domecq and colleagues (2020a) we support the continuation of studies that critically empirically analyse women's entrepreneurship with a feminist – or beyond that – an intersectional lens. Polices may have different outcomes in an economic and socio-cultural context when gender relations are considered (Möller, 2012). In patriarchal societies, such as Nepal, women encounter institutionalised gender restrictions that perpetuate inequalities across cultural, economic, social, and political spheres. Concerns arise about violence against women and their exploitation, particularly in women-only businesses, where different social classes intersect. Policies addressing gender relations must consider the complexities of intersecting identities. Gender-focused interventions may fall short in addressing the diverse challenges women face due to their varying socio-cultural contexts, race, class, ethnicity, and sexuality. Cultural norms further impact women's opportunities in business settings. An intersectional policy approach recognises these intricacies, fostering more inclusive and sustainable outcomes for diverse women, but needs to be explored further. Additionally, besides positive aspects of entrepreneurship for women such as wealth creation or an opportunity for emancipation and social change (Calás et al., 2009), studies focusing on the dark side of entrepreneurship should not be neglected (Dannreuter, 2020, p. ix).

Given the robustness and volumes of the research reported on factors that contribute to the failure or success of women entrepreneurs in H&T (see Table 17.3), we believe this study leads us to three key focus areas that need further research, which we also portray as key takeaway points.

First, there is a sense that women are in a challenging position given that they are often the gender that cares for childcare and family household chores in traditional and in some instances in advanced societies. While there are no easy solutions to this situation given the nexus of cultural practices, policymakers can consider it as an avenue for better childcare support systems at the same time facilitate education and training for prospective women entrepreneurs. This will help situations where women can take a break from household and family responsibilities to focus on their learning and develop crucial skills to initiate their ideas into reality. Government support is instrumental in providing fair and non-discriminatory policies, where women entrepreneurs are supported to overcome family and societal barriers.

Secondly, considering its wide geographical distribution, women's entrepreneurship in South America, Europe, and Asia presents ample opportunities for extensive research. It is crucial to explore this domain comprehensively, as the empowerment of women entrepreneurs transcends being exclusively a woman's issue. Based on our literature survey, South America, Russia, and Central Asia, as well as conditions in North Africa, are fertile ground for further studies on women entrepreneurship and context-specific situations within those societies. At some point, within these respective societies, men and societies have to come to an understanding and negotiate spaces for women to emerge respectfully as equals in business innovation and enterprise. This geographic prospect also alludes to the need to study instances of women entrepreneurs who have successfully managed to extend their reach globally for their business presence. An opportunity exists to expand the research scope within societies where aspects that motivate women to become entrepreneurs in H&T, the barriers they are confronted with, and how they can be supported by non-governmental and governmental policies are yet to be conducted.

Third, much of the literature points to a lack of comparative studies between factors determining the success or failure of women entrepreneurship in H&T, policies, regions, and women's entrepreneurial activities. Comparative studies of women's entrepreneurship can enhance our understanding of the divergences and convergences, and the lessons that can be learned with a robust and critical lens.

References

Abou-Shouk, M. A., Mannaa, M. T., & Elbaz, A. M. (2021). Women's empowerment and tourism development: A cross-country study. *Tourism Management Perspectives, 37*, 100782. https://doi.org/10.1016/j.tmp.2020.100782

Adelaar, T. (2015). Agency, social capital, and mixed embeddedness among Akha ethnic minority street vendors in Thailand's tourist areas. *Sojourn, 30*, 780–818.

Albors-Garrigos, J., Signes, A. P., Segarra-Oña, M., & Garcia-Segovia, P. (2021). Breaking the glass ceiling in haute cuisine: The role of entrepreneurship on the career expectations of female chefs. *Tourism and Hospitality Management, 27*(3), 605–628. https://doi.org/10.20867/thm.27.3.8

Ali, R. S. (2018). Determinants of female entrepreneurs' growth intentions: A case of female-owned small businesses in Ghana's tourism sector. *Journal of Small Business and Enterprise Development, 25*(3), 387–404. https://doi.org/10.1108/jsbed-02-2017-0057

Amrein, T. (2013). Looking back at the parcours Arianna in the Val d'Anniviers: Critical commentaries on a development programme destined for the women of the Swiss alpine Valleys. *Revue de Geographie Alpine, 101*(1). https://doi.org/10.4000/rga.1980

Arbarini, M., Desmawati, L., & Budiartati, E. (2019). Gender equality and women's participation in the development of the tourism village in the era of industrial revolution 4.0. *International Journal of Innovation, Creativity and Change, 5*(5), 311–323.

Ashrafi, A., & Hadi, F. (2019). The impact of tourism on developing Shiraz rural women entrepreneurship. *Revista Universidad Y Sociedad, 11*(4), 72–76. http://rus.ucf.edu.cu/index.php/rus

Ayala, J. C., & Manzano, G. (2014). The resilience of the entrepreneur. Influence on the success of the business. A longitudinal analysis. *Journal of Economic Psychology, 42*, 126–135. https://doi.org/10.1016/j.joep.2014.02.004

Bakas, F. E. (2017). Community resilience through entrepreneurship: The role of gender. *Journal of Enterprising Communities, 11*(1), 61–77. https://doi.org/10.1108/jec-01-2015-0008

Bensemann, J., & Hall, C. M. (2010). Copreneurship in rural tourism: Exploring women's experiences. *International Journal of Gender and Entrepreneurship, 2*(3), 228–244. https://doi.org/10.1108/17566261011079224

Bernhard, I., & Olsson, A. K. (2020). Network collaboration for local and regional development – The case of Swedish women entrepreneurs. *International Journal of Entrepreneurship and Small Business, 41*(4), 539–561. https://doi.org/10.1504/ijesb.2020.111578

Briner, R. B., & Denyer, D. (2012). *Systematic review and evidence synthesis as a practice and scholarship tool*. Oxford University Press.

Calás, M. B., Smircich, L., & Bourne, K. A. (2009). Extending the boundaries: Reframing "entrepreneurship as social change" through feminist perspectives. *Academy of Management Review, 34*(3), 552–569. http://www.jstor.org/stable/27760019

Carvalho, I., Costa, C., Lykke, N., & Torres, A. (2018). Agency, structures and women managers' views of their careers in tourism. *Women's Studies International Forum, 71*, 1–11. https://doi.org/10.1016/j.wsif.2018.08.010

Chong, A., & Velez, I. (2020). Business training for women entrepreneurs in the Kyrgyz Republic: Evidence from a randomised controlled trial. *Journal of Development Effectiveness, 12*(2), 151–163. https://doi.org/10.1080/19439342.2020.1758750

Çiçek, D., Zencir, E., & Kozak, N. (2017). Women in Turkish tourism. *Journal of Hospitality and Tourism Management, 31*, 228–234. https://doi.org/10.1016/j.jhtm.2017.03.006

Cikic, J., Jovanovic, T., & Nedeljkovic, M. (2018). Business and/or pleasure – Gender (in)equalities in rural tourism in Vojvodina. *Journal of Agricultural Science and Technology, 20*(7), 1341–1352.

Costa, C., Breda, Z., Bakas, F. E., Durão, M., & Pinho, I. (2016). Through the gender looking-glass: Brazilian tourism entrepreneurs. *International Journal of Gender and Entrepreneurship, 8*(3), 282–306. https://doi.org/10.1108/ijge-07-2015-0023

Costa, J. C., Shah, H., & Korgaonkar, K. (2014). From grassroots to success: A case study of a successful Goan woman entrepreneur. *Prabandhan: Indian Journal of Management, 7*(2), 40–46. https://doi.org/10.17010//2014/v7i2/59261

Dannreuter, C. (2020). Foreword. In A. Örtenblad (Ed.), *Against entrepreneurship: A critical examination* (pp. v–ix). Springer.

del Mar Alonso-Almeida, M. (2012). Water and waste management in the Moroccan tourism industry: The case of three women entrepreneurs. *Women's Studies International Forum, 35*(5), 343–353. https://doi.org/10.1016/j.wsif.2012.06.002

Ertac, M., & Tanova, C. (2020). Flourishing women through sustainable tourism entrepreneurship. *Sustainability (Switzerland), 12*(14), 17. https://doi.org/10.3390/su12145643

Favre, C. C. (2017). The small2mighty tourism academy: Growing business to grow women as a transformative strategy for emerging destinations. *Worldwide Hospitality and Tourism Themes, 9*(5), 555–563. https://doi.org/10.1108/whatt-07-2017-0034

Figueroa-Domecq, C., de Jong, A., & Williams, A. M. (2020a). Gender, tourism and entrepreneurship: A critical review. *Annals of Tourism Research, 84*. https://doi.org/10.1016/j.annals.2020.102980

Figueroa-Domecq, C. F., Williams, A., de Jong, A. D., & Alonso, A. (2020b). Technology is a woman's best friend: Entrepreneurship and management in tourism. *e-Review of Tourism Research, 17*(5), 777–792. https://journals.tdl.org/ertr/index.php/ertr/article/view/560

García-Machado, J. J., Barbadilla Martín, E., & Gutiérrez Rengel, C. (2020). A PLS multigroup analysis of the role of businesswomen in the tourism sector in Andalusia. *Forum Scientiae Oeconomia, 8*(2), 37–57. https://doi.org/10.23762/fSo_VoL8_no2_3

Gomez-Perez, M., & Jourde, C. (2021). Islamic entrepreneurship in Senegal: Women's trajectories in organizing the Hajj. *Africa Today, 67*(2), 104–126. https://doi.org/10.2979/africatoday.67.2_3.06

Halim, M. F., Barbieri, C., Morais, D. B., Jakes, S., & Seekamp, E. (2020). Beyond economic earnings: The holistic meaning of success for women in Agritourism. *Sustainability (Switzerland), 12*(12). https://doi.org/10.3390/su12124907

Hallak, R., Assaker, G., & Lee, C. (2015). Tourism entrepreneurship performance: The effects of place identity, self-efficacy, and gender. *Journal of Travel Research, 54*(1), 36–51. https://doi.org/10.1177/0047287513513170

Handaragama, S., & Kusakabe, K. (2021). Participation of women in business associations: A case of small-scale tourism enterprises in Sri Lanka. *Heliyon, 7*(11). https://doi.org/10.1016/j.heliyon.2021.e08303

Hikido, A. (2018). Entrepreneurship in South African township tourism: The impact of interracial social capital. *Ethnic and Racial Studies, 41*(14), 2580–2598. https://doi.org/10.1080/01419870.2017.1392026

Hillman, W., & Radel, K. (2021). The social, cultural, economic and political strategies extending women's territory by encroaching on patriarchal embeddedness in tourism in Nepal. *Journal of Sustainable Tourism*, 1–22. https://doi.org/10.1080/09669582.2021.1894159

Hlanyane, T. M., & Acheampong, K. O. (2017). Tourism entrepreneurship: The contours of challenges faced by female-owned BnBs and guesthouses in Mthatha, South Africa. *African Journal of Hospitality, Tourism and Leisure, 6*(4), 1–17.

ILO. (2020). *Tourism poverty reduction and gender equality*. International Labour Organisation (ILO).

Iwu, C. G., & Nxopo, Z. (2015). Determining the specific support services required by female entrepreneurs in the South African tourism industry. *African Journal of Hospitality, Tourism and Leisure, 4*(2), 13.

Jaafar, M., Rasoolimanesh, S. M., & Lonik, K. A. T. (2015). Tourism growth and entrepreneurship: Empirical analysis of the development of rural highlands. *Tourism Management Perspectives, 14*, 17–24. https://doi.org/10.1016/j.tmp.2015.02.001

Katongole, C., Ahebwa, W. M., & Kawere, R. (2013). Enterprise success and entrepreneur's personality traits: An analysis of micro- and small-scale women-owned enterprises in Uganda's tourism industry. *Tourism and Hospitality Research, 13*(3), 166–177. https://doi.org/10.1177/1467358414524979

Kimbu, A. N., & Ngoasong, M. Z. (2016). Women as vectors of social entrepreneurship. *Annals of Tourism Research, 60*, 63–79.

Kimbu, A. N., Ngoasong, M. Z., Adeola, O., & Afenyo-Agbe, E. (2019). Collaborative networks for sustainable human capital management in women's tourism entrepreneurship: A role of tourism policy. *Tourism Planning and Development, 16*(2), 161–178. https://doi.org/10.1080/21568316.2018.1556329

Koc, E. (2020). Do women make better in tourism and hospitality? A conceptual review from a customer satisfaction and service quality perspective. *Journal of Quality Assurance in Hospitality and Tourism, 21*(4), 402–429. https://doi.org/10.1080/1528008x.2019.1672234

Kurtege Sefer, B. (2020). A gender- and class-sensitive explanatory model for rural women entrepreneurship in Turkey. *International Journal of Gender and Entrepreneurship, 12*(2), 191–210. https://doi.org/10.1108/ijge-07-2019-0113

Kwaramba, H. M., Lovett, J. C., Louw, L., & Chipumuro, J. (2012). Emotional confidence levels and success of tourism development for poverty reduction: The South African Kwam eMakana homestay project. *Tourism Management, 33*(4), 885–894. https://doi.org/10.1016/j.tourman.2011.09.010

Lapan, C., Morais, D. B., Wallace, T., & Barbieri, C. (2016). Women's self-determination in cooperative tourism microenterprises. *Tourism Review International, 20*(1), 41–55. https://doi.org/10.3727/154427216x14581596799022

Lim, S. E. Y., & Bouchon, F. (2020). The effects of network hospitality on women empowerment. *International Journal of Tourism Cities, 7*(1). https://doi.org/10.1108/ijtc-07-2019-0112

Lindberg, M., Lindgren, M., & Packendorff, J. (2014). Quadruple helix as a way to bridge the gender gap in entrepreneurship: The case of an innovation system project in the Baltic Sea Region. *Journal of the Knowledge Economy, 5*(1), 94–113. https://doi.org/10.1007/s13132-012-0098-3

Maksimovic, G., Ivanovic, T., & Vujko, A. (2019). Self-employment of women through associations in the rural areas of Sirinicka Zupa. *Ekonomika Poljoprivreda-Economics of Agriculture*, 66(1), 251–263. https://doi.org/10.5937/ekoPolj1901251M

Maliva, N., Bulkens, M., Peters, K., & Van Der Duim, R. (2018). Female tourism entrepreneurs in Zanzibar: An enactment perspective. *Tourism, Culture and Communication*, 18(1), 9–20. https://doi.org/10.3727/109830418x15180180585149

Möller, C. (2012). Gendered entrepreneurship in rural Latvia: Exploring femininities, work, and livelihood within rural tourism. *Journal of Baltic Studies*, 43(1), 75–94. https://doi.org/10.1080/01629778.2011.634103

Morgan, M. S., & Winkler, R. L. (2020). The third shift? Gender and empowerment in a women's ecotourism cooperative. *Rural Sociology*, 85(1), 137–164. https://doi.org/10.1111/ruso.12275

Moswete, N., & Lacey, G. (2015). "Women cannot lead": Empowering women through cultural tourism in Botswana. *Journal of Sustainable Tourism*, 23(4), 600–617. https://doi.org/10.1080/09669582.2014.986488

Movono, A., & Dahles, H. (2017). Female empowerment and tourism: A focus on businesses in a Fijian village. *Asia Pacific Journal of Tourism Research*, 22(6), 681–692. https://doi.org/10.1080/10941665.2017.1308397

Othman, R., Halim, S. F. A. A., Hashim, K. S. H. Y., Baharuddin, Z. M., & Mahamod, L. H. (2015). The emergence of the Islamic spa concept. *Advanced Science Letters*, 21(6), 1750–1753. https://doi.org/10.1166/asl.2015.6187

Pallarès-Blanch, M., Tulla, A. F., & Vera, A. (2015). Environmental capital and women's entrepreneurship: A sustainable local development approach. *Carpathian Journal of Earth and Environmental Sciences*, 10(3), 133–146.

Panta, S. K., & Thapa, B. (2018). Entrepreneurship and women's empowerment in gateway communities of Bardia National Park, Nepal. *Journal of Ecotourism*, 17(1), 20–42. https://doi.org/10.1080/14724049.2017.1299743

Perez, K. T., & Bui, T. L. H. (2010). Closing doors and opening windows: Opportunities for entrepreneurship in an emerging Asian country for a seasoned woman professional. *International Journal of Entrepreneurship*, 14, 45–49.

Priyadarshini, B. N. P., & Ramakrishnan, L. (2016). A study on the problems faced by women entrepreneurs in the service sector in Chennai. *IUP Journal of Entrepreneurship Development*, 13(3), 25–55. https://ssrn.com/abstract=2983728

Qureshi, D., & Ahmed, M. L. (2012). Strengthening women entrepreneurs through tourism employment and entrepreneurship of the rural women in Aurangabad. *Journal of Hospitality Application & Research*, 7(1), 54–66.

Rao, Y., Xie, J., & Lin, X. (2021). The improvement of women's entrepreneurial competence in rural tourism: An action learning perspective. *Journal of Hospitality and Tourism Research*. https://doi.org/10.1177/10963480211031032

Ribeiro, M. A., Adam, I., Kimbu, A. N., Afenyo-Agbe, E., Adeola, O., Figueroa-Domecq, C., & de Jong, A. D. (2021). Women entrepreneurship orientation, networks and firm performance in the tourism industry in resource-scarce contexts. *Tourism Management*, 86, 104343. https://doi.org/10.1016/j.tourman.2021.104343

Savage, A. E., Barbieri, C., & Jakes, S. (2020). Cultivating success: Personal, family and societal attributes affecting women in agritourism. *Journal of Sustainable Tourism*. https://doi.org/10.1080/09669582.2020.1838528

Song-Naba, F. (2020). Entrepreneurial strategies of immigrant women in the restaurant industry in Burkina Faso, West Africa. *Journal of Developmental Entrepreneurship*, 25(3). https://doi.org/10.1142/s1084946720500181

Suminar, T., Budiartati, E., & Anggraeni, D. (2019). The effectiveness of a women's empowerment model through social entrepreneurship training to strengthen a tourism village program. *International Journal of Innovation, Creativity and Change*, 5(5), 324–338.

Surangi, H. A. K. N. S. (2016). The role of female entrepreneurial networks and small business development: A pilot study based on Sri Lankan migrant entrepreneurs of the tourism industry in London. *International Journal of Business & Economic Development*, 4(1), 56–70.

Surangi, H. A. K. N. S. (2018). What influences the networking behaviours of female entrepreneurs?: A case for the small business tourism sector in Sri Lanka. *International Journal of Gender and Entrepreneurship*, *10*(2), 116–133. https://doi.org/10.1108/ijge-08-2017-0049

Tajeddini, K., Ratten, V., & Denisa, M. (2017a). Female tourism entrepreneurs in Bali, Indonesia. *Journal of Hospitality and Tourism Management*, *31*(1), 52–58. https://doi.org/10.1016/j.jhtm.2016.10.004

Tajeddini, K., Walle, A. H., & Denisa, M. (2017b). Enterprising women, tourism, and development: The case of Bali. *International Journal of Hospitality & Tourism Administration*, *18*(2), 195–218. https://doi.org/10.1080/15256480.2016.1264906

Trupp, A. (2015). Agency, social capital, and mixed embeddedness among Akha ethnic minority street vendors in Thailand's tourist areas. *Sojourn (Journal of Social Issues in Southeast Asia)*, *30*(3), 780–818. https://doi.org/10.1355/sj30-3f

Tshabalala, S. P., & Ezeuduji, I. O. (2016). Women tourism entrepreneurs in KwaZulu-Natal, South Africa: Any way forward? *Acta Universitatis Danubius: Oeconomica*, *12*(5), 19–32.

UN. (2015). *Transforming our world: The 2030 agenda for sustainable development (Resolution adopted by the General Assembly on 25 September 2015)*. United Nations (UN).

Untari, S., & Suharto, Y. (2021). The development of youth and woman entrepreneurship program in village tourism through partnership. *Geojournal of Tourism and Geosites*, *33*(4), 1538–1544. https://doi.org/10.30892/gtg.334spl14-605

UNTWO. (2019). *Global report on women in tourism* (2nd ed.). The World Tourism Organization (UNWTO).

Vukovic, D. B., Petrovic, M., Maiti, M., & Vujko, A. (2021). Tourism development, entrepreneurship and women's empowerment – Focus on Serbian countryside. *Journal of Tourism Futures*, *10*. https://doi.org/10.1108/JTF-10-2020-0167

WEF. (2019). *Global gender gap report 2020*. Geneva. The World Economic Forum (WEF).

Wilson-Youlden, L., & Bosworth, G. R. F. (2019). Women tourism entrepreneurs and the survival of family farms in North East England. *Journal of Rural and Community Development*, *14*(3), 125–145.

WTTC. (2020). *Travel & tourism: Driving women's success*. World Travel & Tourism Council (WTTC). Retrieved 06 January 2020, 2020, from https://www.wttc.org/economic-impact/social-impact/driving-womens-success/

Yoopetch, C. (2020). Women empowerment, attitude toward risk-taking and entrepreneurial intention in the hospitality industry. *International Journal of Culture, Tourism, and Hospitality Research*, *15*(1). https://doi.org/10.1108/ijcthr-01-2020-0016

Zapalska, A. M., & Brozik, D. (2014). Female entrepreneurial businesses in tourism and hospitality industry in Poland. *Problems and Perspectives in Management*, *12*(2), 7–13.

Zapalska, A. M., & Brozik, D. (2015). The life-cycle growth and development model and leadership model to analyzing tourism female businesses in Poland. *Problems and Perspectives in Management*, *13*(2), 82–90.

Zapalska, A. M., & Brozik, D. (2017). Māori female entrepreneurship in tourism industry. *Tourism*, *65*(2), 156–172. https://hrcak.srce.hr/183650

Zapalska, A. M., Brozik, D., & Zieser, N. (2015). Factors affecting success of small business enterprises in the Polish tourism industry. *Tourism*, *63*(3), 365–381.

Zhang, C. X., Kimbu, A. N., Lin, P., & Ngoasong, M. Z. (2020). Guanxi influences on women intrapreneurship. *Tourism Management*, *81*. https://doi.org/10.1016/j.tourman.2020.104137

Conclusion

18

WE ARE BREAKING THE GLASS CEILING, BUT CAN WE SHATTER THE CONCRETE ROOF

What Is Next for Gender Studies in Tourism?

Elaine Chiao Ling Yang, Albert Nsom Kimbu,
Wenjie Cai, and Magdalena Petronella (Nellie) Swart

Abstract

In breaking the glass ceiling, this chapter summarises how this handbook contributes to the discourse on gender in tourism from a pluralistic and gender-aware perspective. With 16 chapters covering diverse cultural contexts and geographical regions, it sheds light on the complexities of gender in tourism practice, research, and education. Conversations with thought leaders highlight the potential of tourism to advance gender equality and offer recommendations for achieving this goal. It emphasises the need for inclusive learning spaces, feminist research approaches, and policies that support women's empowerment in the tourism workforce and entrepreneurship. Additionally, these leaders advocate for safer and friendlier environments for female travellers and underscore the significance of intersectionality and gender transformative approaches in tourism research and practice. According to these experts, the impact of a metric-driven academic culture and an underrepresentation of men/masculinity and youth voices are areas in need of research attention. The handbook concludes by proposing gender mainstreaming in research and teaching, adopting intersectionality as a critical praxis, and promoting a gender transformative approach for a more equitable and sustainable future to shatter the concrete roof.

Keywords

Gender equality, Gender mainstreaming, Intersectionality, Gender transformation, Feminist research approach

Introduction

This handbook set out to contribute to the current discourse about gender in tourism. Taking a pluralistic gender-aware perspective, the collection, consisting of 16 chapters, revealed the existence of multiple complexities revolving around practising, researching,

DOI: 10.4324/9781003286721-26

and teaching gender in tourism. The diverse cultural contexts and geographical regions covered in this handbook contribute a multifarious understanding of gender issues pertinent to tourism consumption, experience and knowledge production, and education in the 21st century. As the field of gender studies in tourism matures, evidenced by the growing body of literature and increasing emphasis on gender equality/equity at national and intergovernmental levels (for example, the publication of gender mainstreaming guidelines by the United Nations of World Tourism Organisation in 2022), we asked ourselves, where to from here. The work on gender in tourism is far from being done. In the interest of charting a forward-looking agenda for future tourism practitioners, researchers, and educators, we sought advice from eight prominent gender researchers on their visions and advice for advancing gender equality in tourism. As we conclude the handbook, this final chapter outlines the key recommendations from the collection, offers our editorial reflection, and suggests directions for continuing gender research in tourism if the concrete roof is ever to be shattered.

Summary of Key Takeaways from the Chapters

The chapters in this handbook constitute a full knowledge production and application cycle, ranging from discussions of philosophy, pedagogy, methodology, research ethics, policy, and practice. Albeit having a diverse range of research enquiries, foci, and contexts, the chapters collectively highlight the potential of tourism to advance gender equality and offer recommendations to realise this aspiration. This section summarises some of the key learnings and aggregated recommendations offered by the chapters, organised according to the six sections in the handbook.

The *Teaching and Learning Gender* in Tourism section calls for the creation of inclusive learning spaces. A gender-conscious tourism curriculum is essential in equipping our students, who are the future tourism professionals with the awareness and knowledge to tackle gender issues in tourism. Educators are encouraged to revisit their teaching philosophy, engage with critical pedagogy, and co-create a collective and inclusive learning space with students that enables gender conversations. There is also a call for more cultural awareness and representation in tourism curricula.

The *Researching Gender in Tourism* section offers insightful recommendations for researching marginalised voices (including non-binary and LGBTQIA+ communities) within gender studies in tourism. The collections in this section argue for a feminist research approach that emphasises positionality, reflexivity, and intersectionality to challenge hegemonic and traditional ways of knowing and allows for a more nuanced and inclusive understanding of tourism and tourist behaviour. Trust and respect are to be placed at the forefront when researching marginalised communities. Researchers are also urged to be sensitive and mindful of the vulnerability of their research participants.

The chapters in the *Tourism Development* section engaged with gender issues at three levels – philosophy, policy, and practice. At the philosophical level, indigenous worldviews could offer an alternative approach to inclusive and sustainable tourism development. Ubuntu, an African philosophy that prioritises collective well-being and community development, was used as an example to illustrate the potential of adopting indigenous worldviews in fostering gender equality in tourism development, if well understood and applied. At the policy level, there is a need for clearer alignment, interpretation, implementation, and monitoring of the effects of extant policies that address these gender and racial inequalities. At the practical level, more capacities and skills development opportunities, improved access to leadership and decision-making opportunities in both the public and private sectors, and greater acknowledgement and respect from men are critical to women's empowerment through tourism.

The *Gendered Tourism Workforce* section continues to highlight the importance of skills development for empowerment, which is seen as a political process to support the autonomy and self-determination of marginalised groups. Unique to the contexts of investigation, the chapter on India, for example, calls for strategies to address concerns of career prospects, workplace safety, and flexible work arrangement among female undergraduate students in tourism, whereas the chapter on Brazil adopts an intersectionality lens to examine the double oppressions faced by black women working in the aviation sector and advocates for more efficient complaint channels, the promotion gender and racial literacy, and the review of hiring initiatives.

The penultimate section of the book focuses on *Gendered Mobilities*. The chapters investigated two forms of female travel – solo travel and business travel. Both chapters call for a safer and friendlier environment for female travellers, highlighting the persistent gendered constraints for women to access tourism space. While the recommendations largely echo existing literature on female travel, chapters in this section add to this growing body of knowledge by expanding the enquiry of solo female travel to include Spanish-speaking context and data and, in so doing, expanding our understanding of post-pandemic female business travellers.

Finally, the section on *Gender and Entrepreneurship* highlights the need for better institutional, societal, and cultural support systems and policies to support women in overcoming the challenges they face as tourism entrepreneurs. In particular, education, networks, and family support were identified as some of the critical success factors for women entrepreneurs. Collectively, the chapters in this section call for more research to unpack the intersectional dynamics of the women tourism entrepreneurship phenomenon across different geographical and cultural contexts, especially in emerging destinations of the global south. A multistakeholder gender transformative approach is proposed to address the underlying gender inequalities prevalent in the communities before the impacts of any entrepreneurial policies and initiatives could be felt.

Editorial Reflection

This handbook project embarked with an aspiration to provide an extensive coverage of gender issues relevant to tourism education, research, and praxis. Nonetheless, the impact of metrics-driven culture was felt strongly in our editorial journey. On several occasions, we faced some challenges in getting contributions from colleagues in certain universities or countries. Although these potential contributors were interested in our book, they were discouraged from spending time on book chapters; instead, their institutions expected them to devote research time to developing papers for "prestigious" journals. Under pressure to produce a large number of research outputs in highly regarded journals (e.g., high impact factors, or ABS 3-4*), many early career researchers (ECRs) and mid-career researchers (MCRs) only submit their scholarly work in specific journals, which could advance their career. As a result, many ECRs and MCRs do not consider alternative platforms for publishing. In addition, we are disappointed to see that many peers have resorted to judging a paper's research quality by where the paper was published. This reflects not only how the metric-driven academia has created a vicious circle for knowledge production but also the problematic assessment of research outputs, which significantly impacts universities' research funding, and its influence on individual practices. Relatedly, time constraints as an influencing factor in the number of submissions received. Many academics supported and/or spoke favourably about the book initiatives but could not contribute chapters due to time constraints, preferring to focus on meeting the requirements of their institutions, especially if they did not consider book chapters in academic appraisals.

We also want to emphasise that the metric-driven neoliberal culture is very much reinforcing structural inequalities and disadvantages women academics and academics from non-Western backgrounds. The pandemic revealed that women academics continue to assume more caring responsibilities (Crook, 2020), which could render them experiencing greater time deficits in the academic race. For scholars who are non-English speakers, they face language barriers to effectively articulate their work and publish in the few highly regarded tourism journals, which are exclusively in English language. Often scholars from the Global South are playing catch up to the research agenda and standards set by Western scholars, and this only widens the gap (Yang & Ong, 2020). As a result, the metric-driven culture, compounded by the Western dominance in knowledge production, disadvantages a more creative, liberal knowledge creation and discredits alternative forms of knowledge production.

Being non-native English speakers ourselves, we tried to take a more empathetic and developmental approach in the process of review and editing. Regrettably, some submissions did not make it to the book. We noticed the issue of "gender as tokenism" among these submissions. With the growing emphasis on the gender equality agenda, articulated through the United Nations Sustainable Development Goals (UNSDG) 5, there has been increasing interest in considering the gender aspect in tourism research. However, based on our observation from editing this handbook and reviewing manuscripts for journals, a lot of tourism studies remain at the "add women and stir" and "gender differences" stages (Henderson, 1994), where descriptive gender comparison takes precedence over meaningful gender discussions that consider the complexities of gender relations. This observation is most noticeable in submissions from developing countries. While the growing research attention to gender issues is commendable, there is still a lot of work to be done to close the gap in tourism gender research between the Global North and Global South – while recognising the Global North has the first mover advantage in setting the research agenda. We hope this handbook contributes to closing such gap to some extent through the developmental approach of editing. We also like to extend our thoughts to scholars who were unable to proceed with the book due to health issues, life events, and natural disasters. We thank them for their involvement in this project and wish them well in this epoch of vulnerabilities.

Reflecting on the collection we have, there are some missing pieces that we wish to acknowledge. Despite our initial intention as outlined in the call for chapters and targeted approach in sourcing submissions, men/masculinity still remained vastly underrepresented. Gender research continues to be mistaken as women studies. While the handbook comprises two chapters on non-binary and transgender communities, further work is warranted to see gender as a spectrum to truly transform gender relations and address gender-based inequalities. Further, the youth voice is rather silent in this collection. Apart from chapters on tourism students, there is a need for more attention to understanding the impact of gender on the empowerment of young participants in tourism as entrepreneurs, tourists, and workers.

Conversations with Thought Leaders

We reached out to eight esteemed tourism scholars specialised in the area of gender research for their thoughts, visions, and advice for advancing gender equality in tourism. Table 18.1 summarises the profiles of these thought leaders. A combination of email and online interviews were conducted in March and April 2023. Ethical clearance was obtained from the University of South Africa (2022_CRERC_096 [FA]) to maintain the research integrity of our investigation. The scholars were asked to envision what gender equality looks like to them. They were also asked to evaluate the current status of gender equality in tourism in relation to research, teaching, and practice and provide advice for future tourism researchers, educators, and practitioners.

Table 18.1 Thought leaders profiles

Name	Title	Affiliation	Years of research career	Main areas of expertise
Ana María Munar	Associate professor	Copenhagen Business School, Denmark	20	Gender and tourism academia, gender as social phenomenon, philosophy
Can Seng Ooi	Professor	University of Tasmania, Australia	30	Tourism experiences, cultural studies, male researchers in the field
Catheryn Khoo	Professor/ gender expert	Torrens University, Australia/ UNWTO	12	Women in tourism, marginalised voices in tourism
Donna Chambers	Professor	Northumbria University, UK	23	Heritage/culture and gender in tourism, critical theories (e.g., postcolonial, decolonial, and black feminist theories)
Heather Jeffrey	Assistant professor	University of Birmingham, UAE	9	(Re)presentation of women in tourism, gender-based violence, tourism gender education
Nigel Morgan	Professor	University of Surrey, UK	35	Social sustainability, gendered tourism marketing, gender-based violence, female travel
Stroma Cole	Reader/ director	University of Westminster, London, UK/ Equality in Tourism	28	Tourism, water, and gender nexus
Vanessa GB Gowreesunkar	Associate professor	Anant National University, India	15	Tourism management and marketing, social entrepreneurship, island tourism, and sustainability

The Visions: What Gender Equality Looks Like

Justice, freedom, and transformation underpin the thought leaders' views and visions for gender equality. As Ana emphatically stated, *"It (gender equality) is a fight that is never-ending. ... It's a fight for justice. And a fight for freedom."*

Nigel and Stroma situated gender equality at the centre of wider social and ecological transformations. Nigel's vision for gender equality is *"a world in which patriarchy and prejudice have been dismantled everywhere and where planetary boundaries are respected, allowing regenerative ecological and human flourishing."*

Relatedly, Stroma conceived gender equality as the *"transformation of relations"*. She explained,

> *The survivability as humans requires a massive transformation and gender equality has got to be a key ingredient in that. And I think if you are able to address gender inequality and put that at the centre, everything else flows out. It will lead to the transformation that is required for the planet. You care about people, you care about the planet, you care about each other. If you value care.*

Taking a reflective stance, Can Seng ruminated on how his view of gender equality has evolved over time and astutely pointed out the pitfalls of the gender-blind stance:

> *I used to hold the view that equality will only take place when we are gender-blind. I now realise that we can be "gender-blind" by not talking about gender. However, inequality and injustices can still be perpetuated.*

Catheryn succinctly summarised gender equality as *"when we don't even need to have this conversation anymore"* and called for attention to intersectional inequalities arising from race and ethnicity. Recognising the existence of diverse experiences and injustices, Donna also emphasised the need for gender equity beyond gender equality: *"I believe that even more important is the need for gender equity which means recognising that in practice, even if we provide the same opportunities to all, some will find it more difficult to succeed than others."*

The Realities: Status of Gender Equality in Tourism

The scholars then proceeded to undertake a critical accounting of the current status of gender equality in the tourism industry and academia. There was general consensus on the progress of gender equality in tourism, evidenced by more women occupying thought leadership and management positions, albeit progress being *"desperately slow"* (Stroma). Some highlighted the need to consider geographical, contextual, and socio-economic differences when evaluating the status of gender equality. Acknowledging that while *"there is a commonality in that gender matters"*, Ana cautioned the importance of considering contextual differences, stating, *"This is very contextual, we cannot take a universal view on this [status of gender equality]"*. Stroma further called for attention to take account of socio-economic differences in gender equality, especially in light of the impending digital revolution:

> *There is progress, but if you look at the digital divide and the robotic shift, these are all things that are going to have negative impacts on women. They're going to disproportionately affect women, and they're gonna disproportionately affect poorer women.*

The Realities of the Tourism Industry

As an industry, tourism is plagued by inherent characteristics that are detrimental to achieving gender equality. Heather encapsulated these characteristics in her comment: *"The industry is built out of a patriarchal society which benefits from an army of low-paid and low-skilled workers who, for the most part, are women."* Nigel identified some of the key barriers concerning women in tourism, stating, *"Vertical and horizontal segregation remains a key barrier to women's advancement, as do unsafe and unfair employment practices. Workplace gender-based harassment remains a major issue for many women, especially for migrants and women of colour."*

For women who were able to surmount vertical segregation and occupy leadership positions, their representation remains marginal in the tourism industry, and there is concern that the progress is reversing in some places. As Nigel observed, *"Whilst women are generally present in C-suites in more numbers, their overall representation remains pitifully low at senior levels and, by many indices, has barely shifted; indeed, in some instances in some countries has shown a reverse in recent years."* Noting the increasing presence of white women juxtaposing the chronic lack of local women in decision-making positions in Asia, Cateryn shrewdly remarked, *"It's not enough that every woman has a seat on the table. We need the right women, not only white women."* These views further highlight the intricacy of gender equality in relation to contexts and intersectional inequalities.

The Realities of Tourism Academia

Reflecting on where they started, many thought leaders acknowledged the positive changes in tourism academia in terms of the number of women as professors and journal editors and the growing scholarship on gender and tourism, within which the scholarship on LGBTIQ+ is also emerging. However, Nigel cautioned about the interlocking systems of inequalities, which would require transformations beyond equal gender representation to truly achieve equity:

> *Academia reflects our unequal societies, with all their prejudices and intersectional inequalities ... there is so much more to be done to achieve equity given the extent and multifaceted and embedded nature of the gender gap. Even appointing an equal number of women and men to key roles, it will take us far too many years to achieve equity.*

A similar sentiment was echoed by Vanessa, who voiced her doubts about women's presence in leadership positions as genuine progress. She observed, *"In tourism, women are seen in the positions as tourism thought leaders and editorial board members. However, the question is whether they are really assuming the role, or they have just been given the role to satisfy a condition."* Reflecting on her own experience working in India, Vanessa felt that women in those positions still do not have the liberty to assert their opinion and questioned if men are truly supportive of gender equality. This raises questions about *femwashing* practices (Khoo-Lattimore et al., 2019) and the lack of genuine engagement of gender equality in tourism academia.

While gender research in tourism has been gaining traction in the past decade, there are concerns about tourism gender researchers being disadvantaged in the metric-driven culture underpinned by the neoliberal model. Funding and citation are two critical concerns.

Drawing on his extensive experience as advisor and panellist in international promotion and research evaluation exercises, Nigel observed that *"gender research is undervalued by funders"* and *"continues to be seen as a 'soft' or 'niche' topic that only affects and interests women."* He commented, *"To work in gender is to attract fewer citations."* Catheryn shared a similar view, stating, *"Gender researchers in tourism are very disadvantaged if we measure success by H-index and publications."* Stroma also underlined the challenges in getting research fundings and reviewers for gender topics and lamented *"...we could advance the field with so much more care and cooperation within the academe."*

When questioned about what we can do to be the change we want, Heather insightfully pointed out that *"by playing the game, we are the game [that] we are reproducing every day. There are examples of academics who go against the grain, but quite often at a cost, we are not all willing or able to pay."* The generation gap and work-life balance were raised as some of the barriers to breaking out of the current system. Heather believed that dialogues between different generations of academics and reflection on our own complicity in the system are pivotal to initiating changes.

Priority Research Agendas for Gender in Tourism

Table 18.2 summarises the collective views of the scholars regarding the priority research agendas for gender studies in tourism. Research addressing pressing real-world challenges with gender implications is urgently needed. In particular, the fourth industrial revolution (e.g., robotics, automation, and artificial intelligence) disproportionately impacts less privileged women and warrants research that fast-tracks gender transformation. Gender equity is also essential to sustainability, which is another pressing issue that requires further research attention. As highlighted earlier, gender-based violence is a persistent problem in tourism, while safe and fair employment remains a concern, with tourism jobs being highly gendered and women dominating low-paid positions. Many of these issues are interconnected and thus require *"joined-up thinking"* (Stroma) from researchers to *"situate gender and sexual discrimination in the wider context of social justice, inequality and inequity in society"* (Can Seng). There are also calls for more participatory and feminist methodologies and deeper engagement with theories and critical scholarship.

Above all, intersectionality was foregrounded by almost all thought leaders as the most crucial lens cutting across the research agendas. As Heather explained, *"intersectionality allows us to unpack these power asymmetries, it better enables understanding of context-specific nuances and is a more realistic (re)presentation of the world that we live in where we are not all the same."* Also calling for more work on gender intersectionality, Donna elucidated the importance of giving voice to marginalised groups: *"Issues to do with Black, indigenous and women of colour need to be explored and especially from the perspectives of these women themselves. It is important that marginalised voices speak for themselves and not be constantly objectified."* Ana accentuated solidarity through intersectional research and advocated for *"more support for each other in the fight of emancipatory forms of knowledge."*

Teaching Gender in Tourism

All eight scholars recognised the importance of integrating gender topics in tourism education and have done so in their own teaching practices. However, they also identified the institutional barrier where gender is often sidelined in tourism programmes and taught in

Table 18.2 Priority research agendas for gender in tourism

Focus areas	Research agendas
Technological change	• Labour policies, corporate governance, and ethics research to address the masculine and Western approach to technology development and implications on equality, diversity, and inclusion issues in tourism • Digital gender gap • Skill transfer for women in tourism • Social media as a form of digital activism
Sustainability	• Eco feminist approach to understand the relationship between gender and the environment • Decolonised feminist ethics of care that emphasises collective mobilisation of humanity and structural shift towards equity
Gender-based violence	• Men and (toxic) masculinities research to change the discourse of gender-based violence
Safe and fair employment	• Gendered tourism work and pay gap • Gender representations in leadership positions – genuine equality or condition to be satisfied
Methodology	• Feminist participatory action research • Feminist epistemologies • Methodologies that go beyond specific language
Theory	• Intersectionality theory • Indigenous and non-Western knowledge • Critical tourism scholarship • Scholarship that addresses material conditions, distribution of social welfare, and ongoing exploitation of people under neoliberal governance
Others	• Non-binary and transgender research • Issues related to sexuality and maternity in tourism

isolation. To address this, there are parallel views of embedding gender in wider tourism curricula and having gender-specific modules. Some believed that gender equality is fundamental to United Nations Sustainable Development Goals (UNSDGs) and, therefore, should be mainstreamed and normalised in the curriculum, as we have done with sustainability. Nigel further called to *"demasculinise the curriculum just as we are beginning to decolonise it."* There was also support for bespoke gender modules to delve deeper into the topic. As Heather explained, *"It takes a long time to un-learn gender and to explore privilege, one class or one session is not enough."* However, the scholars also pointed out the lack of appetite for such modules from the institutions, even though the younger generation appears to be more embracing of gender discussion.

The scholars have mixed experience teaching gender topics. Several factors come into play, including gender (of the educators and students) and culture. Donna has had positive student engagement with gender discussions but observed that most of her students were women. Nigel, on the contrary, has often been disappointed by the lack of willingness of students to engage with these topics. Catheryn who has taught in both Asian and Western contexts observed the cultural difference in students' responses. She remarked,

"*In Asian countries, men seem to be more hesitant. It's harder to get through to them.*" Vanessa made a similar observation, stating "in conservative societies and countries, including gender as a module in the school programme triggers lots of discussion and debate as it is generally believed that girls are being brain-washed at school to become anti-men."

Ana spoke of the "*experience of silence*" when teaching gender but encouraged educators to not fear those silences in classrooms because silences can be productive spaces for change where students are challenged to reflect on the status quo. Or in Ana's words, "*We are provoking crisis of thought in the sense that we are provoking a crisis of the understanding of the symbolic order.*" She also underlined the "*double vulnerability*" faced by young female academics due to their bodies falling short of the traditional image of a professor. This demands "*double courage*" for them to approach gender discussions, which are often open spaces to questions rather than having definite answers. Ana espoused that high-quality scholarship on gender, including those drawing on indigenous knowledges will speak for itself, create a better curriculum and empower educators. She also emphasised the importance of collective strength through community building, stating, "*We need a movement in academia that represents this inclusive knowledge. And you'll feel very different to be in the class feeling like part of a movement. ... we don't feel alone, [although] we may be alone physically in the class.*"

Advice for Future Generations

The thought leaders' advice for future generations of tourism scholars, educators, and practitioners revolves around *Community*. This entailed "*find[ing] your tribe*" (Heather) and "*fight[ing] for your community*" (Can Seng). At its core, community is about care and love, from which power in numbers could be attained. Critiquing the hyper-pace and utility-driven culture, Stroma called for more "*care, cooperation and collaboration,*" while Ana accentuated the importance of friendship, stating "*I have one advice, friendship, the form of love that you can develop with colleagues that work on the same fight as you.*" Heather advised, "*Reach out to others and seek support, you would be surprised at the experiences others have had, there is strength in numbers, and you are part of an army.*" She also urged, "*When you make it don't forget what you have experienced and remember the same experience is felt differently by different women.*"

For women in the tourism sector, Donna suggested "*forming more collectives – there is strength in numbers, and this might also help with financing,*" while Vanessa advocated for women to "*take part in the success of other women and celebrate and appreciate them.*" Catheryn encouraged women to "*speak up and take your space*" and not be afraid to do things differently. She also advised threading gender strategically when engaging with the industry: "*Don't label yourself as a gender expert. Label yourself as whatever expertise you have but incorporate gender into that work. Make it an underlying feature in your work.*"

There were also calls for men in tourism to become allies by practising self-reflection, listening, speaking up, and stepping back. Can Seng encouraged men to "*be reflexive and self-critical.*" Echoing this view, Donna added, "*Men need to educate themselves about gender inequality and inequity in tourism and not leave all the heavy lifting up to women alone.*" Heather urged the opposite gender to "*call out inequality when you see it, empower women colleagues to take the lead. Acknowledge privilege and dismantle your bias.*"

Finally, Nigel encouraged future generations working in this space to *"remain passionate, bold and yet caring – for people, planet and self."* He stated,

> To research and advocate for gender equity is to produce meaningful research with impact. As academics, you may at times feel that the deck is stacked against you and that your topic, methods and relevance are questioned, especially by managers and colleagues in academic collectives focused on business and management. Yet, I truly believe that understanding and promoting gender equity is so important that researching it is crucial to the flourishing of humanity on this planet.

Where to from Here? Shattering the Concrete Roof!

Guided by the collective wisdom from the chapters in this handbook and conversations with thought leaders, here we offer our concluding thoughts for continuing the gender work in tourism to shatter the concrete roof.

Gender Mainstreaming in Research and Teaching

There is a clear need for gender to be more widely embedded in tourism education, research, policy, and practice. The question is how we could normalise gender discussions, making them more inclusive instead of divisive. Some recommendations from the collection and thought leaders include creating an inclusive space for conversations, embracing silence (trusting it to be a productive space for change), producing and drawing on high-quality gender scholarship, and making gender an underlying feature of our work.

Intersectionality as Critical Praxis

Intersectionality is foregrounded across the chapters as well as the interviews, marking the profound need to consider the intersections of multiple forms of social inequalities in shaping the gendered tourism experience. Since Henderson and Gibson (2013) advocated for more intersectionality research in the leisure context ten years ago, this stream of research has gained some traction but remains marginalised in tourism scholarship. In this handbook, intersectionality has been proposed as a theory, feminist paradigm, and analytical framework to investigate gender issues in tourism. The converging views in this handbook suggest the adoption of intersectionality as a critical praxis moving forward, through the practical application of the theory in addressing intertwined inequalities in tourism.

Gender Transformative Approach

Gender equality plays a critical role in social and ecological transformations. This is especially pertinent at this juncture of human history, where our younger generations are plagued by mass disempowerment and existential crisis, battling with a future marked by climate change and the fourth industrial revolution, from which gender and other social inequalities will be further intensified. A gender transformative approach is warranted to transform gender relations and, more broadly, our relations with people and the planet, with the feminist ethics of care.

References

Crook, S. (2020). Parenting during the covid-19 pandemic of 2020: Academia, labour and care work. *Women's History Review*, 29(7), 1226–1238. https://doi.org/10.1080/09612025.2020.1807690

Henderson, K. A. (1994). Perspectives on analysing gender, women, and leisure. *Journal of Leisure Research*, 26(2), 119–137.

Henderson, K. A., & Gibson, H. J. (2013). An integrative review of women, gender, and leisure: Increasing complexities. *Journal of Leisure Research*, 45(2), 115–135.

Khoo-Lattimore, C., Yang, E. C. L., & Je, J. S. (2019). Assessing gender representation in knowledge production: A critical analysis of UNWTO's planned events. *Journal of Sustainable Tourism*, 27(7), 920–938. https://doi.org/10.1080/09669582.2019.1566347

Yang, E. C. L., & Ong, F. (2020). Redefining Asian tourism. *Tourism Management Perspectives*, 34, 100667, https://doi.org/10.1016/j.tmp.2020.100667

INDEX

Note: **Bold** page numbers refer to tables and *italic* page numbers refer to figures.